P. Brosche J. Sündermann (Eds.)

Earth's Rotation from Eons to Days

Proceedings of a Workshop
Held at the Centre for Interdisciplinary Research (ZiF)
of the University of Bielefeld, FRG
September 26-30, 1988

With 89 Figures
and 25 Tables

Springer-Verlag
Berlin Heidelberg New York
London Paris Tokyo
Hong Kong Barcelona

Professor Dr. PETER BROSCHE
Observatorium Hoher List der Universitätssternwarte
Universität Bonn, 5568 Daun
Federal Republic of Germany

Professor Dr. JÜRGEN SÜNDERMANN
Institut für Meereskunde, Universität Hamburg
Troplowitzstraße 7, 2000 Hamburg 54
Federal Republic of Germany

ISBN-13:978-3-642-75589-7 e-ISBN-13:978-3-642-75587-3
DOI: 10.1007/978-3-642-75587-3

Library of Congress Cataloging-in-Publication Data
Earth's rotation from eons to days: proceedings of a workshop held at the Centre for Interdisciplinary Research (ZIF) of the University of Bielefeld, FRG, September 26–30, 1988 / P. Brosche, J. Sündermann (eds.). p. cm.
Includes bibliographical references and index.
ISBN-13:978-3-642-75589-7 (U.S.: alk. paper)
1. Earth–Rotation–Congresses. 2. Tides–Congresses.
I. Brosche, P. (Peter) II. Sündermann, Jürgen. III. Universität Bielefeld. Zentrum für Interdisziplinäre Forschung.
QB633.E25 1990 525'.35--dc20 90-9900 CIP

Softcover reprint of the hardcover 1st edition 1990

Typesetting: K + V Fotosatz GmbH, Beerfelden
2132/3145(3011)-543210 – Printed on acid-free paper

In Memory
of Karl Friedrich Küstner
(1856-1936)

Preface

These proceedings are the result of the third workshop „*Tidal Friction and the Earth's Rotation*" which took place from 26 until 30 September 1988 at the Center for Interdisciplinary Studies at the University of Bielefeld. The participants express their gratitude to the board of directors for the grant and to the staff for indefatigable daily support.

The workshop at the same time was a meeting of the Special Study Group 5.99 of the International Association of Geodesy; this Study Group has the same target, and one of the convenors (P.B.) is its president.

In comparing the present situation with the one at our first workshop (1977), two statements can be made: (a) nearly all kinds of data and especially the numerical values became distinctly more precise, (b) the complexity of the Earth's rotation at present forbids its use for the determination of tidal friction parameters. Instead, our accurate values today stem from the motion of our natural satellite, the moon, and, most recently, from some artificial satellites. This is an unsatisfying situation. Therefore, we have decided to extend our work to all problems which have to be solved in order to reduce the observed values for the Earth all disturbing effects. Thus our activity in the future should be titled „Long-term changes in the Earth's rotation". Even this, of course, is not strictly true. The very tides which cause the secular changes have periods of half a day and must be checked at such a time scale as well. It is for this reason that we have selected the actual title of this book.

One hundred years ago, K.F. Küstner proved the existence of (short-term) variations in the terrestrial location of the Earth's axis of rotation. Tidal friction was misunderstood for a long time as a purely lunar phenomenon. Therefore, Küstner's observations can be considered as the final breakthrough towards the acceptance of the Earth's rotation as an empirically founded and scientifically challenging field of research. We dedicate this volume to his memory.

Daun and Hamburg, March 1990

PETER BROSCHE
JÜRGEN SÜNDERMANN

Contents

Earth's Rotation Measured by VLBI
H. SCHUH (With 8 Figures) 1

1 Determination of the Earth Rotation Parameters by VLBI 1
1.1 Project IRIS 1
1.2 DSN VLBI Observations at JPL 3
1.3 The GJRO Campaign 3
1.4 The VLBI Campaigns with HartRAO 3
2 Analysis of the Earth Rotation Parameters Observed by VLBI 4
2.1 Analysis of the UT1 Series 4
2.2 Analysis of the IRIS Pole Series 6
3 Relation Between the Fortnightly and Monthly Tidal Variations and the Variations of the Pole? 9
4 Conclusions 11
References 11

Relativity and the Earth's Rotation
E. GILL, M. SOFFEL, H. RUDER, and M. SCHNEIDER 13

1 Introduction 13
2 Post-Newtonian Effects: Free Rotation of the Earth 14
3 Post-Newtonian Effects: Forced Precession and Nutation 16
4 Relativity Experiments and the Earth's Rotation 18
References 20

Astrometric Observations of Hevelius and Derived Values of $\triangle$T (Dynamical Time − Universal Time)
J. WÜNSCH (With 2 Figures) 21

1 Introduction 21
2 The Solar Observations 21
3 The Lunar Observations 24
References 25

Historical Chinese Astronomical Observations
LIU CIYUAN 27

1 Historical Chinese Astronomical Observations (1) 27
2 Historical Chinese Astronomical Observations (2) 30
2.1 Stars 30
2.2 Special Astronomical Phenomena 30

Application of Early Chinese Records of Lunar Occultations and Close Approaches
LIU CIYUAN and KEVIN K. C. YAU (With 5 Figures) 33

1 Introduction 33
2 Observational Criteria and Method of Analysis 33
3 Data Analysis 35
4 Discussion 37
5 Conclusions 39
References 39

Ancient Central Eclipse Records of China and Variation of the Earth's Rotation
HAN YANBEN, LI ZHISEN, and YANG XIHONG (With 1 Figure) 40

1 Introduction 40
2 General Principle 41
3 Data Analysis and Results 42
4 Discussion 44
References 44

Ancient Eclipse Records of East Asia and the Earth's Rotation
LI ZHISEN, HAN YANBEN, and ZENG ZHIFANG (With 1 Figure) 46

1 Introduction 46
2 Valuable Time Records of Eclipses 46
3 Importance of Central Eclipse 48
4 Use of the East Asian Series of Ancient Records 49
References 50

Tidal Acceleration of the Moon
XX NEWHALL, J.G. WILLIAMS, and J.O. DICKEY 51

References 51

The Variation in J_2 and in the Moments of Inertia:
Satellite Results and Consequences for the Angular Momentum Budget of the Earth-Moon-Sun System
M. Burša 52

1 Introduction 52
2 Recent Values of Quantities Needed for the Angular Momentum Budget 52
2.1 Secular Variation in the Moon's Mean Motion Due to Tidal Dissipation 52
2.2 Secular Variation in J_2 and Related Quantities 53
2.3 Secular Variations in the Orbital Elements of the Moon's and Earth's Orbit Due to Tidal Dissipation 53
3 Secular Tidal Variation in the Orbital Angular Momentum of the Earth-Moon-Sun System 54
4 Secular Tidal Variations in the Angular Velocity of the Earth's Rotation 55
5 Residuum in the Total Angular Momentum Budget 56
6 Conclusions 56
References 57

Celestial Mechanics of Present Tidal Friction
J. Kostelecký 58

1 Introduction 58
2 Equations of Motion 58
3 Effect of Gravitational Field 59
4 Earth's Body Tides 60
5 Ocean Tides 60
6 Effect of Tides on the Motion of Artificial Satellites, Solution of the Inverse Problem: Determination of Tidal Parameters 61
7 Effect of Tides on the Moon's Motion 62
8 Parameters of Ocean Tides – Satellite Solutions 63
9 Parameters of Body Tides 63
10 Secular Tidal Variation of Parameters of the Moon's Orbit 65
11 Conclusion 67
References 67

The Consideration of Solid Earth Effects in Ocean Tide Modeling
W. Zahel (With 7 Figures) 69

1 Introduction 69
2 The Computation of Free and Forced Oscillations 69
3 Free and Forced Hemispherical Oscillations 71

4 Near-Resonance Semi-Diurnal Oscillation Systems 75
5 Conclusions 79
References 80

Variations of the Angular Momentum Budget for Tides of the Present Ocean
U. Seiler (With 7 Figures) 81

1 Introduction 81
2 Simulation of Ocean Tides 81
3 The Angular Momentum Budgets 83
4 Variations of the Earth's Rotation 87
5 Summary and Conclusions 91
References 94

The Pole Tide and the Damping of the Earth's Free Nutation
H. Lenhardt (With 2 Figures) 95

1 Introduction 95
2 A Glance at the History 95
2.1 At the Beginning: Klein and Sommerfeld (1910) 96
2.2 A Milestone: Bondi and Gold (1955) 97
3 A Spherical Harmonic Model of the Pole Tide 99
3.1 The Algorithm 99
3.2 The Comparison with Observed Values 102
4 Some Aspects of the Modern Theory 104
5 Conclusions 106
References 106

The Seasonal Angular Momentum of the Thermohaline Ocean Circulation
A. Frische and J. Sündermann (With 16 Figures) 108

1 Introduction 108
2 Calculation of Thermohaline Currents 109
2.1 Basic Equations 109
2.2 Thermal Wind Equations 110
2.3 The Hydrographic Data 111
3 Oceanographic Results 111
3.1 The Global Circulation 111
3.2 Seasonal Variations 114
3.3 Transport Calculations 118
4 Angular Momentum Contained in Thermohaline Currents 120
4.1 Calculations Based on Geostrophic Currents 120

4.2 Calculations Based on a General Circulation Model . . . 123
4.3 Implications for Length of Day . . . 125
5 Conclusions . . . 125
References . . . 126

Atmospheric Effects on the Earth's Rotation
H. VOLLAND . . . 127

1 Introduction . . . 127
2 Outline of the Theory of Atmospheric Large-scale Motions . . . 127
3 Internal Friction . . . 128
4 Surface Friction . . . 129
5 Atmospheric Angular Momentum . . . 130
6 Polar Motion . . . 133
7 Solar Gravitational Tidal Torque . . . 135
8 Magnetospheric Friction . . . 136
References . . . 140

The Solar Torque – A Leak for the Angular Momentum
of the Earth-Moon System
P. BROSCHE and J. WÜNSCH . . . 141

1 Introduction . . . 141
2 The Ratio . . . 141
2.1 The Schematic View . . . 141
2.2 Periodic Variations . . . 142
2.3 Average Torques . . . 142
3 The Balance . . . 144
References . . . 145

Tectonic Consequences of the Earth's Variable Rotation
on Geological Time Scales
C. DENIS and P. VARGA (With 8 Figures) . . . 146

1 Introduction . . . 146
2 Outline of the Theory of Equilibrium Figures . . . 147
3 Tidal Despinning of the Earth . . . 149
4 Kinetic Parameters of the Earth Throughout Geological Time . . . 152
5 Lithospheric Stresses Caused by Tidal Despinning . . . 155
6 Tidal Shrinkage . . . 158
7 Tectonic Evolution of the Earth Due to Tidal Despinning . . . 159
References . . . 161

Secular Variations of the Earth's Moment of Inertia and Related Quantities
P. VARGA and C. DENIS 163

1 Introduction 163
2 Rotational Deformation and Trace Changes of the Inertia Tensor 165
3 Variations of the Polar Moment of Inertia and Related Quantities Caused by Changes of l.o.d. 167
4 Some Causes of Secular Change of the Polar Moment of Inertia 169
5 Long-Term Variations of the Geopotential Coefficient J_2 171
6 Conclusion 174
References 174

Global Consequences of the Tidal Secular Deceleration for the Solid Earth and its Fluid Core
J. HINDERER and H. LEGROS 177

1 Introduction 177
2 Secular Deceleration of the Core-Mantle System 177
3 Geodynamic Consequences Due to the Change in the Rotational Potential 179
3.1 Radial Part 179
3.2 Zonal Part 180
4 Resonance in the Paleowobbles 181
5 Secular Deceleration and Magnetic Scaling Laws 184
6 Conclusion 185
References 186

Astronomical Frequencies in Paleoclimatic Data and the Dynamical Ellipticity of the Earth
M.F. LOUTRE, V. DEHANT, and A. BERGER (With 3 Figures) 188

1 Introduction 188
2 Climate Changes at the Astronomical Frequencies 188
2.1 Climate Variations 188
2.2 Milankovitch Astronomical Theory of Paleoclimates 190
2.3 Classical Astronomical Frequencies in Diverse Geological and Insolation Data all over the World 190
3 Earth's Orbital Parameters 192
4 Astro-Climatic Elements 193
5 Influence of the Variation of the Lunar Orbit 194
6 Impact of the Ice Sheets During the Quaternary 197
7 Conclusion 199
References 199

Quaternary Sea-Level Changes in the North Sea,
an Analysis of Amplitudes and Velocities
H. STREIF (With 6 Figures) 201

1 Introduction 201
2 Geological Evidence and Processes 201
3 Conclusions 212
References 213

Tidal Rhythms in the Shell Secretion of Living Bivalves
C.A. RICHARDSON (With 10 Figures) 215

1 Introduction 215
2 Experimental Approach 216
3 Effects of Tidal Emersion 218
4 Effects of Continuous Immersion 222
References 225

Implications of Some Recent Sedimentological Studies
to the History of the Earth-Moon System
J.D.A. PIPER (With 1 Figure) 227

1 Introduction 227
2 Sedimentary Rhythmites 228
References 232

Continental Configurations and Mantle Reference Frames
over Geological Time
J.D.A. PIPER (With 12 Figures) 234

1 The Palaeomagnetic Constraint and the Dipole Assumption 234
2 Configurations and Movements of the Continental Crust 236
3 True Polar Wander and the Hotspot Frame 243
4 The Geomagnetic Field Source and the Core-Mantle Boundary Zone . . . 248
References 253

Earth's Rotation Measured by VLBI

H. Schuh

1 Determination of the Earth Rotation Parameters by VLBI

On 1 January 1988 the International Earth Rotation Service (IERS) started operation replacing the International Polar Motion Service (IPMS) and the Earth rotation section of the Bureau International de l'Heure (BIH). Initially, it relies on three observing methods: Very Long Baseline Interferometry (VLBI), Lunar Laser Ranging (LLR), and Satellite Laser Ranging (SLR).

VLBI plays an important role in the determination of the Earth rotation parameters by IERS because it is the only one of modern observational techniques that can accurately measure all of the parameters which define the orientation of the Earth (UT 1, polar motion and nutation). In this chapter all such VLBI programs are presented as are mainly dedicated to monitor variations in the rotation of the Earth.

1.1 Project IRIS

The *I*nternational *R*adio *I*nterferometric *S*urveying (IRIS) activities consist of VLBI observing sessions designed to monitor UT 1 and polar motion. At present, two independent IRIS VLBI networks exist (IRIS-A and IRIS-P) which both contribute to the IERS. A third network (IRIS-S) has been used to conduct pilot observing campaigns with a station in the southern hemisphere. Figure 1 shows the three IRIS networks.

a) The IRIS-A Network

The IRIS-A observations normally involve three stations in the United States (the Westford radio telescope in Massachusetts, the George R. Agassiz station in Texas[1] and the Richmond Observatory in Florida) and the 20-m radio telescope of the Fundamental Station Wettzell in the Federal Republic of Germany. One session per month also includes the Onsala Space Observatory in Sweden. The experiments are 24 h in duration at 5-day intervals.

The observational tapes from the IRIS-A sessions are processed at correlators either of the US Naval Observatory in Washington D.C. (before 1986 at the correlator of the Haystack Observatory, Mass.) or of the Max-Planck-Institute for Radioastronomy in Bonn, FRG. The output of the correlation is distributed to several organizations for

---

Geodätisches Institut der Universität Bonn, Nussallee 17, D-5300 Bonn 1, FRG.

[1] Replaced by Mojave in California since 22 June 1989.

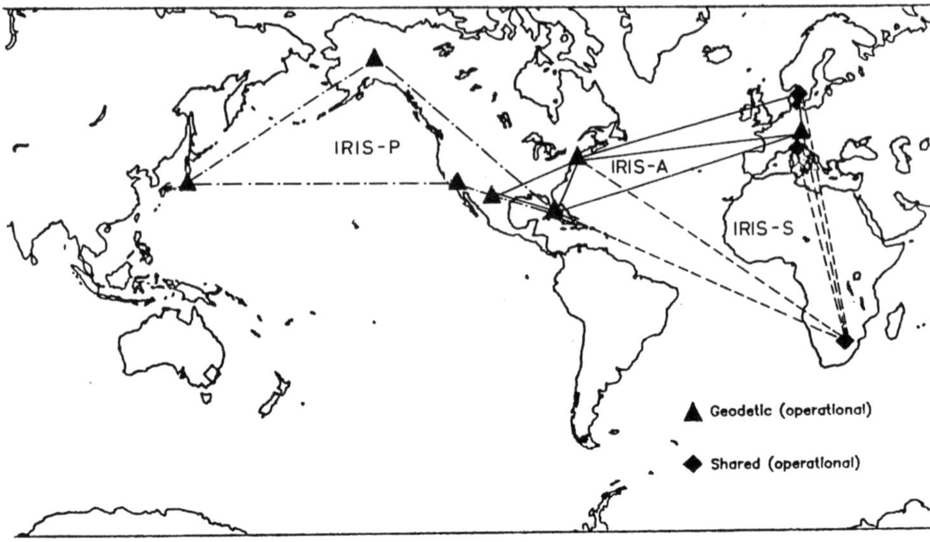

Fig. 1. The VLBI networks involved in project IRIS; *IRIS-A:* Westford/Mass., Fort Davis/Texas; Richmond/Florida; Wettzell/FRG; Onsala/Sweden. *IRIS-P:* Mojave/Calif.; Fairbanks/Alaska; Richmond/Florida; Kashima/Japan. *IRIS-S:* Hartebeesthoek/South Africa; Westford/Mass.; Richmond/Forida; Fort Davis/Texas; Wettzell/FRG; Onsala/Sweden; Bologna/Italy

analysis. The IRIS-A experiments provide all of the Earth orientation parameters (UT 1, polar motion and nutation), baseline components, extragalactic radio source positions and parameters of geophysical and astronomical models.

Thus, since 5 January 1984, the date at which the Wettzell observatory began regular operations, the IRIS-A system has been routinely providing the x and y components of polar motion with an accuracy of ± 1 to ± 2 marcsecs, and UT 1 to ± 0.05 to ± 0.1 ms of time.

At present, due to improvements of the VLBI hardware and software, the formal error of x_p, y_p is about ± 0.7 marcsecs for each component, and that of UT 1 about ± 0.03 ms of time. Since April 1985 short VLBI experiments (~ 45 min) on the baseline Westford to Wettzell have been performed ("Intensive" series) in order to monitor UT 1 every day except every fifth day when it is superseded by the regular IRIS-A session. This series produces values of UT 1 with an accuracy of better than ± 0.1 ms. More details of the "IRIS" and of the "Intensive" sessions are given in the IRIS Earth Orientation Bulletin A (Campbell et al. 1988) and by Carter et al. (1988 a).

b) The IRIS-P Network

In 1987 a second IRIS network was established called IRIS-P (Pacific) in order to get an independent comparison of the IRIS-A results. Since April 1987 IRIS-P is running once per month with Kashima in Japan, Fairbanks in Alaska, Mojave in California and Richmond in Florida. First results of IRIS-P and a comparison of the determined Earth rotation parameters between IRIS-P and IRIS-A VLBI experiments are presented by Yoshino et al. (1987).

c) The IRIS-S Network

In January/February 1986 and 1987 and in December 1987/January 1988 VLBI campaigns were made including the Hartebeesthoek Radio Observatory in South Africa and stations in the USA and Europe. This network of four, respectively five observatories spanning three tectonic plates (Africa, Eurasia and North America) was called IRIS-S (South). A detailed description of the first two campaigns and a comparison between the results of IRIS-S and IRIS-A are given by Carter et al. (1988 b).

1.2 DSN VLBI Observations at JPL

The JPL Time and Earth Motion Precision Observations (TEMPO) project makes VLBI earth orientation measurements using the DSN (Deep Space Network) radio telescopes in California, Spain and Australia since 1978. These short-duration (3 h) VLBI observations are conducted approximately once per week using the JPL Block I VLBI system. Additional irregular 18-to-24-h duration VLBI observing sessions for source catalog maintenance have been conducted since late 1978 on the DSN baselines. Results of a combined homogeneous reduction of the VLBI data are presented by Morabito et al. (1988). The JPL-Deep Space Network also contributes to the new International Earth Rotation Service (IERS). In 1988 the DSN radio telescopes were equipped with the Mark III system which is today's most accurate VLBI technique.

1.3 The GJRO Campaign

At the end of 1985, a first joint experiment between Japan and Germany was performed (GJRO, *G*erman *J*apanese Earth *R*otation *O*bservations). Daily VLBI observations were conducted between Kashima and Wettzell from 23 November to 6 December, 1985 to monitor some very short time scale variations of UT 1. The baseline vector from Kashima to Wettzell (Fig. 2) should provide good UT 1 determinations due to its considerable length (8500 km) and its east-west orientation. One of the most important goals of this campaign was to provide an independent standard of comparison for the daily IRIS "Intensive" VLBI observations between Wettzell and Westford. The GJRO campaign consisted of 12 short experiments (100 min), each with 16 observations, and one long experiment (22 h, 203 observations). The starting time of each short experiment has been set close to the daily sessions between Wettzell and Westford in order to compare these two independent UT 1 determinations. UT 1 was determined on each session with fixed station positions and fixed pole coordinates, but solving for clock and atmospheric parameters. Both UT 1 series (GJRO and IRIS "Intensive") agree within 0.2 ms if the regular (5-day) IRIS pole positions are used, but there is still a remaining offset of 0.1 ms. More details of the GJRO campaign are given by Yoshino et al. (1986).

1.4 The VLBI Campaigns with HartRAO

Additionally to the IRIS-S experiments (Sect. 1.1 c) the Hartebeesthoek Radio Astronomy Observatory (HartRAO) in South Africa and the Wettzell radio telescope

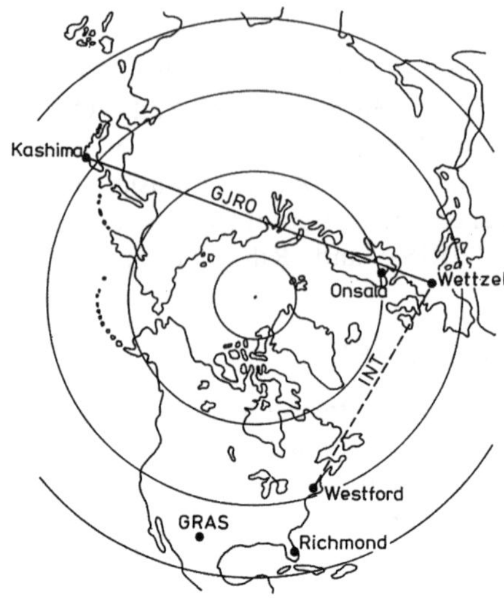

Fig. 2. GJRO and IRIS-Intensive baselines, and IRIS-A stations

performed a series of Mark III VLBI experiments. In January and February 1986 27 single baseline observing sessions of 2-h duration were scheduled additionally to the six IRIS-S sessions observed within multi-station experiments of 24-h duration. The main purpose of the short daily sessions was to demonstrate the potential of the VLBI technique to monitor polar motion by relatively short and inexpensive experiments on a north-south baseline. Furthermore, this first daily pole position monitoring project of more than one-month duration should permit to investigate short period fluctuations of the pole which have been recently considered by several authors (Kolaczek and Kosek 1985; Brzezinski 1987; Eubanks et al. 1988; Ballani 1989). The measurements yielded accuracies of about ± 2 marcsecs for the x pole component and ± 1 marcsec for the y pole component. The intensive series of pole positions agrees very well with the 5-day IRIS pole positions and indicates that periodic pole path fluctuations may exist. More details of the first campaign with HartRAO are given by Nothnagel et al. (1987).

This campaign was repeated in January and February 1987 with an improved observing schedule and more redundancy. First results of this second campaign with HartRAO show that the formal errors of the least squares fits are even lower and that the geometrical configuration is more stable then for the first experiment series.

2 Analysis of the Earth Rotation Parameters Observed by VLBI

2.1 Analysis of the UT1 Series

The UT1 series, which have been observed within project IRIS (5-day "IRIS-A" and daily "Intensive" series) since 5 January 1984, have been subjected to an iterative spectral analysis. In each iteration of this analysis the power spectrum is computed and the frequency of the maximum peak is extracted. Then a least squares fit is carried out, entering all periods which have been determined until the current iteration and solving for their amplitudes and phases. The residuals are used to compute the power spectrum of the next iteration. Of course, the residuals cannot be rigorously considered as uncorrelated observations, but with this data type the effect is negligible (Schuh 1981). In that reference the method which as previously been proved to be very effective for a spectral decomposition of the BIH UT1 data, has been described in detail.

The results for the detected periods between 2.5 months and 1 year agree very well with those of an earlier analysis of the BIH data. In addition, various significant oscillations with periods less than 2 months could be detected in the IRIS UT1 data. Most of the periods between 18 and 63 days obviously correspond to short period variations in global atmospheric data reported by several authors. These are in particular the periods of about 50 days, 32–36 days, 25 days and 21 days. The other significant variations with periods shorter than 35 days can be clearly identified as being due to the influence of zonal tides on the rotation of the Earth. The results of the spectral analysis have been compared with the theoretical periods and phases and with the amplitudes given by Yoder et al. (1981). The spectral analysis reveals very clearly the monthly (27.6 days) and the fortnightly (13.7 days) terms and also the smaller variations at 14.77 days and 9.13 days. A similar analysis of eight subsets, each with 2 months of the daily UT1 determinations, indicates that the small 7.10-day term can already by extracted from the VLBI observations. Concerning the scaling factor k/C of the tidal-induced UT1 variations, the present VLBI data confirm the theoretical value of 0.94 recommended by Yoder et al. (1981) within its error range of ±0.04. However, there are some indications that it could be slightly larger at the fortnightly range and a little smaller for the monthly periods. A detailed description of this analysis and a discussion of the results are given by Schuh (1987, 1988).

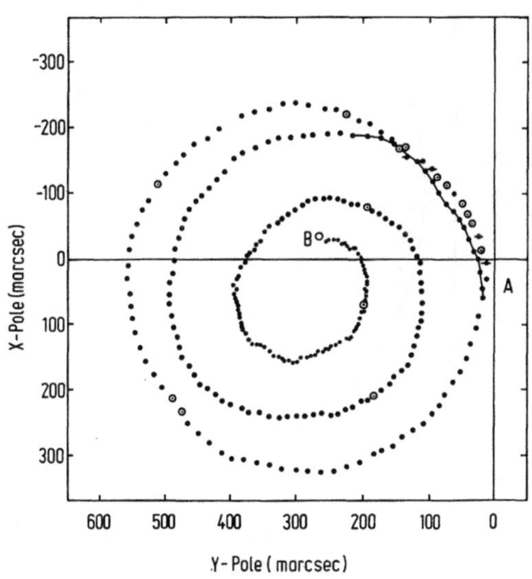

Fig. 3. Pole coordinates observed within project IRIS from 17 November 1983 (*A*) to 16 September 1987 (*B*) at 5-day intervals (⊙: *y* from SLR)

In the results of the GJRO campaign (see Sect. 1.3) the fortnightly period (13.7 days) induced by the Earth tides was detected with an increase of the theoretical amplitude by 5%–10% (Yoshino et al. 1986).

2.2 Analysis of the IRIS Pole Series

The pole coordinates observed within project IRIS (IRIS-A network) from 17 November 1983 until 16 September 1987 are shown in Figure 3. Of course, this time interval is not long enough to investigate either the mean frequencies in polar motion like the Chandler wobble (~1.2 y), the annual, semiannual and biannual periods, or the longer periods like a spurious oscillation of about 30 years ("Markowitz wobble"), or the secular drift of the mean pole. However, already today we are able to analyze the pole data observed by VLBI in the high frequency range, i.e., for periods of less than 2 months.

According to recent publications variations between 10 and 60 days due to atmospheric excitation can be expected in the motion of the pole (Brzezinski 1987; Eubanks et al. 1988) and also have been detected by analysis (Kolaczek and Kosek 1985; Ballani 1989) although their amplitudes are relatively small.

For the investigation of the short periods of the polar motion the pole path was analyzed in the alongtrack and crosstrack components. In order to obtain these variations we used a simple model by fitting a circle through half-year portions of the pole path and computing the residuals Δr to the mean radius r_m and the differences $\Delta\varphi$ from the average angle φ_m which has been covered in the time interval of 5 days.

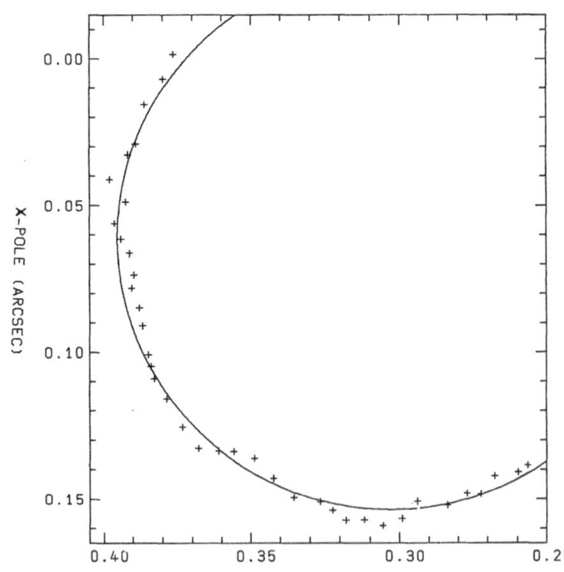

Fig. 4. Pole coordinates observed within project IRIS from 12 August 1986 until 13 February 1987 and a *circle* fitted to these data

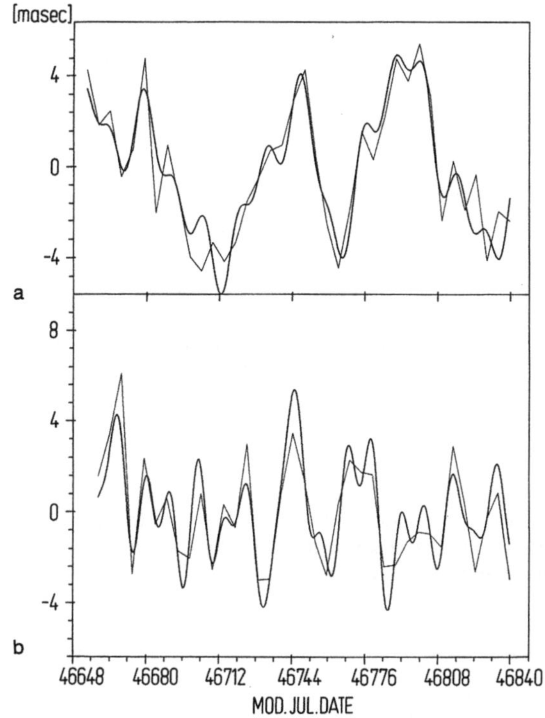

Fig. 5a. Crosstrack components of the pole path Δr (*thin line*) from 12 August 1986 until 13 February 1987 and the curve constituted by the variations found by spectral analysis (*bold line*), periods: 137.0, 60.8, 46.2, 32.5 and 13.7 days. **(b)** Alongtrack components of the pole path $\Delta\varphi$ (*thin line*) from 12 August 1986 until 13 February 1987 and the curve constituted by the variations found by spectral analysis (*bold line*), periods: 89.3, 35.0, 22.0, 18.6, 12.6, and 11.0 days

As an example, the pole coordinates observed by IRIS from 12 August 1986 until 13 February 1987 and the circle fitted to these data are plotted in Figure 4. Figure 5, the upper part showing the residuals Δr in direction of the center of the circle fitted to these points, i.e., the crosstrack ($\triangleq$ radial components) of the pole path, and the curve constituted by the variations detected by the spectral analysis. Figure 5, lower part, shows the differences $\Delta\varphi$ from the average angle φ_m, where $\Delta\varphi$ is given in geocentric arclength (marcsec) (positive $\Delta\varphi$ means that the pole was "faster" than on the average during those 5 days), i.e., the alongtrack ($\triangleq$ tangential) components. Figures 6 and 7 show all oscillation periods below 35 days and with amplitudes ≥ 0.8 marcsecs which have been detected in 13 subsets of the IRIS data from 17 November, 1983 until 16 September 1987 (Fig. 6: crosstrack components; Fig. 7: alongtrack components) each of them covering 6 months and overlapping by 3 months. Alongtrack and crosstrack motions show variations with periods of ~13.7 days. This could point to a relationship with the fortnightly tides (Earth tides and/or ocean tides) with a period of 13.7 days which will be considered in Sect. 3. In the alongtrack components the largest variations were found with an average period of ~10.8 days with an amplitude of 1.6 marcsecs. This confirms the assumption made by Brzezinski (1987) and

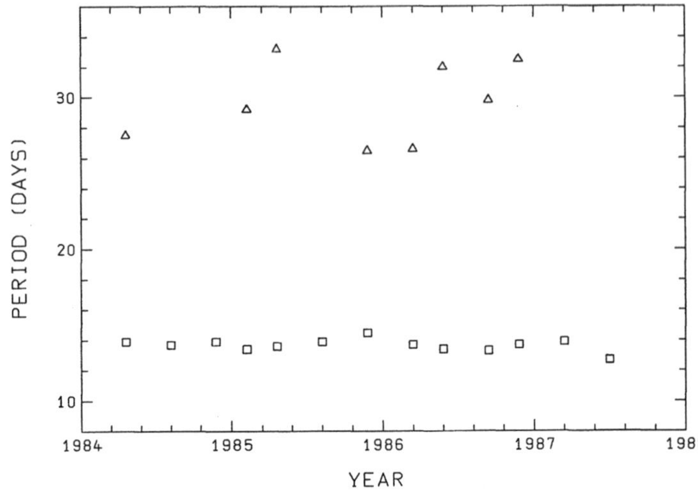

Fig. 6. Periods < 35 days with amplitudes ≥ 0.8 marcsecs detected in 6-month subsets of the crosstrack components of the pole path; □ average period: 13.7 days, average amplitude: 1.0 mas; △ average period: 29.6 days, average amplitude: 1.2 mas

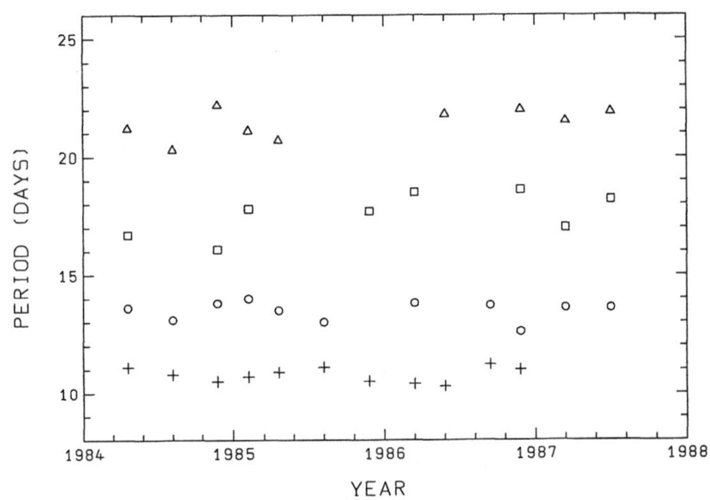

Fig. 7. Periods < 35 days with amplitudes ≥ 0.8 marcsecs detected in 6-month subsets of the alongtrack components of the pole path; + average period: 10.8 days, average amplitude: 1.6 mas, ○ average period: 13.7 days, average amplitude: 1.3 mas, □ average period: 17.5 days, average amplitude: 1.2 mas, △ average period: 21.4 days, average amplitude: 1.2 mas

Eubanks et al. (1988) about the influence of the short period variations in global atmospheric data on the motion of the pole. The authors estimate the amplitudes of the 10-to-12-day pole variations to be between 1 and 2 marcsecs. This agrees with the amplitudes of the variations observed directly by VLBI in that period range. Additionally, significant variations were found with periods around 17.5 days and 21.4 days.

3 Relation Between the Fortnightly and Monthly Tidal Variations and the Variations of the Pole?

For the investigation of a possible relation between the observed pole variations and the fortnightly and monthly tidal variations crosscorrelation methods can be used. For a first intercomparison the least squares fits for each of the 6-month subsets were repeated but now with the theoretical periods of the main fortnightly tide (M_f with a period of 13.66 days) and the main monthly term (M_m with a period of 27.56 days) entered. All other periodic fluctuations in the data have been determined as usual by spectral analysis and have been removed from the data. After solving again for phases and amplitudes we compared the observed phases Φ_{obs} of the two variations (13.66 and 27.56 days) referred to the beginning of each data set obtained by the least squares fits [Φ_{obs} from $A_{obs} \cdot \sin(\omega \cdot t + \Phi_{obs})$] with the theoretical phases φ_{th} of the variations of the tidal potential [$K \cdot \cos(\omega \cdot t + \Phi_{th})$]. Thus, if both effects are connected in phase, this should result in an offset $\Delta\Phi (\Delta\Phi = \Phi_{th} - \Phi_{obs})$ of $-90°$ between the sine-term and the cosine-term [because $\sin(\alpha) = \cos(\alpha - 90°)$]. Whereas, if the two effects are completely independent, we should get phase differences $\Delta\Phi$ varying arbitrarily from $-180°$ to $+180°$ and thus having an expectation value of zero degrees. For the 13.66-day term observed in the crosstrack pole components most of the phase differences $\Delta\Phi$ appeared to be negative and the weighted mean phase difference was determined to $-56.6 \pm 41.3°$ (weighted rms error[2]. For the weighted mean phase difference between the theoretical tidal variation and the 13.66-day term in the alongtrack components we found $+7.2 \pm 76.8°$ and for the phase difference between the theoretical 27.56 day term and the corresponding periods in the crosstrack pole components we got $-1.4 \pm 99.2°$. Both latter results show that those tidal variations cannot be detected in the IRIS pole observations.

However, for the 13.66 day variation of the crosstrack components, both the phase offset of almost $-57°$ which is significantly different from $0°$ and the relatively small rms error are a strong indication for a relation between the two effects, i.e., between the fortnightly pole variation and the M_f Earth tide (and/or ocean tide with the same period). Thus, it seems that the pole variation lags by about $33°$ (or 1 day) behind the M_f tidal variation. Figure 8 shows the differences $\Delta\Phi (\Delta\Phi = \Phi_{th} - \Phi_{obs})$ for the 13.66-day crosstrack pole components. We can see that the spreading around the mean phase difference of $-57°$ is relatively small, except for some outliers with high formal errors.

In a recent new solution the phases of the observed *combined* variations of the two main fortnightly terms (13.66 and 13.63 days) were determined and were compared with the theoretical phases of the *combined* variations. The weighted mean phase difference was determined to $-63.5 \pm 30.4°$, again showing a lag of about 1 day, now even with a smaller rms error than in the solution shown before. In comparison, all other phase differences $\Delta\Phi$ (for the 13.66-day alongtrack pole variation and for the

---

[2] weighted rms error σ_i

$$\sigma_i = \pm\sqrt{([p_i \cdot (l_i - \bar{l})^2])/([p_i] \cdot n)}$$

p_i – weights; l_i – observations; $\bar{l}$ – weighted mean ($\bar{l} = [p_i l_i]/[p_i]$); n – number of observations.

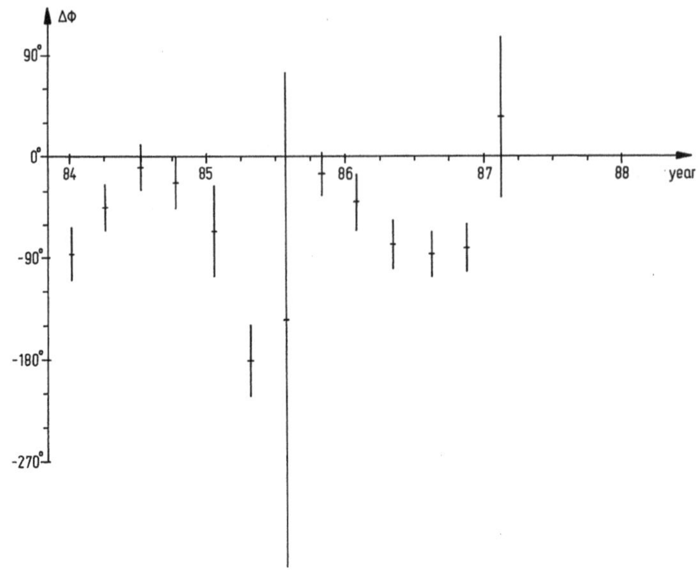

Fig. 8. Phase differences $\Delta\Phi$ between Φ_{obs} of the 13.66-day sine-variation of crosstrack pole components and Φ_{th} of the theoretical 13.66 day cosine-variation (M_f) of the tidal potential ($\Delta\Phi = \Phi_{th} - \Phi_{obs}$)

monthly terms, both alongtrack and crosstrack) vary randomly from $-180°$ to $+180°$.

Of course all these results need further empirical verification. First, it has to be checked whether the detected relation between the fortnightly crosstrack pole variations and the M_f-tide is not due to errors of the VLBL analysis models. Those errors could be caused by insufficient models for the tidal deformation of the fixed Earth, for the tidally induced variations of UT 1 or for the nutation parameters, all of which contain fortnightly and monthly terms. Due to the mathematical correlations in the VLBI solutions between all parameters related to the orientation of the Earth (pole position, UT 1, nutation) there could be a kind of second order influence of those errors on the pole parameters. However, if such an indirect effect can be excluded, the theoretical geophysical models for the interaction between the tides and polar motion will have to be extended because the influence of the solid Earth tides on polar motion has been estimated as considerably smaller (i.e., Yoder et al. 1981; Scherneck, pers. commun.) than what was found by our analysis. Indeed, very new results by Seiler (1988, this volume) indicate that the influence of the M_f ocean tide can cause a fortnightly variation of the pole curve which is in the range that has been observed by VLBI.

4 Conclusions

The VLBI approach to monitoring the variations in the Earth's rotation appears to be one of the most efficient, both on the conceptual side and as regards the accuracy achieved. Currently, several independent VLBI networks play an important role in the determination of the Earth rotation parameters by the new International Earth Rotation Service (IERS). However, for the study of long-term effects, such as polar wander and long-term Earth's rotation, it is highly desirable to achieve global coverage with VLBI networks, in order to be able to discriminate between changes in Earth orientation and plate tectonic motions.

For the investigation of long-term effects of the Earth's rotation it is necessary to determine the short-period variations first. The positive results of the spectral analysis of the Earth rotation parameters observed within project IRIS show that the chosen method is quite effective and holds good promise for future investigations. These are needed to confirm the complex relations between the Earth rotation parameters and the driving forces in the short-period range such as the tidal, atmospheric and hydrospheric excitation. After the division of polar motion into its alongtrack and crosstrack components different variations have been found in the two components which could facilitate the geophysical interpretation of the results.

References

Ballani L (1989) Polar motion fine structure by MERIT data. Proc VII Int Lohrmann Colloq Geodetical Astrometry, March 1988, Karsdorf/Dresden, GDR. Wiss Z Techn Univ Dresden 38 (2) pp 53–55

Brzezinski A (1987) Statistical investigations on atmospheric angular momentum functions and on their effects on polar motion. Manuscripta Geodaetica 12:268–281

Campbell J et al. (1988) IRIS BULLETIN A. Issued monthly by the Subcommission IRIS, NGS/NOAA, Rockville, USA

Carter WE, Robertson DS, Fallon FW (1988a) Polar Motion and UT1 Time Series Derived from VLBI Observations. Bur Int l'Heure Annu Rep Paris 1987:D-23

Carter WE, Robertson DS, Nothnagel A, Nicolson GD, Schuh H, Campbell J (1988b) IRIS-S: Extending Geodetic VLBI Observations to the Southern Hemisphere. J Geophys Res

Eubanks TM, Steppe, JA, Dickey JO (1988) The Atmospheric Excitation of Rapid Polar Motion. In: Babcock AK, Wilkins GA (eds) Proc IAU Symp 128, Coolfont, Oct. 20–24, 1986. Reidel, Dordrecht, pp 365–371

Kolaczek B, Kosek W (1985) Analysis of Short Periodical Variations of Pole Coordinates Determined by Different Techniques in the MERIT Campaign. Proc Int Conf Earth Rotation and the Terrestrial Reference Frame, Columbus, Ohio, Vol. 2, Dep Geod Sci Surv, Ohio State Univ, Columbus Ohio 43210-1247, pp 505–523

Morabito DD, Steppe JA, Eubanks TM (1988) Earth Orientation Results from VLBI at JPL. Bur Int l'Heure Annu Rep Paris 1987:D-13

Nothnagel A, Nicolson GD, Schuh H, Campbell J, Kilger R (1987) Daily Pole Positions Monitored by Very Long Baseline Interferometry. IUGG XIX General Assembly, Vancouver, Canada, Aug 9–22, 1987

Schuh H (1981) Zur Spektralanalyse von Erdrotationsschwankungen. Veröff Bayer Komm Int Erdmess Bayer Akad Wiss Astron-Geod Arb, München 41:176–193

Schuh H (1987) Analysis of UT1 Observations by VLBI for the Determination of the Love Number k. Proc 5th Working Meeting on European VLBI for Geodesy and Astrometry. Wettzell, Nov 7–8, 1986. Mitt Geodätischen Inst Rhein Friedr-Wilhelm Univ Bonn, 71:61–72

Schuh H (1988) Analysis of UT1 Determinations by VLBI within Project IRIS. In: Babcock AK, Wilkins GA (eds) Proc IAU Symp 128, Coolfont, Oct. 20–24, 1986. Reidel, Dordrecht, pp 171–180

Yoder CF, Williams JG, Parke ME (1981) Tidal Variations of Earth Rotation. J Geophys Res 86:881–891

Yoshino T, Hama S, Shiomo T, Campbell J, Cloppenburg H, Schuh H, Kilger R (1986) First VLBI Experiments between Kashima and Wettzell for Monitoring UT1. Adv Space Res 6 (9):13–16

Yoshino T et al. (1987) The IRIS-P VLBI Experiment for the Earth Rotation Measurement and Time Comparison. 4th Asia-Pacific Regional Meeting of the IAU, Beijing, Oct 1987

Relativity and the Earth's Rotation

E. Gill[1], M. Soffel[1], H. Ruder[1], and M. Schneider[2]

1 Introduction

It is now widely appreciated that relativity plays an increasing role in the fields of astrometry, celestial mechanics and geodesy (see, e.g., Soffel 1989). Typical orders of magnitude associated with special and general relativistic effects are given by $(v/c)^2$ and $(GM/c^2 r)$ respectively, where v is a typical velocity of a body in an appropriate coordinate system, c is the speed of light; G is the gravitational constant, M the mass of a gravitating body and r the (coordinate) distance to it. E.g., for the motion of the Earth about the Sun $(v_{\oplus}/c)^2 \sim (GM_{\odot}/c^2\,\mathrm{A.U.}) \sim 10^{-8}$, whereas for the Earth's rotation velocity $(v_r/c)^2 \sim 10^{-12}$.

Considering the consequences in astrometry, e.g., the angle of light deflection in the gravitational field of the Sun is of order

$$\delta\varphi \simeq 2\,\frac{GM_{\odot}}{c^2\,\mathrm{A.U.}} \simeq 4'' \times 10^{-3}$$

for light rays incident at about 90° from the Sun, a relatively large effect for modern VLBI standards. For precise geodetic VLBI measurements, used, e.g., for the determination of the Earth's rotation parameters, even the light deflection in the gravitational field of the Earth should be taken into account. More precisely, in VLBI it is not directly the light deflection that is measured but instead the associated effect in the time domain, called the gravitational time delay. For a baseline of ~6000 km this time delay caused by the Earth itself has a maximum amplitude of ~0.02 ns.

In celestial mechanics modern numerically produced ephemerides of our solar system, such as the international reference ephemeris DE200 from JPL, are based on Einstein's equations of motion for an N-body system in the so-called post-Newtonian approximation, i.e., they solve the Einstein-Infeld-Hoffmann equations of motion.

In geodesy relativistic effects are taken into account in the problem of time comparison (e.g., by clock transport, VLBI or satellite techniques) or time dissemination (e.g., GPS system). According to the gravitational redshift the rate of some clock located at a height h above mean sea level differs from that of a clock on the geoid by a factor

[1] Lehrstuhl für Theor. Astrophysik, Auf der Morgenstelle 12, D-7400 Tübingen, FRG.
[2] Technische Universität München, Arcisstraße 21, D-8000 München 2, FRG.

$$1+gh/c^2 \simeq 1+(h/\mathrm{km})\times 10^{-13} \ ,$$

where g denotes the Earth's gravity.

In a relativistic framework problems related to the Earth's rotation have to be treated by means of the so-called space-time *metric.* This metric tensor $g_{\mu\nu}$ generalizes the Newtonian potential U and accounts for the effects from special relativity. As a tensor of second rank it contains more degrees of freedom; it not only describes electric-type gravity related to U but also magnetic-type gravity due to gravito-magnetic forces induced by moving (e.g., rotating) matter. In the (parametrized) post-Newtonian approximation in standard PN-coordinates $x^0 = ct$, $x^1 = x$, $x^2 = y$, $x^3 = z$ the components of the metric tensor for an isolated, rotating body can be written as $(i,j = 1,2,3)$

$$g_{00} = -1+2\frac{U}{c^2}-2\beta\left(\frac{U}{c^2}\right)^2 \tag{1a}$$

$$(g_{0i}) = -(\gamma+1)G\cdot\frac{\mathbf{J}\times\mathbf{x}}{c^3|\mathbf{x}|^3} \tag{1b}$$

$$g_{ij} = \left(1+2\gamma\frac{U}{c^2}\right)\delta_{ij} \ . \tag{1c}$$

Here, all indices are space-time indices. The post-Newtonian parameters β and γ, that take the value of unity in Einstein's theory of gravity, describe the non-linearity of gravity and space curvature produced by unit rest mass, respectively. Those parts containing the Newtonian potential U arise from the static mass distributions; they lead to electric-type gravitational effects. In contrast to this, the g_{0i}-terms are induced by the Newtonian angular momentum $\mathbf{J}$ of the field generating source and give rise to magnetic-type gravity. In this geometrical framework of describing gravitational fields the tidal forces responsible for the external torques acting on the Earth's angular momentum are contained in the (Riemann) curvature tensor $R^{\mu}_{\ \nu\lambda\sigma}$ $(\mu = 0,1,2,3;\text{etc.})$ associated with the metric $g_{\mu\nu}$. In the Newtonian limit the only non-vanishing components of this curvature tensor are given by

$$R^i_{\ 0j0} = -\frac{\partial^2 U}{\partial x^i \partial x^j} \ . \tag{2}$$

Hence, to lowest order the curvature tensor of space-time agrees with the usual Newtonian tidal force tensor. To higher orders the Riemann tensor contains all kind of corrections due to electric- or magnetic-type gravity effects.

2 Post-Newtonian Effects: Free Rotation of the Earth

Let us now turn to some problems related with the Earth's rotation. In the absence of external torques the angular momentum vector of the Earth would be parallel for all times just as in the Newtonian case. However, the parallelity now refers to the space-time metric $g_{\mu\nu}$ with the consequence that the spin-vector $\mathbf{S}_1$ of an individual body in a system of N ($N>1$) extended and rotating bodies will not stay constant in time. If the body (Earth, mass M_1) under consideration is revolving about some other body (Sun, mass M_2) then $\mathbf{S}_1$ will precess due to a spin-orbit coupling with angular velocity

$$\boldsymbol{\Omega}^{(1)}_{\text{S-O}} = \frac{GM}{c^2|\mathbf{x}|^3}\left[(\gamma+1)+\left(\gamma+\frac{1}{2}\right)\frac{M_2}{M_1}\right]\mathbf{x}\times\mathbf{v}\ , \tag{3}$$

(Barker and O'Connell 1970, 1975; $M = M_1 M_2/(M_1+M_2)$, $\mathbf{x} = \mathbf{x}_1 - \mathbf{x}_2$, etc.) which is directed perpendicular to the orbital plane (ecliptic). In (3), the γ-dependent terms indicate the general relativistic contributions, the γ-independent terms arise due to the (special) relativistic kinematics. For the Earth in orbit about the Sun this spin-orbit coupling leads to precessional motion of the Earth's spin axis, with the time-dependent accumulated angle $\boldsymbol{\Phi}^{\oplus}_{\text{S-O}}$ being given by ($M_1 = M_\oplus \ll M_2 = M_\odot$) (Voinov 1988)

$$\boldsymbol{\Phi}^{\oplus}_{\text{S-O}} = \int \Omega^{\oplus}_{\text{S-O}}dt = \left(\gamma+\frac{1}{2}\right)\frac{GM_\odot}{c^2 a(1-e^2)}(f+e\sin f)\ . \tag{4}$$

Here, a, e and f denote the semi-major axis, eccentricity and true anomaly of the Earth's orbit, respectively. Note, that (3, 4) contain the so called geodetic precession with frequency

$$\langle\boldsymbol{\Omega}^{\oplus}_{\text{S-O}}\rangle = \left(\gamma+\frac{1}{2}\right)\frac{GM_\odot n}{c^2 a(1-e^2)} \tag{5}$$

($n^2a^3 = GM_\odot$) as well as nutational motion. For the Earth in orbit about the Sun the "geodetic precession" amounts to $\sim 2''$/century; in the IAU precession matrix (Lieske 1979) this geodetic precession has already been taken into account. The same precession applies for the angular momentum vector of the Earth-Moon system leading to a secular drift in the lunar perigee and node of $2''$/century. Shapiro et al. (1988) have experimentally determined this precession from lunar laser-ranging data. Their result confirms the prediction from Einstein's theory of gravity within an estimated error of $0.04''$/century.

To study the problem of Earth rotation in the post-Newtonian approximation the choice of the reference systems (coordinates) becomes particularly important. For many purposes it is convenient to solve the relevant equations of motion in the barycentric (standard post-Newtonian) coordinate system (x^0, x^i). However, especially for the problem of Earth's rotation it is much more useful to employ a "good" geocentric coordinate system. If "simple" geocentric coordinates, defined as differences of barycentric coordinates, are used then the final result for the Earth's rotational motion would contain many relativistic terms that disappear, when a coordinate system is employed whose grids are moving together with the geocenter (Brumberg and Kopejkin 1988, 1989). A transition from barycentric coordinates (x^0, x^i) to a "good" (in accordance with the principle of equivalence) geocentric reference system (GRS) with coordinates $(X^0 = \tau, X^i)$ is obtained to linear order in X^i by (Voinov 1988)

$$x^{a} = x^{a}_{\oplus}(\tau) + e^{a}_{(m)} X^{m} + O(X^{i} X^{i}) \ , \tag{6}$$

where the tetrad vectors $e^{\mu}_{(\alpha)}$ are parallel-transported along the geocenter, just like the Earth's spin axis *in the absence of torques.* Here, $e^{a}_{(m)}$ describe the local inertial axes and the "good" geocentric coordinates as given by (6) are non-rotating with respect to the local inertial frame. To sufficient accuracy the inertial axes are given by (Voinov 1988)

$$e^{0}_{(i)} = v^{i}_{\oplus}$$

$$e^{i}_{(j)} = \Gamma^{m}_{j} \tilde{e}^{i}_{(m)}, \ \tilde{e}^{i}_{(m)} = \delta^{i}_{m}[1 - U(\mathbf{x}_{\oplus})] + \frac{1}{2} v^{i}_{\oplus} v^{m}_{\oplus} \ , \tag{7}$$

where Γ^{m}_{j} represents a rotation about the $\mathbf{e}_{(z)}$ axis with angle $\mathbf{\Phi}^{\oplus}_{\mathrm{S-O}}$ and accounts for the geodetic precession. The term containing the Newtonian potential U is due to normalization of the tetrad with respect to the metric, whereas the velocity-dependent terms arise from the transition to the comoving frame and are therefore of special relativistic origin.

In analogy to the corresponding situation in atomic physics there is also a spin-spin coupling due to the gravito-magnetic interaction of $\mathbf{S}_1$ with the spin $\mathbf{S}_2$ of a second body with precession frequency (Barker et al. 1975, $\mathbf{n}_2 = \mathbf{S}_2/|\mathbf{S}_2|$)

$$\mathbf{\Omega}^{(1)}_{\mathrm{S-S}} = \frac{(\gamma+1)}{2} \frac{GS_2}{c^2 |\mathbf{x}|^3} \left(\frac{3\mathbf{x}(\mathbf{n}_2 \cdot \mathbf{x})}{|\mathbf{x}|^2} - \mathbf{n}_2 \right) . \tag{8}$$

For the solar spin interacting with the Earth's spin the amplitude of this precession

$$\langle \Omega_{\mathrm{S-S}} \rangle \sim 0.03 \left(\frac{GM_{\odot}}{c^2 R_{\odot}} \right) \left(\frac{R_{\odot}}{\mathrm{A.U.}} \right)^3 \omega_{\odot} \tag{9}$$

is about 10^{-5}″/century and will not be measurable in the foreseeable future.

3 Post-Newtonian Effects: Forced Precession and Nutation

Recently new mathematical techniques have been developed that can be employed to start with the formulation of a post-Newtonian theory of forced precession and nutation. Roughly speaking, one constructs the metric tensor in the vicinity of the body under consideration (Earth) and matches this local tensor with some appropriate global metric resulting from the other bodies in the system (Moon, Sun and planets)[3]. Using these arguments one can show that the time dependence of $\mathbf{S}_1$ (neglecting the spin of other bodies) quite generally should be given by an equation of the form (Misner et al. 1973; Thorne and Hartle 1985; Voinov 1988):

---

[3] The same technique has been used to construct a "good" geocentric coordinate system (Brumberg and Kopejkin 1988, 1989).

$$\frac{DS^{(i)}}{D\tau} = \varepsilon^{ijk} \hat{I}_{(j)(m)} W_{(m)(k)} \tag{10}$$

where higher mass- and all current-multipole moments have been neglected. Here, $\hat{I}_{(j)(m)}$ is a generalized, symmetric and trace-free inertia-tensor ("reduced quadrupole moment") and $W_{(m)(k)}$ is the tidal force tensor as given by the curvature of space-time produced by the Moon, Sun and planets

$$W_{(m)(k)} = R_{\alpha\beta\gamma\delta} e^{\alpha}_{(0)} e^{\beta}_{(m)} e^{\gamma}_{(0)} e^{\delta}_{(k)} \ . \tag{11}$$

$e^{\mu}_{(0)} = u^{\mu} = dx^{\mu}/d\tau$ and τ is the (proper-) time a geocentric clock would indicate. To lowest order (10) takes the familiar Newtonian form [$\mathbf{x}$ and $\mathbf{v}$ refer to the center of mass of body (1)]:

$$\begin{aligned}\frac{dS_1^i}{dt} &= \varepsilon_{ijk} \int_{(1)} \varrho x^j \frac{dv^k}{dt} d^3x = \varepsilon_{ijk} \int_{(1)} \varrho x^j \frac{\partial U}{\partial x^k} d^3x \\ &\simeq \varepsilon_{ijk} \int_{(1)} \varrho x^j x^l d^3x \times \left.\frac{\partial^2 U}{\partial x^k \partial x^l}\right|_{CM} = \varepsilon_{ijk} \hat{I}_{jl} R^l_{\ 0k0} \ .\end{aligned} \tag{12}$$

In (10) relativity enters at various places:

- clock rates (rate of corotating clock $\neq$ rate of non-rotating clock; gravitational red-shift effects, etc.),
- corrections to the components of the inertia-tensor, e.g., due to rotational energy or gravitational energy,
- corrections to the Newtonian tidal force tensor.

Electric-type corrections to the tidal acceleration produced by some mass M at a (coordinate) distance r typically are of order $\eta = GM/c^2 r$ times the Newtonian value, which is $\sim 50''$/year for the luni-solar precession. In the vicinity of the Earth $\eta_{\odot} \simeq 10^{-8}$ ($\eta_{☾} \simeq 10^{-13}$) and leads to a residual luni-solar relativistic precession of order $\simeq 10^{-6}{}''$/year, whereas the nutation amplitudes are $< 10^{-6}{}''$. Corrections from magnetic type gravity due to the angular momenta of the tide raising bodies can safely be neglected in the solar system.

Whereas in the preceding discussion a Newtonian definition of the angular momentum (spin) of an individual body was sufficient, this is no longer the case for a discussion of the motion of body-fixed axes. In the full relativistic framework such a definition is highly problematic; however, in the (first) post-Newtonian approximation there is general agreement of how the spin of an individual body can be defined in suitably chosen coordinates. Other definitions important for a theory of Earth rotation, such as rigidity, require some care to restrict the propagation velocity of a signal to the speed of light. In standard (PN) inertial coordinates a body is called quasi-rigid if there is another coordinate system rotating with angular velocity $\mathbf{\Omega}$ with respect to the rest-frame in which the material particles of the body are at rest, as in the Newtonian case. In inertial coordinates the motion of particles of some quasi-rigid body is then simply given by

$$\mathbf{v} = \mathbf{\Omega} \times \mathbf{x} \ . \tag{13}$$

For the free rotation of a quasi-rigid model Earth the (PN) Euler-equations formally agree with the Newtonian version and one finds the usual Poinsot-motion, the components of the inertia-tensor only renormalized by post-Newtonian "corrections terms" due to the fact that all kinds of energy (rotational energy, gravitational energy, etc.) contribute to the inertial mass (Schastok et al. 1987).

4 Relativity Experiments and the Earth's Rotation

Not only might relativity play some role in future accurate descriptions of the Earth's rotation but the Earth's rotation can be used for various relativity experiments involving the Earth's gravito-magnetic field. Because of the spin-spin coupling to the Earth's angular momentum and its corresponding gravito-magnetic field the spin axis of a (torque-free) gyroscope in orbit should precess with respect to the fixed stars. For the spin axis of such a gyroscope in a low altitude polar orbit oriented perpendicular to the orbital plane this precession amounts to $\sim 0.5''$/year. Measuring this effect is one goal of the well-known Stanford gyroscope experiment (NASA's gravity probe B, GPB; see, e.g., Everitt 1974; Lipa and Everitt 1978). The same effect leads to a precession of satellite orbits in space; for the LAGEOS's orbit the corresponding nodal drift is of order $2'' \times 10^{-5}$/revolution, roughly comparable with the effect from the $l = 12$ multipole of the Earth's gravity field.

An interesting idea to measure this effect is by means of a laser gyroscope. In such a device we find two laser beams traveling around a closed circuit in opposite sense and some means to extract an interference pattern from the system. For an Earth fixed laser gyroscope the Sagnac phase shift is given by (Scully et al. 1981; Soffel 1989)

$$\Delta\Phi \simeq \frac{8\pi}{\lambda}\,\frac{\mathbf{A}\cdot\mathbf{\Omega}}{c}\,, \tag{14}$$

where λ is the mean wavelength of the laser beams, $\mathbf{A}$ denotes the 3-vector of oriented area enclosed by the laser beams and

$$\mathbf{\Omega} = \mathbf{\Omega}_{\oplus} + \mathbf{\Omega}_{\mathrm{T-P}} + \mathbf{\Omega}_{\mathrm{S-O}} + \mathbf{\Omega}_{\mathrm{S-S}} \tag{15}$$

with

$$\mathbf{\Omega}_{\mathrm{T-P}} = -\frac{1}{2}\,\mathbf{v}\times\mathbf{a}/c^2 \tag{16}$$

$$\mathbf{\Omega}_{\mathrm{S-O}} = \left(\gamma + \frac{1}{2}\right)\mathbf{v}\times\nabla U/c^2 \tag{17}$$

$$\mathbf{\Omega}_{\mathrm{S-S}} = -\frac{(\gamma+1)}{2}\,\nabla\times\left(\frac{G}{c^2}\,\frac{\mathbf{J}_{\oplus}\times\mathbf{x}}{r^3}\right)\,. \tag{18}$$

The $\mathbf{\Omega}_{\oplus}$-term results from the classical Sagnac effect and the first term containing the non-gravitational acceleration $\mathbf{a}$ describes the Thomas precession, that is known for the electronic spin in the atom. The other post-Newtonian corrections result from spin-orbit and spin-spin coupling, respectively. These are about 10^{-10} times smaller than the classical part of the Sagnac phase shift.

It is unlikely that such high accuracies needed for relativity experiments will be achieved in the near future. However, even in the near future the use of several Sagnac devices for the determination of the Earth's rotation parameters with a short ($\ll$day) integration time looks promising. One experiment in this direction performed at the Max-Planck institute for quantum optics (Garching, FRG) (Leuchs, Ruder, Schneider) is presently in the phase of testing. It uses an Argon ion laser with a power of $P \sim 4$ W, mounted on a steel plate that is allowed for pivoting and swivelling up to 45° to get a maximum signal of the Earth's rotation. This happens when the plate is parallel to the Earth's equator. Together with several high-quality mirrors the laser constitutes an active resonator enclosing an area of $\sim 5\,\mathrm{m}^2$. High measuring accuracies of $\boldsymbol{\Omega}$ requires

- misalignment control due to thermal stresses ($\delta T \sim 10^{-3}$ K)
- mirror reflection coefficients $R \geq 0.9995$
- a vacuum environment to avoid perturbations due to air density fluctuations.

The Earth's rotation causes a beating of the counterpropagating laser beams with frequency of ~ 100 Hz. However, a signal of a few kHz is observed due to the plasma drift in the laser tube. This plasma drift has the advantage of avoiding the frequency locking of the two laser modes, but the irregular drift, on the other hand, produces an unstable signal, giving rise to calibration problems. It is planned to replace the active laser gyroscope by a passive one in the near future.

The relative measuring accuracy of the component of the Earth's angular momentum vector perpendicular to the Sagnac platform at the shot noise limit is determined by the reflection coefficient R, the enclosed area A, the integration time τ and the laser power P:

$$\frac{\delta\Omega}{\Omega} \simeq 10^{-3}(1-R)\frac{1}{A\,[\mathrm{m}^2]}\frac{1}{\sqrt{\tau[\mathrm{s}]}}\frac{1}{\sqrt{P[\mathrm{W}]}}\,. \tag{19}$$

For $A \sim 5\,\mathrm{m}^2, \tau \sim 100$ s and $P \sim 1$ W a relative accuracy of

$$\frac{\delta\Omega}{\Omega} \sim 10^{-8} \triangleq 1\ \mathrm{ms/d}$$

should be achievable in the test experiment. In a future Sagnac station with $A \sim 100\,\mathrm{m}^2, \tau \sim 400$ s and $P \sim 4$ W

$$\frac{\delta\Omega}{\Omega} \sim 10^{-10} \triangleq 0.01\ \mathrm{ms/d}$$

might be feasible. Another idea to construct a large Sagnac device for the measurement of the Earth's rotation parameters is to modify the planned gravity wave detector at the MPQ, with physical arm lengths of ~ 3 km. Closing the envisaged interferometer geometry to a closed triangle would lead to a powerful Sagnac instrument with $A \sim 5 \times 10^6\,\mathrm{m}^2, \tau \sim 100\,\mathrm{s}$ and $P \sim 4\,\mathrm{W}$ with a measuring accuracy of about

$$\frac{\delta\Omega}{\Omega} \sim 10^{-14} \triangleq 1\,\mathrm{ns/d} \ .$$

For such a device relativity effects would play an important role.

References

Barker BM, O'Connell RF (1970) Derivation of the equations of motion of a gyroscope from the quantum theory of gravitation. Phys Rev D2:1428

Barker BM, O'Connell RF (1975) Gravitational two-body problem with arbitrary masses, spins, and quadrupole moments. Phys Rev D12:329

Brumberg VA, Kopejkin SM (1988) Reference systems. In: Kovalevsky J, Mueller II, Kolaczek B (eds) Relativistic theory of celestial reference frames. Reidel, Dortrecht, p 115

Brumberg VA, Kopejkin SM (1989) Relativistic reference systems and motion of test bodies in the vicinity of the earth. Nuovo Cimento 103B:63

Everitt CWF (1974) Experimental gravitation (Proc of Course 56 of the Int School of Physics "Enrico Fermi"). In: Bertotti B (ed) The gyroscope experiment I: general description and analyses of gyroscope performance. Academic Press, New York

Lieske JH (1979) Precession matrix based on IAU (1976) system of astronomical constants. Astron Astrophys 73:282

Lipa JA, Everitt CWF (1978) The role of cryogenics in the gyroscope experiment. Acta Astronautica 5:119

Misner CW, Thorne KS, Wheeler JA (1973) Gravitation. Freeman, San Francisco

Scully MO, Zubairy MS, Haugan MP (1981) Proposed optical test of metric gravitation theories. Phys Rev A24:2009

Schastok J, Soffel MH, Ruder H, Schneider M (1987) The post-Newtonian rotation of the Earth: a first approach. In: Wilkins G, Babcock A (eds) The Earth's rotation and reference frames for geodesy and geodynamics. Proc of the IAU Symp No 128

Shapiro II, Reasenberg RD, Chandler JF, Babcock RW (1988) Measurement of the de Sitter precession of the Moon: a relativistic three-body effect. Phys Rev Lett 61:2643

Soffel MH (1989) Relativity in Astrometry, Celestial Mechanics and Geodesy. Springer, Berlin, Heidelberg, New York, Tokyo

Thorne KS, Hartle JB (1985) Laws of motion and precession for black holes and other bodies. Phys Rev D31:1815

Voinov AV (1988) Motion and rotation of celestial bodies in the post-Newtonian approximation. Celest Mech 42:293

Astrometric Observations of Hevelius and Derived Values of ΔT (Dynamical time – Universal time)

J. Wünsch

1 Introduction

About 1500 meridian altitudes of the sun observed by Johannes Hevelius (1611–1687) at Danzig in the years 1652–1679 and about 1160 distances of fixed stars from the lunar limb obtained in 1658–1679 as well as 48 occultations of stars by the moon were analyzed with the aim to obtain a value of the time difference $\Delta T = ET - UT$ between ephemeris time and universal time for the time of Hevelius. This time difference is a measure of the "clock error" of the rotation of the Earth caused mainly by secular deceleration due to tidal friction.

The observations of Hevelius mentioned above are listed by Hevelius (1673, 1679, 1685). Figures of the instruments (see Fig. 1) are also reproduced by Repsold (1908). Ephemeris time (or, from 1984 onwards, dynamical time TD) is a theoretically uniform measure of time, while UT is measured relative to the slightly non-uniform rotational phase of the Earth. Accurate observations which can be used for the determination of ΔT in the seventeenth century are very rare. For comparison with the observations the ephemeris DE 102 – LE 51 of the Jet Propulsion Laboratory (JPL) was used, which is described by Newhall et al. (1983). The differences (observed – calculated) = (O–C) in the measured coordinates were formed and graphically displayed as a function of time T, right ascension and declination for the sun; and as a function of time, position angle and measured distance for the lunar distances and occultations.

2 The Solar Observations

The solar observations obtained with the instrument Quadrans Azimuthalis were corrected for daily parallax as well as for refraction with monthly mean temperatures for Gdingen, near Danzig, according to: *Brockhaus Texte und Tabellen* (1983). A value of the geographical latitude of Hevelius' observatory of $\varphi = +54°\ 21'\ 3''$ was used, after Rybka (1984). The equation of time was taken into account carefully.

Solar observations of Hevelius have been analyzed earlier by Tupman (1900) who used only a very small sample and by Wittmann (1984) for the determination of the obliquity of the ecliptic. The definition of ΔT and general aspects of the determination of ΔT from solar observations are reviewed, e.g., by Lambeck (1980).

Universitäts-Sternwarte Bonn, Auf dem Hügel 71, D-5300 Bonn 1, FRG.

Fig. 1. The backside of the great quadrant "Quadrans Azimuthalis" is depicted in this figure from "Machina Coelestis" (Hevelius 1673). A diagonal arm holding the vertical plane and a counterweight below are seen. Strings to move the alhidade and a plumb line are visible near the hand of the observer. The instrument was used mainly to measure meridian heights of stars and the sun. This reproduction is from the copy held by the Astronomisches Institut der Universität Bonn, FRG

The O−C in δ were analyzed by a variety of least squares calculations with different data sets and assumptions. The well-known equation (e.g., Schmeidler 1965) for the orbit of the sun is:

$$\Delta\delta = -D + \sin\varepsilon\cos\alpha\ \Delta L' + \sin\alpha\ \Delta\varepsilon + 2\cos\alpha\sin\delta\ \Delta h' - 2\sin\varepsilon\cos^2\alpha\cos\delta\ \Delta k'$$

where α and δ are right ascension and declination of the sun, ε is the obliquity of the ecliptic. The five unknowns D, $\Delta L'$, $\Delta\varepsilon$, $\Delta h'$ and $\Delta k'$ have the following meaning:

D is a correction to geographical latitude and the direction to the zenith, $\Delta L'$ is the correction to the mean longitude of the sun which corresponds to the time difference ΔT:

$$\Delta T\,[\mathrm{s}] = 24.35\ \Delta L'\,['']\ ;$$

the factor 24.35 arises from 365.2422/15.

$\Delta\varepsilon$ is a correction to the obliquity of the ecliptic, $\Delta h'$ and $\Delta k'$ are corrections to the elements of the orbit of the Earth.

A calculation with 1490 data gives $\Delta L' = -1''\!.6 \pm 4''\!.3$. If the yearly mean of the $\Delta\delta$ is subtracted from each $\Delta\delta$, one obtains $\Delta L' = -0''\!.4 \pm 3''\!.8$. This value was selected as the final result from the solar observations. It is equivalent to a time difference of $\Delta T = -9^{\mathrm{s}}\!.7 \pm 93^{\mathrm{s}}$.

The standard deviation of a single measurement of solar altitude is $\pm 45''$. From series of several solar altitudes measured near the meridian on the same day, it was found that the instrument Quadrans Azimuthalis had an azimuth error. It was turned $-0^\circ\!.47 \pm 0^\circ\!.13$ from the exact direction to the South (true azimuth = measured azimuth + 0°.47). This was taken into account in the reduction of the solar altitudes. The differences $\Delta\delta$ = (observed − calculated declination) when studied as a function of right ascension α contain besides the effect which is sought for, with a period of 1 year (unknowns $\Delta L'$ and $\Delta\varepsilon$), further variations with a period of half a year (formally corresponding to $\Delta h'$ and $\Delta k'$) which are interpreted as a seasonal effect of the instrument caused by temperature variations. Thus, for the solar observations of Hevelius, the unknowns $\Delta h'$ and $\Delta k'$ are merely a conventional representation of systematic errors. Solutions with only two unknowns, D and $\Delta L'$, were computed also, but we consider the above the best result, because semiannual variations are filtered out. The result for ΔT has, regrettably, a very great standard deviation, so that a reliable numerical value for the secular decrease of the angular velocity of rotation of the Earth $\dot{\omega}/\omega$ cannot be found from these observations. However, as shown, interesting and valuable insights into seventeenth-century astrometry have been gained.

The subdivision of the solar observations into seven groups of 4 years each gives very different results for different groups. If the data are distributed in two groups only, a decrease of ΔT from the first to the second group is indicated:

group 1, 1652 – 1664, N = 665, $\Delta T = +114^{\mathrm{s}}\!.4 \pm 136^{\mathrm{s}}$
group 2, 1665 – 1679, N = 825, $\Delta T = -172^{\mathrm{s}}\!.9 \pm 136^{\mathrm{s}}$.

Calculations with a seasonal variation of azimuth error were also tried but did not give more reliable results.

Rybka (1984), in his Table 14, gives division errors of the instrument Quadrans Azimuthalis which he derived from the star catalog of Hevelius. The declinations of the stars were determined with the same instrument as the solar altitudes. If corrections for these division errors are applied, the results for the unknowns change appreciably and a $\Delta L'$ of $+0''\!.3 \pm 4''\!.2$ is obtained. However, this does not improve the accuracy significantly.

3 The Lunar Observations

The lunar distances were measured with the instrument Sextans Magnum Orichalcicus. For the reduction of the lunar distances, daily parallax of the moon and refraction were considered and an apparent star place was calculated. Hevelius checked the readings of his pendulum clocks by measuring typically four star altitudes per night. The standard deviation of the resulting clock correction is about $\pm 20^s$.

The residuals (O−C) of the lunar distances show an interesting difference between the measurements to the near and far lunar limb which is interpreted as the effect of a filar breadth. Different subsets of the data were used (1652−1668, 1674−1679, near limb, far limb) to find the term $x_c = -\Delta L$ depending on the cosine of the position angle. A small sine term obtained in addition was interpreted as flexure of the sextant perpendicular to its plane. From the lunar distances, an apparent difference ΔL of the moon along its orbit of $\Delta L = +7''.0 \pm 9''.6$ relative to the ephemeris is found from a subset of 300 data; this is equivalent to $\Delta T = +12^s.7 \pm 17^s.5$ (factor 1.821 for the lunar orbit). The mean post-fit residual of a single distance is about $\pm 120''$, which is relatively high, though the internal scatter of a series of measurements of the same star is only about $\pm 30''$. Furthermore, one has to remember that the calculated places of the moon depend on the secular acceleration of the moon $\dot{n}$. The ephemeris DE 102−LE 51 is based on a numerical value of $n = -26''.21/cy^2 \pm 2''.0/cy^2$ (Newhall et al. 1983, p. 157). The most recent value of n is $-24''.9/cy^2 \pm 1''.0/cy^2$ (Newhall et al., this Vol.).

In general, one has $\Delta L = n\Delta T + 0.5\,(\Delta\dot{n})\,(T_{1975})^2$ where $\Delta\dot{n}$ is a correction to the value of $\dot{n}$ used, n is the mean motion of the moon ($0''.549/s$) and T_{1975} is the time in Julian centuries measured from 1975.0.

In a diagram of the (O−C) versus measured distance (between 0° and 60°) no clear effect of division errors of Sextans Magnum Orichalcicus could be recognized which should appear as a unique curve with only a small scatter around it.

The reduction of the 48 lunar occultations which were timed by telescopic observation and have been studied earlier by Newcomb (1878) and Martin (1969), gives a value of $\Delta T = +22^s.8 \pm 6^s.9$. The mean of the three values (sun, moon, occultations) is $+8^s.6 \pm 16^s.6$, a weighted mean $+21^s.7 \pm 4^s.7$, which is the final result.

The result for the lunar occultations and the weighted mean lie on the smoothed curve for $\Delta T(T)$ by Stephenson and Morrison (1984) and about 40 s below extrapolated values for the curve by Krasinsky et al. (1985). The result from the lunar distances is about one standard deviation smaller than the smooth curve by Stephenson and Morrison (1984). It is very likely that ΔT in 1669 was about 15 s smaller than the value given by these authors (see Fig. 2).

The weighted mean shows a negative fluctuation of about 40 s relative to a mean parabola as a function of T:

$$\Delta T = (T_{1820})^2\, 25^s.5/cy^2$$

as given by Stephenson and Morrison (1984, page 57 and their Figure 5).

The difference is not proved with certainty, but has a strong likelihood. It is further supported by the fact that it continues the trend of the plot of the fluctuations in ΔT by Spencer Jones (1939, Figure 1) before 1680.

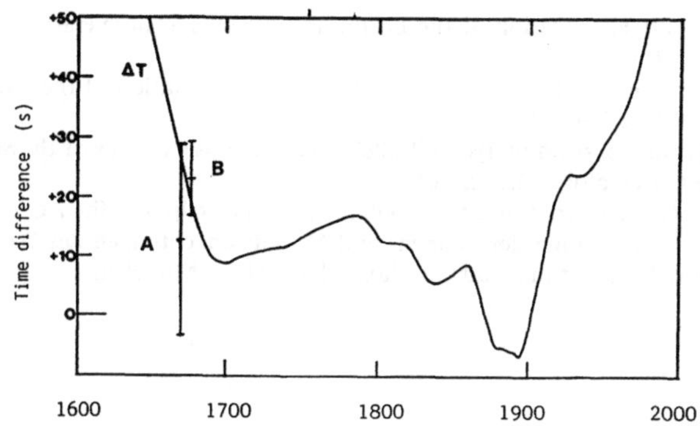

Fig. 2. Smoothed curve of the time difference $\Delta T(T)$ in [s], derived mainly from timings of lunar occultations (Stephenson and Morrison 1984); abscissa is the year; results of the present work are shown with *error bars: **A***, lunar distances; ***B***, occultations

For full details see Wünsch (1987).

It may be remarked that Hevelius also observed distances of planets from fixed stars which might be interesting to study and to compare with a modern ephemeris to obtain the standard deviation of (O – C) as a first step.

Acknowledgments. I thank Dr. M.E. Standish (Jet Propulsion Laboratory) for providing a copy of the ephemeris DE 102 – LE 51, Prof. R.P. Kudritzki for support and Prof. F. Schmeidler, who guided this work, for helpful discussions. Comments are also due to Dr. K.P. Hartmann, Dr. K.P. Simon and O. Montenbruck. This work was supported by the Ludwig-Maximilians-Universität München from the Friedrich-Geissendörfer-Stiftung and the Stiftung Przemysler-Przemyslav and by the Bayerischen Staatsministerium für Unterricht und Kultus with a Graduierten-Stipendium.

References

Brockhaus Texte und Tabellen (1983) Länder und Klima. Europa, UdSSR. Zsstellung d. Tab.: Grünewald G, Höller E, Stranz D, Wiesbaden

Hevelius J (1673, 1679) Machina Coelestis, Danzig. Nachdruck: Schmeidler, Osnabrück 1969

Hevelius J (1685) Annus Climactericus. Danzig

Krasinsky GA, Saramonova EY, Sveshinkov ML, Sveshnikova ES (1985) Universal time, lunar tidal deceleration and relativistic effects from observations of transits, eclipses and occultations in the XVIII – XX centuries. Astron Astrophys 145:90

Lambeck K (1980) The earth's variable rotation. Geophysical causes and consequences. Cambridge University Press, Cambridge

Martin CF (1969) A study of the rate of rotation of the Earth from occultations of stars by the Moon 1627 – 1860. Ph.D. thesis, Yale Univ

Newcomb S (1878) Researches on the motion of the moon. I. Astron Met Observ Washington

Newhall XX, Standish EM Jr, Williams JG (1983) DE 102: a numerically integrated ephemeris of the Moon and planets spanning forty-four centuries. Astron Astrophys 125:150

Repsold JA (1908) Geschichte der astronomischen Meßwerkzeuge I. W. Engelmann, Leipzig

Rybka P (1984) Katalog Gwiazdowy Heweliusza. Monografie z dziejów nauki i techniki, Tom CXXXII. Wrocław, Warszawa

Schmeidler F (1965) Methods in meridian astronomy. Vistas Astron 6:69

Schmeidler F (1969) Nachdruck von: Hevelius, Machina Coelestis. Verlag Otto Zeller, Osnabrück

Spencer Jones H (1939) The rotation of the Earth, and the secular accelerations of the Sun, Moon and Planets. M N R Astron Soc 99:541
Stephenson FR, Morrison LV (1984) Long-term changes in the rotation of the Earth: 700 BC to AD 1980. Phil Trans R Soc Lond A 313:47
Tupman GL (1900) A comparison of Tycho Brahe's meridian observations of the Sun with Leverrier's solar tables. The Observatory 23:132; 165
Wittmann A (1984) On the variation of the obliquity of the ecliptic. Mitt AG 62:201
Wünsch J (1987) Die Auswertung der Sonnen- und Mondbeobachtungen des Danziger Astronomen Johannes Hevelius. Dissertation, Ludwig Maximilians Univ, München

Historical Chinese Astronomical Observations

Liu Ciyuan

1 Historical Chinese Astronomical Observations (1)

Astronomy was one of the earliest sciences in human history. People who were in the early stages of civilization needed to determine seasons and time by using the stars. For example, the ancient Egyptians predicted the flood of the Nile by Sirius' appearing in the east sky in the morning.

In ancient China, people paid special attention to astronomy. To issue the calendar was the sign of highest authority. It is well known that the solar calendar is quite simple. It can be determined by the number of days in 1 year. The Julian Calendar is fairly accurate over several centuries (in it 1 year is equal to 365.25 days), while the Gregorian Calendar nearly solved the whole problem (1 year = 365.2425 days). However, the Chinese calendar is a luni-solar one. It employs a solar year and a lunar month, and requires solar eclipses to take place on the first day of the month. This means that accurate calculation of solar and lunar positions is necessary. Therefore, to regulate the Chinese calendar was a quite difficult work. Some hundred kinds of calendars were offered in the past 2000 years (half of them were used in practice). The calendar problem was solved thoroughly only by the spreading of modern Western knowledge of astronomy to China.

Another special characteristic of traditional Chinese astronomy is its complete and abundant records of celestial phenomena. This was closely related to superstitious ideas of the ancient people. They thought seriously that sky phenomena were predictions of human social affairs, particularly related to the rise and fall of states and the fate of families of emperors and important officials. The following story is an example:

There were three states in China in the third century. The king of Wei state asked one of his officials: "Which is the orthodox state?" The official replied: "You should judge this by astronomy". Mars was stationary near the Fire star (α Sco. which was regarded as a bad omen for the emperors). Afterwards, the king of Wei died, while nothing amiss took place in the other two states. From this we can see that Wei was the orthodox state. It is such superstitious ideas that enabled astrology to prosper for a long time and made Astronomers Royal record important celestial phenomena, report them to the emperors, write them down in the Histories and leave us a piece of valuable heritage.

At the very beginning of Chinese civilization, astronomy has already played an important role. The *Shujing* (written in the seventh century B.C.) says: "The King Yao

Shaanxi Astronomical Observatory, Academia Sinica, P.O. Box 18 Lintong, Xian, China.

(living approximately in the twentieth century B.C.) appointed Xi and He to honor the sacred sky, observe the Sun, the Moon and the stars and tell people the time and the seasons. The *Zuozhuan* says: "The fire official Yanbo of King Yao who lived at Shangqiu consecrated the Fire star to keep time". At that time, there had already been special officials for astronomy. In ancient times, the status of astronomers was very high. They were in charge of making documents and regulations, and of recording important events in addition to doing astronomical work. At that time, the office of astronomers was usually hereditary from generation to generation in some families because this occupation was considered as profound and mysterious. Later, their work gradually became concentrated on determining calendars and observing the sky. They were then no longer the most important officials, although they still had high positions.

The *Jiutangshu* recorded the organization of the astronomical bureau in the seventh century: There were 66 officials and 726 staff in the royal observatory, including 90 observers who were monitoring the sky day and night. The director of the observatory was an official third in ranking and being equal to a minister of the government.

In ancient China, astronomy was an imperial scholarship; astronomers served the emperors directly. The emperors usually prohibited the common people from learning astronomy seriously and sometimes even did not allow astronomers to have contact with the common people. The emperors were afraid that someone who was in possession of the knowledge might be in the service of a person who would want to overthrow the government. Also, if the common people knew astrology, they would be able to know the fate of the government and this would lead to unrest. However, amateur astronomers always existed. During the long wars and chaos after the birth of a new dynasty the emperor issued decrees to assemble amateur astronomers to serve him. In the sixth century, Zhang Zixin escaped to an island to avoid the war. He carried on astronomical observations for 30 years and discovered the unevenness of the Sun's motion. Other amateur astronomers also made some contribution, but their work was lost.

At the imperial observatory, observation was carried on day and night for thousands of years. Several observers kept watch on different parts of the sky. They paid particular attention to the time when certain stars transited the meridian, measured the length of the shadow of the dial, sent out time signals by bells and drums and recorded unusual sky phenomena. Whenever an important phenomenon took place, the director of the observatory reported it to the emperor, and it would be written down by the official who was in charge of recording history. When the official history of the dynasty was then compiled in a later dynasty, the astronomical record could be preserved in it. We are sure that ancient observers kept indeed detailed records, but only rarely can their original manuscripts be found.

The records on the supernova of 1572 A.D. provide an example showing how the original observation was entered in official history:

The *Mingshilu* (the veritable records of the Ming dynasty in which national events and the emperor's activities were recorded) says: "Longqing 6th year, 10th month, 3rd day, i.e., the day Bingchen (1572, 11, 8) a guest star appeared in the northeast sky, like a bullet. It was near the constellation Gedao belonging to Bi lunar mansion. It became brighter and brighter gradually for 19 days. On the night Renshen (1572, 11, 24) the star was red-yellow, like a lamp, very splendid . . . . In the 12th month, day

Jiaxu (1573, 1, 25) the Ministry of Rites reported to the emperor . . . . Since the 10th month, a guest star can be seen in day-light, unusually splendid. By Wanli 1st year, 2nd month (1573, 3, 4–4, 1) it became faint gradually; 2nd year, 4th month (1574, 4, 21–5, 19) it disappeared".

The *Yifeng xianzhi* (a local chronicle of Yifeng county compiled during Wanli years, 1573–1610 A.D.) says: "Longqing 6th year Renshen, winter, 10th month every latter half of night a ghost star appeared in the northeast sky, splendid like the moon".

The *Zhejiang xuzhigao* (A draft of the local chronicle of Zhejing province) records: "Longqing 6th years, a guest star appeared in daylight for 1 month, then it became faint gradually; it disappeared after 16 months".

The *Mingshigao* (the draft of history of the Ming dynasty) says: "Longqing 6th year, winter, 10th month, the day Bingchen, a broom star appeared in the northeast sky. Wanli 2nd year, 4th month it disappeared".

The Mingshi (the formal history of the Ming dynasty) states: "In the winter of Longqing 6th year a ghost star appeared".

We have not found the original records on this event but it is not difficult to imagine that it would be far more vivid and detailed. From this example we can see that most of the information has been lost in the process of compiling from the original records the report for the emperor, then the veritable records, the draft of the history, and finally the formal history.

The earliest astronomical instrument perhaps was the "Guibiao" (gnomon) which is a vertical post with a horizontal narrow platform toward the north. It was employed to determine the length of the shadow at the Sun's transit so as to decide the dates of the 24 terms of the seasons. The "Hunyi" (armillary) is the most typical Chinese astronomical instrument. It has existed at least since the "Warior states" period (fifth to third century B.C.). It was an equatorial device which had a polar axis and a declination axis as well as horizontal and ecliptic circles. The instrument was used to determine the positions of celestial bodies. Observers sighted an object through the tube and read the numbers on the circles. Another instrument called "Simplified instrument" had separate equatorial and horizontal devices and was very much like a modern equatorial.

The "Louhu" (clepsydra or water clock) was used to keep time and indeed a complete time service was set up. Unfortunately timed astronomical records only seldom have been preserved.

The astronomical observation was something very important in ancient China. The observers were astronomers with abundant experience. Mistaken astronomical calculation or observation might be punished by death. Solar eclipses were considered warnings from Heaven to the emperors. When one took place, they would reduce their food, withdraw from the main palace, even issue announcements in self-ansure. Therefore, astronomers were wath to give false information about celestial events unless they had a special motive. Studies of many hundreds of lunar and planetary records show that nearly all are genuine and reliable. Mistakes were usually made in the process of repeated copying rather than through dishonesty on purpose. For example, there are some records that the moon was near a certain star – at the west or the east. If such predictions were recorded they would be recorded as "occultations".

There are a few such records as "the sun should have been eclipsed but it did not take place", or "the eclipse was not seen because of clouds". These were considered fortunate and the emperors would accept the congratulation of the officials. Such a prediction would not be recorded as a "warning from Heaven".

Of course, since Chinese records extended over more than 2000 years, there might be different situations in different periods. This problem needs further research. But, generally speaking, ancient Chinese astronomical records are very reliable. They have made a significant contribution to modern astronomy and will continue to do so.

2 Historical Chinese Astronomical Observations (2)

2.1 Stars

There had been the star names "fire, bird, Shang" on the oracle bones in the Shang dynasty (sixteenth to eleventh century B.C.). By the "Spring and Autumn" period (eighth to fifth century B.C.), the traditional 28 lunar mansion system had already been formed basically. The *Shiji tianguanshu* (first century B.C.) systematically recorded Chinese stars and described some one hundred constellations. By the Jin dynasty, the Chinese constellation system had been formed and in the main included 283 constellations, a total of 1464 stars. There was little obvious change until the end of the Ming dynasty when western missionaries came to China.

The Chinese measured star positions with the equatorial parameters: Ruxiudu was the difference in right ascension between the star and the determinative star of that lunar mansion; Qujido was the distance from the celestial star. The differences in right ascension between the 28 determinative stars were measured, while the winter solstice was measured as being at certain degrees in some mansion. So there was a complete equatorial coordinate system in ancient Chinese astronomy. Its unit was the Du (degree). There were 365.25° in a full circle so that the Sun travels 1° per day.

The earliest star catalog which is still extant is the *Shishi xingjing* which was written during the period of the "Warring states" (fifth to third century B.C.) and is partially preserved in an astrology book, the *Kaiyuan zhanjing* of the Tang dynasty (seventh to eighth century B.C.). There are Ruxiudu and Qujidu of 120 stars in this catalog. Unfortunately, few ancient catalogs still exist; there is not a complete catalog up to the Song dynasty (tenth century).

The earliest star map which is still extant is a mural in a tomb of the Western Han dynasty (second to first centory B.C.) unearthed last year in Xian, on which 28 lunar mansions were painted. Another typical Chinese star map is the *Dunhuang juanzi* of the eighth century, which is now in the British Museum; this includes 1350 stars. The *Suzhou stele* of the Song dynasty is an all-sky star map. It includes the celestial equator, ecliptic, Milky Way and some 1400 stars. It is thought that this chart may even show the site of the supernova of A.D. 1054.

2.2 Special Astronomical Phenomena

The earliest astronomical records appeared on the oracle bones of the Shang dynasty and are the earliest Chinese writing. Systematical written history began in the *Spring and Autumn* period (722–481 B.C.). Since then, many astronomical phenomena have been recorded with exact dates. In the *Hanshu* and most of the later official histories, there are special astronomical chapters in which a large number of astronomical records are preserved.

a) Eclipses and Solar Activity

The *Shujing* records the story of a solar eclipse during the Xia dynasty (twenty-first century B.C.): The astronomer royal failed to give a prediction for a solar eclipse because of drunkenness and this led to terrible unrest. So the king executed the astronomer. Reliable eclipse records have been found on oracle bones of the Shang dynasty. However, the first record with an exact date is as follows: "King Luyun 3rd year, spring, 2nd month, Jisi day, there was a solar eclipse (22/2/720 B.C.)". It has been confirmed from this record that the sexagesimal cycle of dates has been correctly continued up to now. There are more than a 1000 records of solar eclipses in Chinese history. The general form of solar eclipse records is as follows: "In certain year, month, day, there was a solar eclipse." A few records are more detailed, e.g., "total eclipse", "sky became dark and stars appear", "like a sickle". Occasionally, a careful estimate of the magnitude, time and direction and so on of the contacts is also given. The main problem is that usually the place where an eclipse was seen is not specified.

It is recorded on an oracle bone of the Shang dynasty that "three flames had eaten the sun; there was a big star". It is perhaps a report of prominences when a total eclipse took place.

The earliest lunar eclipse records can be seen on the oracle bones. The ancient Chinese considered lunar eclipses to be regular phenomena, so there are few early records until the Tang dynasty. Still, about 2000 lunar eclipse records in all are extant. Some of these reported the color of the obscured moon, time of contacts, magnitude, and the constellation in which the moon was situated. For example: "Jingming 2nd year, 5th month, the day Dingyou, the moon was in lunar mansion Dou. It rose eclipsed to the extent of 12/15" (25/6/503 A.D.).

b) Comets and Novae

Various celestial bodies, which newly appeared, are recorded in Chinese histories. There were dozens of names for them and they were often confused. Generally, "broom star" or "Bo star" meant comet and "guest star" was used for both. Therefore, a currect judgement should be made depending on the context.

The earliest exact record about a comet can be found in the *Spring and Autumn*: "King Luwen 14th year, autumn, 7th month, a Bo star entered the constellation Beidou" (613 B.C.). This may also be the earliest record about Halley's comet. From 240 B.C. to 1910 A.D. almost each return of Halley's comet can be found in Chinese histories. In all, there are about 500 records of comets. They are more detailed than records of eclipses. Sometimes shape, length and direction of the tail and position of a comet are mentioned on several dates.

The earliest record of a nova is on an oracle bone of the fourteenth century B.C.: "A new big star near the fire star". To identify novae is a difficult task because of the ambiguous records. About a 100 records are considered to be candidates for historical novae or supernovae.

c) Meteors, Meteoric Showers and Meteorites

The records concerning meteors are rich and colorful. Many records mentioned brightness, color, shape, speed, sound, position in the sky, trail, the place where the meteorite fell and the state of the meteorite. There are thousands of such records and they are usually quite detailed.

About 100 meteor showers were recorded. The earliest comes from the nineteenth century B.C. in the *Bamboo Annals*, while the record in *Spring and Autumn* is quite reliable: "King Luzhuang 7th year, summer, 4th month, Xinmao day, the stars fell down like rain at night" (16/3/687 B.C.).

Meteors are of course local phenomena and there are many valuable records in local histories.

d) The Moon and Planets

The ancient Chinese considered that there were corresponding relations between the stars in the sky and persons or places on the Earth, while the movement of the moon and planets was controlled by Heaven. So the moon or planets covering or approaching certain stars meant disaster for some persons or places. Such records are so numerous that few people have kept track of them seriously, but they must run into tens of thousands. Regular records of the movement of the moon and planets were started during the Han dynasty. They can be classified into two types: occultation and conjunction of stars and planets by planets, and occultation and conjunction of stars and planets by the moon.

Another type of records deal with the appearance of planets in daylight. They were reported usually for Venus, sometimes for Jupiter, and even occasionally for stars. The *Kaiyuan zhanjing* of the eighth century said that Gande, living in fourth-third century B.C., found "there is a little red star near Jupiter; that is its alliance". Experiments have shown that the satellites of Jupiter can really be detected by the naked eye under favorable conditions.

Polar lights, halo, clouds with special shapes, other unidentified sky phenomena were also recorded as astronomical events.

Historical Chinese astronomical records are a huge treasure house. Much information in it has not been seriously dealt with; a lot of work is waiting for us.

Application of Early Chinese Records of Lunar Occultations and Close Approaches

Liu Ciyuan[1] and Kevin K. C. Yau

1 Introduction

One of the ways in which the Earth's rate of rotation in the last few thousand years may be investigated is by analyzing early observations of eclipses of both the Sun and the Moon. From the eighth to the first century B.C. there are records of timed lunar eclipses from Babylon. Around the nineth to the eleventh century, there are Arabian records of timed solar and lunar eclipses. In addition, records of total solar eclipses from Medieval Europe prove of value. Apart from these records, numerous Chinese records of total eclipses can be found from the eighth century B.C. to the fall of the last imperial dynasty at the beginning of the present century, together with a number of Korean and Japanese records of total solar eclipses from the eighth century onwards. All of these records are very useful in calculating the rate of the Earth's rotation in the past. However, from the second to the eighth century there is a lack of records of both lunar and solar eclipses to enable the rate of the Earth's rotation to be investigated. Fortunately, there are several thousand records of occultations and close approaches by the Moon of both planets and stars contained in Chinese texts which may remedy the deficiency of eclipse records.

2 Observational Criteria and Method of Analysis

A typical Chinese record of lunar occultation or close approach of a planet or star follows a fairly standard pattern. It begins with the year of accession of the reigning emperor, followed by the lunar month and day number of the sexagesimal cycle. Then comes the text describing the circumstances of the occultation or close approach, which is in a form similar to the following: "The Moon occulted (*yan*) or invaded (*fan*) a planet or star". Only excerpts of the original observation carried out by imperial astronomers stationed at the capital at the time are now preserved, other details are unfortunately lost. As these records contain very little detail and no timing information, researchers working on the problem of the Earth's rate of rotation in the past have tended to opt for the easy way of employing only records of eclipses.

The length of time the Moon and its occulted object are visible to the naked eye is often limited (see details in Liu 1988a, b). Two of the basic criteria would be: 1. a darkened sky; and 2. that the Moon and planet or star be situated above the horizon.

Department of Physics, University of Durham, South Road, Durham DH1 3LE, UK.
[1] On leave from Shaanxi Observatory, Lintong, Shaanxi, China.

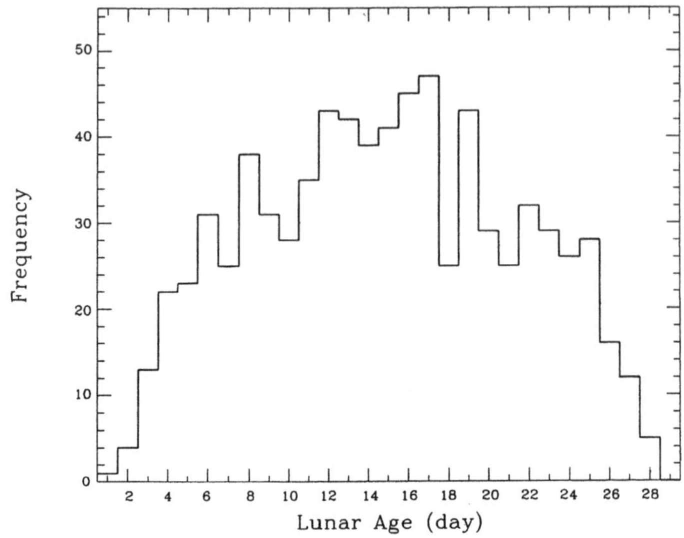

Fig. 1. The distribution of the age of the Moon for the 800 records of occultations and close approaches contained in the *Yuan-shi*

For instance, at the beginning of each lunar month, the new Moon can only be observed for a short time above the horizon before it sets again. It is during this short time span or "time window" that an occultation or close approach by the Moon of a planet or star is most likely to be observed. In general, the beginning and the end of a time window are governed by the time of rise or set of the object involved in the occultation or close approach, and the beginning and the end of twilight. These four time factors are in terms subject to the brightness of the object concerned.

In the case of a star, a simple relation may be derived from practical experience which could represent the magnitude–altitude angle relationship quite well. Although, using a slightly different formula to represent this relation will shift the final result systematically by a small amount. However, if the distribution of observation times is centered symmetrically at midnight, the systematic shift will be compensated for. Figure 1 shows the distribution of the age of the Moon obtained from the 800 records of occultations and close approaches contained in the *Yuan-shi* (History of the Yuan Dynasty A.D. 1279–1368) (see details in Liu and Stephenson 1988). The shape of the distribution may be understood in terms of the changing phases of the Moon. It is not difficult to infer from the symmetrical distribution that the observations were carried out equally in both halves of the night.

Regarding planets, the situation is slightly different. Planets are in general brighter than stars, especially in the case of Venus and Jupiter. It is not uncommon to find records of these two planets (particularly Venus) seen during daytime in Far Eastern texts. In practice, it is very difficult to judge the exact times of the day when these two planets will become visible to the naked-eye. An occultation by the Moon of either of these planets may well occur just after sunrise or before sunset. The brightness of these two planets is such that in fine weather they can be easily seen even

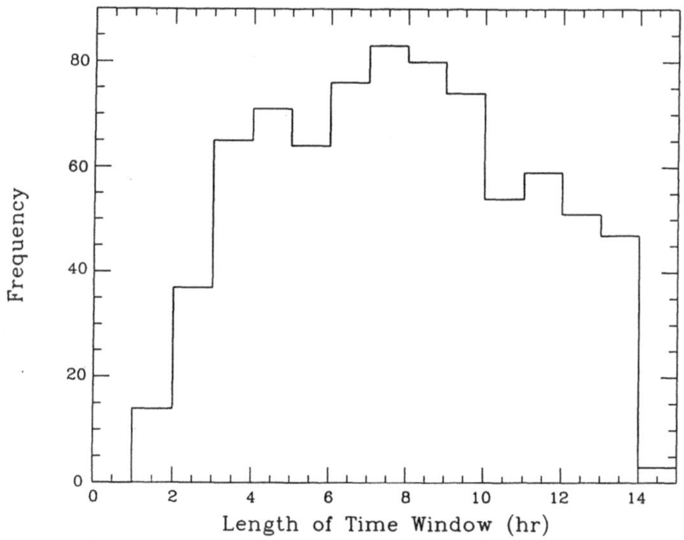

Fig. 2. The distribution of lengths of the time window for the 800 records of occultations and close approaches contained in the *Yuan-shi*

when they are near to the horizon. Therefore, it is necessary to employ different criteria for the time of rise and set of planets. We have chosen these times when their altitudes would be zero. At the same time, the occultation of a bright planet by the Moon should be a more reliable observation than an occultation of a faint star because planets can be seen even very close to the Moon. Thus it is possible to use the time window of an occultation of a planet by the Moon to calculate a more reliable result for the rate of the Earth's rotation.

From a statistical point of view, the center of the time window of visibility is taken arbitrarily to be the observation time and the limits of the time window will give an estimate of the uncertainty in the time of observation. The ET and UT of the observation can be calculated for a number of such records and thus the corresponding ΔT (= ET−UT) for the respective periods. Figure 2 shows the distribution of the length of time windows for the 800 records of occultations and close approaches contained in the *Yuan-shi.* It can be seen that the majority of the records show long time windows, greater than 3 h. Only 51 records have time windows of less than 3 h.

3 Data Analysis

Considering that records of occultations are much more accurate than of close approaches, only the former is considered in the present analysis. About 600 such records in the period before A.D. 1200 are contained in Chinese texts. It can be seen from Figure 3 that the distribution of these records is very irregular. The principal reason for this pattern of distribution is that most of the records have been lost (see Yau 1988).

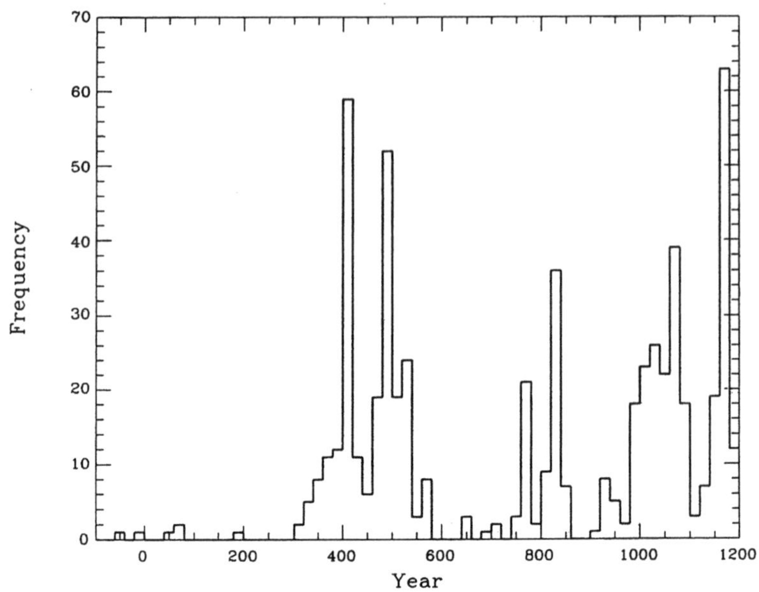

Fig. 3. The decade distribution of records of lunar occultations from 100 B.C. to A.D. 1200

We will illustrate how the long-term changes in the rate of the Earth's rotation may be calculated by analyzing all records of lunar occultations prior to A.D. 600. About 20% of the data was useless due to dating problems. This was due to either the wrong lunar month or cyclical day number being given in the record. These descriptions probably resulted from scribal errors introduced during the compilation of the historical texts. In addition, a significant number of records merely give the name of an asterism without any mention of the star that was occulted. In some cases, it is possible to identify the occulted star from the motion of the Moon. However, for a closely packed star group, the responsible star cannot always be identified. From 200 B.C. to A.D. 600, only 171 records may be used for the purpose of calculation (58 records of planets and 113 records of stars).

We divided the 171 records into four groups according to four different periods: (1) prior to A.D. 300, (2) A.D. 301–400, (3) A.D. 401–500 and (4) A.D. 501–600. We have obtained averaged values of ΔT and the respective acceleration of the Earth's rotation $\dot{\omega}$ for these four periods (see Table 1, where $\Delta T = -\frac{1}{2}\dot{\omega}T^2$ and T is in

Table 1. Some parameters of Earth's rate of rotation obtained from 171 Chinese lunar occultations

Mean epoch	ΔT (s)	$\dot{\omega}$ (s/cy^2)
24	11220	−71.2
372	7970	−78.1
444	5650	−61.2
526	4730	−58.3

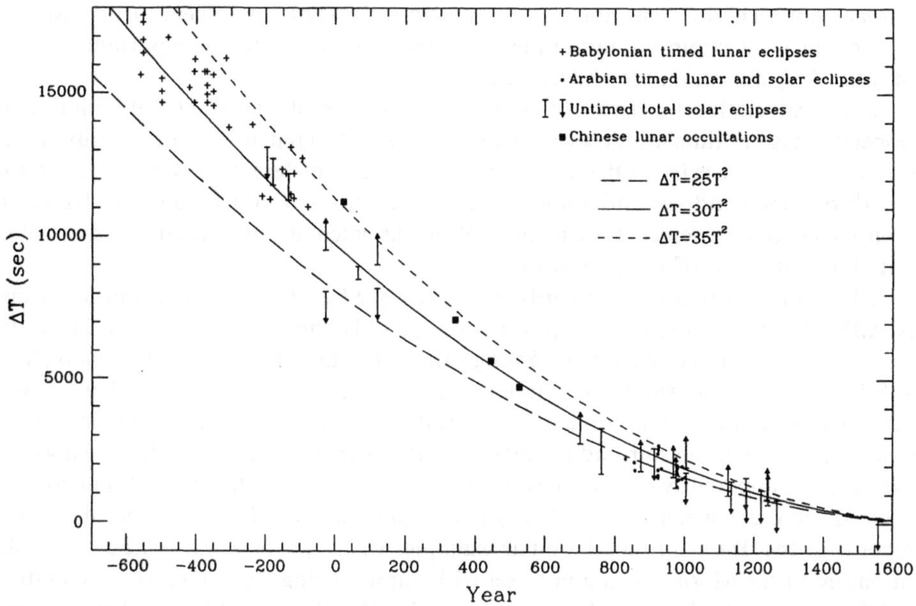

Fig. 4. A comparison of ΔT obtained from Chinese occultations with results given in Stephenson and Morrison (1984). The three curves were obtained with $\dot{\omega} = -50$, -60 and -70 s/cy^2, respectively

Julian centuries since the reference epoch of A.D. 1800). For our computation, we have employed for the lunar tidal acceleration, $\dot{n} = -26'' \text{ cy}^{-2}$. A comparison of the present results with those obtained by Stephenson and Morrison (1984) is shown in Figure 4. By choosing the appropriate grouping of records (for example, solving separately records of stars and planets) and varying some of the parameters concerning the visibility of the objects, the respective calculations show that the results remain virtually unchanged.

4 Discussion

The principal assumption of the Time Window Method is that the records were based on true observations and not on predictions. The following points may summarize our believes that these records are real observations:

1. The main reasons for keeping a watch of the sky both day and night by Chinese astronomers were mainly for astrological purposes. If an event were predictable then the omen value of the event would undoubtedly be very much reduced. Besides, the importance placed by the throne on keeping a careful watch of the sky throughout the whole of Chinese history should be emphasized. If predictions were used then it would not be necessary to employ a large team of diligent sky watchers, which numbered several hundreds during the T'ang Dynasty. However, in a few cases (principally solar and lunar eclipses), the event was described as "should have been seen

but was not seen" or "was not seen due to clouds". These may seem to suggest that predictions were employed, but at the same time they illustrate the importance placed on recording only the observed events.

2. Our calculation has shown that almost all of the observations fall within their respective visible time windows with $\dot{\omega} = -60\ \mathrm{s/cy}^2$. That is, a "concentration" of observation times exists. Of course, by employing a different $\dot{\omega}$ would change the length of time windows and some of the observations would lie outside. However, their concentration remains unchanged. Whereas calculations and predictions would not show this kind of concentration.

3. The much more detailed records in the *Nan-qi Shu* (History of the Nan-qi Dynasty A.D. 479–502) often give the position of stars like "to the east" or "not far to the west" of the Moon. For example, "Young-ming 6th year, 6th month, day *yi-mao* (A.D. 488 Jul 1), the Moon was one *cun* (about 0.1 °) to the east of *Jiao-xing* (α *Vir*). It was regarded as *fan* (a close approach)". According to calculations, the event was an occultation in which the star had already reappeared from behind the Moon. However, the observer apparently did not observe the whole process of the occultation and only noticed the event when the star had already gone past the Moon. Some of the 800 records in the *Yuan-shi* reveal similar examples of actual observation. For example, in one event the Moon was about to set with Jupiter being very close, the occultation would have occurred only a short time away. Yet the observer still recorded the event as a close approach because he did not actually observe the occultation.

4. Figure 5 shows the distribution of the calculated minimum distance between the center of the Moon and planets or stars for the 800 records contained in the *Yuan-shi*. The values of ΔT for this period would be reasonably small, hence any uncertainty in the calculation should be small. According to the above definition, use of the term *fan* means that a planet or star comes to within about 1 ° of the Moon (or 1.25 ° from

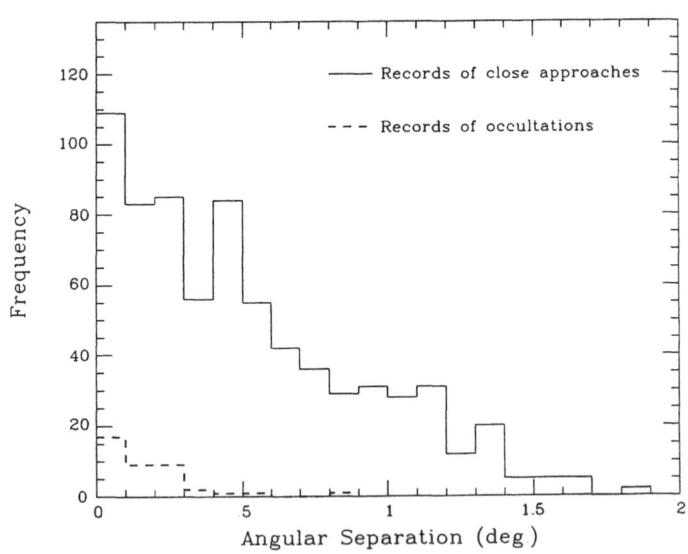

Fig. 5. The distribution of calculated angular distances from the center of the Moon for records of close approaches and occultations contained in the *Yuan-shi*

the center). For an occultation, the calculated minimum distance from the center of the Moon should be less than about 0.25°. However, Figure 5 shows that a sizeable proportion of records of close approaches described actually events of occultations. One explanation for this incompatibility is that the observers did not keep a continuous watch through the night. At the time when they carried out observations most of these events were still at the close approach stage. Another reason may be due to the inability of observers of seeing faint stars near the bright Moon. Under this circumstance and being cautious not to misreport an event, they recorded them as *fan* instead.

5 Conclusions

In conclusion, it is reasonable to believe that in general early Chinese astronomers recorded only those events that they actually observed, and not predictions. In the process of analyzing a set of rough observations, it is not difficult to introduce some forms of systematic errors into the final results. However, systematic errors cannot be reduced by simply increasing the number of observations. In order to estimate the size of systematic errors introduced in our method, a set of relatively recent data of occultations and close approaches around A.D. 1300 from the *Yuan-shi* was analyzed. The advantage of using a set of data collected near the modern epoch is that the uncertainty in the values of ΔT would be fairly small.

A total of 240 records of occultations and close approaches were selected. By taking $\dot{n} = -26'' \mathrm{cy}^{-2}$ and $\Delta T = 33\,T^2$ s (i.e. with $\dot{\omega} = -66$ s/cy^2), the mean difference in longitudes between the occulted star and the center of the Moon, $\Delta\lambda$, was found to be $-0.025° \pm 0.051$. From the expression:

$$\frac{\Delta\lambda}{n} = -\frac{\Delta\dot{\omega}}{2}\,T^2$$

(where n is the mean motion of the Moon, $1.525° \times 10^{-4}$/s and $\Delta\dot{\omega}$ is the corresponding corrections to the assumed $\dot{\omega}$), systematic errors of 160 s in our assumed initial values of ΔT are obtained. It seems likely that the value represents typical systematic errors introduced by our Time Window Method of analysis. For earlier periods, with ΔT being sufficiently large and $\Delta\lambda$ remaining constant for any epoch, records of occultations or close approaches are worthy of a detailed investigation.

Acknowledgments. We would like to thank Dr. F.R. Stephenson for reading the draft and making several valuable suggestions. One of us (Liu Ciyuan) acknowledges the support provided by a K.C. Wong Fellowship of the Royal Society.

References

Liu Ciyuan (1988a) Chin Astron Astrophys 12:169
Liu Ciyuan (1988b) Chin Astron Astrophys 12:260
Liu Ciyuan, Stephenson FR (1988) J Hist Astron (submitted)
Stephenson FR, Morrison LM (1984) Phil Trans R Soc Lond A 313:47
Yau KKC (1988) An Investigation of Some Contemporary Problems in Astronomy and Astrophysics by Way of Early Astronomical Records. PhD Thesis, Univ Durham

Ancient Central Eclipse Records of China and Variation of the Earth's Rotation

Han Yanben, Li Zhisen, and Yang Xihong

Abstract

Tidal friction leads to a long-term deceleration of the rotational speed of the Earth. Some of the ancient fossils and ancient astronomical records may be used to study the long-term variations of the Earth's rotation. There is a lot of ancient astronomical records in Chinese history. In this paper, we briefly describe a method which we use to study the rate of variations of the Earth's rotation with the aid of the records of central solar eclipses. About 200 records in Chinese annals from the Spring and Autumn Period (700 B.C.) to the beginning of the Tang Dynasty (A.D. 700) have been analyzed and calculated, but only the results of 13 eclipses that have definite descriptions about the observed phenomena of central eclipses are given.

1 Introduction

Research of the variations of the Earth's rotation can provide some important information and a scientific basis for the research of some theoretical and practical problems, for example the evolution of the Earth-Moon system, earthquakes and other geophysical problems. Many astronomers and geophysicists are very interested in this research (Runcorn 1970; Li Qibin 1973; Li Shiguang 1973; Monin 1974; Rosenberg et al. 1975; Jeffreys 1976). The research on fossils of ancient organisms give some data describing the general picture of the Earth's rotation in a geological age (Wells 1966). Data measured by astronomical instruments were not available until the eighteenth century (Luo Shifang et al. 1974). However, variations of the Earth's rotation from the times of recorded history to the seventeenth century are still unknown. The records of astronomical events from ancient civilizations contributed much to the study of the Earth's rotation. There is a large number of ancient records of astronomical events in China. Records corresponding to the research of variations of the Earth's rotation are of solar and lunar eclipses, the time of equinox, the time of planetary conjunction, opposition and culmination, the time of New Moon and occultation of the stars, etc.

Chinese ancient records of eclipses have two features, they are age-old and in large numbers. The *Shujing* eclipse may be the earliest one recorded both in China and the world in general. Its title was taken from the recording source, a Chinese historical book, *Shujing.* It is a pity that the date of the eclipse has not been found in any an-

Beijing Astronomical Observatory, Chinese Academy of Sciences, China.

cient books up to now. It appeared probably in the twentyfirst century B.C. through textual research. About 3000 years ago, some records of eclipse began to appear on the oracle bones of China. More and more (over 400) eclipse records with definite date are found in many Chinese annals of the Spring and Autumn Period to the later dynasties of China. The magnitude and time of the eclipses and central eclipses have been contained in the historical records. Most of the recorded eclipses are central eclipses, because the phenomena of these are easily noticed.

2 General Principle

The central line positions of total and annular eclipses on the Earth's surface were obviously affected by the variations of the Earth's rotation, and the effects are shown mainly in the longitude of the observing places. The formula with the Ephemeris time scale and Greenwich mean time at the mean noon of Jan. 0, 1900 as epoch is

$$\Delta\lambda = \lambda_e - \lambda \ , \tag{1}$$

where λ_e is ephemeris longitude of the central line from gravitational theory, λ is geographical longitude of the actual place where the central line falls on the Earth's surface. The western directions of λ_e and λ are both positive. $\Delta\lambda$ is the difference between λ and λ_e caused by the variation of the Earth's rotation.

The relative position of the Greenwich meridian and Ephemeris meridian changes continuously due to the variations of the Earth's rotation. If the difference of these two meridians in longitude is

$$\Delta\lambda^* = 1.002738\Delta T \tag{2}$$

where ΔT is the correction to the UT, and $\Delta\lambda$ would be

$$\Delta\lambda = 1.002738\Delta T \ , \tag{3}$$

then

$$\Delta T = \frac{\Delta\lambda}{1.002738} = \frac{(\lambda_e - \lambda)}{1.002738} \ . \tag{4}$$

The geographical coordinates of the place where the ancient central eclipse was observed would be known if the place in the ancient record is defined by textual research. λ_e is computed by the theoretical formula with ephemeris time scale. Therefore, ΔT can be obtained by formula (4).

Since the observed places of most central eclipses are not clearly noted, λ will be determined with the error $\delta\lambda$. Meanwhile, there is also an uncertainty $\delta\lambda_e$ to the λ_e on account of the width of the central eclipse zone. Then,

$$\Delta T = \frac{\lambda_e - \lambda \pm (|\delta\lambda_e| + |\delta\lambda|)}{1.002738} \tag{5}$$

The works of Meeus (1962, 1966), Brown (1919) Oppolzer (1887), and *Improved Lunar Ephemeris* (1954) have been used to compute the parameters of the eclipses.

3 Data Analysis and Results

Among the 400 eclipse records in ancient Chinese books, about 200 eclipses have been found after preliminary textual research with the central path across ancient China. These 200 eclipses have been computed, but only the 13 eclipses are shown in Table 1 and the definite descriptions about the details of these 13 eclipses are given in Figure 1. The results obviously show the tendency of deceleration of the Earth's rotation.

It is necessary to explain the following three key arguments for the selection of the central eclipse records.

1) Record of the observed phenomena

We are very interested in the descriptions of phenomena about ancient central eclipses in Chinese historical books. These descriptions tell us whether the observed eclipse is a central one. The Chinese character JI is used in seven records. JI means "all" or "finish", it obviously describes the scene of the second contact of a central eclipse. Five records use the Chinese characters ZHOUHUI, XINGJIAN, LIEXIU CANRAN, which mean that daytime becomes night and stars appear. These descriptions are also about the phenomena of the central eclipses. The Chinese charaters JI JIN are used in two records, one stands for an annular eclipse, another for an annular-

Table 1. Records and results of the central eclipses in ancient China

No.	Date	J.D.	Place (Log., Lat.)	ΔT	Type	Remarks
1	−708.07.17	1462659	Qufu (117.0, +35.7)	17.47±0.48	t	(a)
2	−600.09.20	1502171		18.19±0.53	t	(a)
3	−441.03.11	1566052	Xianyang (108.7, +34.4)	14.05±0.79	t	(b)
4	−381.07.03	1582081		16.44±0.45	t	(c)
5	−299.07.26	1612055		12.85±0.36	t	(c)
6	−187.07.17	1652954	Changan (108.9, +34.2)	11.70±0.17	r-t	(d)
7	−180.03.04	1655376		11.23±0.34	t	(a)
8	−79.09.20	1692466		7.16±0.19	t	(a)
9	2.11.23	1722115		8.14±0.02	r-t	(a)
10	65.12.16	1745149	Luoyang (112.5, +34.7)	7.06±0.24	t	(a)
11	360.08.28	1852788	Jiankang (118.8, +31.9)	7.61±0.57	r	(d)
12	454.08.10	1887103	Jiankang (118.8, +31.9)	6.41±0.22	t	(e)
13	523.11.23	1912410	Jiankang (118.8, +31.9)	7.37±0.43	r	(f)

Note: (1) The error $\delta\lambda$ is excluded in the result. (2) The letters in the last column indicate the descriptions of phenomena, they are: a) eclipse, and the second contact of the eclipse; b) eclipse, daytime fades, and stars appeared; c) eclipse, daytime fades; d) eclipse, almost total; e) eclipse, the second contact of the eclipse, and stars twinkling; f) eclipse, Venus was seen at daytime. Type t means total eclipse; r means annular eclipse. ΔT is given in 10^3 s

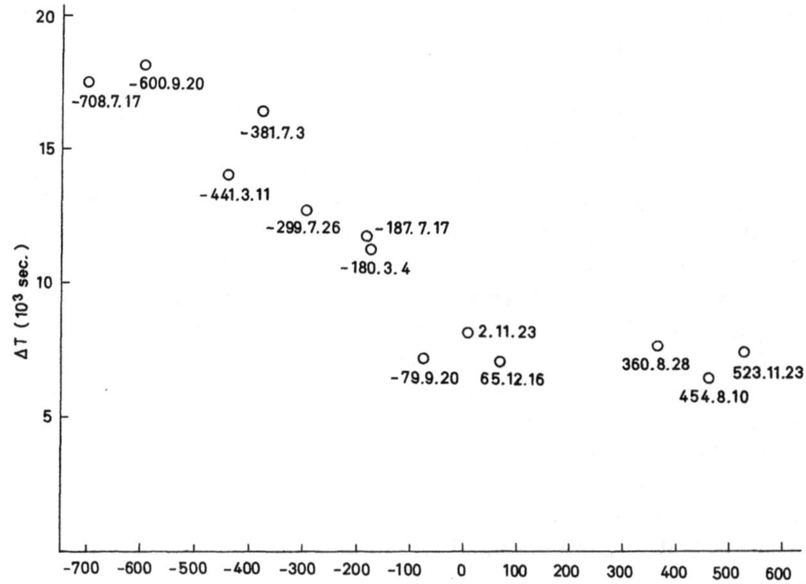

Fig. 1. The distribution of ΔT obtained from the 13 central eclipses in ancient China

total eclipse, and they mean to approach "all" or "finish", i.e., the third contact of an eclipse.

2) Record of the observing place

As the observing places of many eclipses were not recorded in Chinese historical books generally, we had to suppose that the observing places were the capitals. At that time most of the ancient dynasties in China set up the astronomical organizations with officials in their capitals. The records of observing places show that: (a) the observers and/or recorders were in the capital, (b) the recorder in the capital did not see the central eclipse. The information was obtained then from someone who lived in the zone of the central eclipse. Obviously, the records from (a) are better than from (b). In the case of earlier central eclipse records, even if the record is from (b), it does not lead to a gross error, because the ancient counties of dukes or princes under an emperor were very small, and the observer would not have been very far away from the capital. But as for the later eclipses, it might lead to a greater error if the information came from (b), because the counties became larger and information was transmitted easily. Some eclipses recorded in local chronicles might be available in the course of further investigations.

3) The date of the eclipse

The records of the dates in Chinese historical books are generally definite. Only a few errors have been discovered. The errors of date probably originated from the making of the private copies. Because the central eclipse was limited to the observing place and there are only a few eclipses every year in the world, the errors may be corrected by computation and textual research. Two examples are given:

a) *11th March−441 (t)*. The record of this eclipse is found in a historical book of China, SHIJI, and reads: "eclipse, daytime became night, stars were seen in the 34th year of Duke QinLi". This is the year −442nd, but no central eclipse occurred that year in the world (Oppolzer 1887). There were two central eclipses in the 35th year of Duke QinLi (−441st year). An annual eclipse of 3rd September could not be seen by the Chinese as its central path swept over the Southern Hemisphere. The central path of the total eclipse of 11th March passed through the territory of the Qin dynasty. The eclipse recorded in SHIJI should be the total eclipse of 11th March, −441.

b) *26th July, −299*(*t*). This record is also found in SHIJI, with the words: "eclipse, daytime became night in the 6th year of Duke QinZhao". Two central eclipses occurred that year, i.e., −300th year. One of them could be seen in the Southern Hemisphere only. Although another could be seen in the Northern Hemisphere, the distance of its central path from China was too great. The Chinese could not see the phenomenon "that daytime became night". The central zone of the total eclipse which occurred on July 26th, −299 and passed over China should be the one that is recorded in SHIJI.

4 Discussion

The records of central eclipses with definite observed phenomena, place, and date are useful for the research of the variation of the Earth's rotation. However, the three key arguments of most ancient central eclipse records were incomplete, which is the main reason why some large errors may appear and so this method of the central eclipse is naturally of limited usefulness. But the dates of the central eclipse records are earlier than those of the eclipse time records and magnitude records. So it is of advantage for the study of the Earth's rotation to use the record of central eclipses. To suppose the ancient capital as having been the observing place is reasonable in the case of the very early records.

Although only the 13 central eclipse records are given in this chapter, the resulting values of ΔT may be significant for describing the general trend of the Earth's rotation at that time.

References

Brown EW (1919) Tables of the motion of the Moon, vol 3. Yale University Press, New Haven

Improved Lunar Ephemeris (1952−1959). United States Government Printing Office, Washington 1954

Jeffreys H (1976) The Earth, its history and constitution. Cambridge Press

Li Qibin et al. (1973) A preliminary study on the relation betwen the Great Earthquakes of China and the secular variation of the angular velocity of rotation of the Earth. Acta Geophys Sin 16:71−80 (in Chinese)

Li Shiguang (1973) Essential of Geomechanics. Science (in Chinese)

Luo Shifang et al. (1974) Analysis of the periodicity of the irregular rotation of the earth. Acta Astron Sin 15 (1):79−85

Meeus J (1962) Tables of Moon and Sun. Kesselberg Sterrenwacht, Kessel-Lo
Meeus J, Grosjean CC, Vanderleen W (1966) The canon of solar eclipses. Pergamon Press, Oxford
Monin AS (1974) Earth's Rotation and Climate. Inst Ocean USSR Acad Sci, Moscow
Oppolzer Th v (1887) Canon of eclipses. (Canon der Finsternisse). Kaiserl-Königl Hof- und Staatsdruckerei, Wien
Rosenberg GD et al. (eds) (1975) Growth rhythms and the history of the Earth's rotation. Wiley & Sons, London
Runcorn SK (ed) (1970) Palaeogeophysics. Academic Press, London
Wells JW (1966) Paleontological evidence of the rate of the Earth's rotation. In: Marsden BG, Cameron AGW (eds) The Earth-Moon System. Plenum, New York, pp 70–81

Ancient Eclipse Records of East Asia and the Earth's Rotation

Li Zhisen, Han Yanben, and Zeng Zhifang

Abstract

East Asia (China, Japan and Korea) is very rich in ancient eclipse records. This paper presents the analysis and use of them in the study of the Earth's rotation. These astronomical activities of East Asia have the same origin, therefore their records may constitute one combined series. It is the longest and least broken series in the world, which extends from 776 B.C. to A.D. 1630, and includes 512 central eclipses and 388 eclipse timings (139 solar, 249 lunar). It has become important to use the ancient astronomical records for studying the secular variations of the Earth's rotation, but the value of materials of East Asian civilizations is still ignored. This situation should be changed.

1 Introduction

At present, using ancient astronomical records is a unique and effective method to study the historical variation of the Earth's rotation. Since the 1960's, research in this field has been flourishing and repeatedly new advances have been made (Newton 1970, 1972; Brosche et al. 1978, 1982; Stephenson 1978). Great attention has been paid to observations and records of the west, but the observations involved were limited to the central eclipse records. If this attitude is changed, much greater progress could be made.

The ancient civilization of East Asia is a characteristic community of cultures in the world. During the last 2000 years, China has had a long history of contact with Japan and Korea. In the process of cultural development, they promoted, encouraged and cooperated with each other, created jointly the glorious civilization of East Asia. It was unique in the ancient world. Their astronomical activities have much in common, and it is the outstanding feature of East Asian astronomy to devote itself to observation and recording. This paper presents a discussion of the eclipse records of East Asia and their use in the study of the Earth's rotation.

2 Valuable Time Records of Eclipses

Among the astronomical records in antiquity, the most valuable one is the time observations, especially for studying the Earth's rotation. The emergence and development

---

Beijing Astronomical Observatory, Chinese Academy of Sciences, China.

Table 1. Gaps in eclipse timing records in ancient China and their comparison with those from Japan and Korea

No.	Gap	Duration (years)	Number of Japanese and Korean records	Mean frequency (per 10 years)
1	B.C. 144 – 133	111		
2	B.C. 133 – 88	45		
3	B.C. 88 – 33	55		
4	B.C. 27 – A.D. 61	88		
5	A.D. 61 – 175	114	(5)	0.5
6	223 – 311	88	(2)	0.5
7	311 – 429	118	(2)	0.2
8	546 – 680	134	(3)	0.2
9	794 – 840	46	(5)	1.1
10	840 – 940	100	9	0.9
11	959 – 1040	81	9	1.1
12	1302 – 1450	148	33	2.2
13	1450 – 1572	122	43	3.5
14	1785 – 1867	82	92	11.2

Note: The numbers in parenthesis refer to available records of central eclipses from Japan and Korea.

of the time concept is an important mark of human civilization. Ancient China is very rich in time records of celestial phenomena, though many of them have been lost and few have been analyzed (Li Zhisen 1987). From 145 B.C. to A.D. 1636, there are many time observations, 87 for solar and 72 for lunar eclipses. However, the continuity of the records was affected by wars and a decline of astronomical activity. There are many gaps extending for 40 – 50 and even 100 years (see Table 1).

Both Japan and Korea are close neighbors of China, their territories either adjoin each other, or are separated only by a strip of water. Since the fifth or sixth century, they began making contact. In the ninth century (Tang Dynasty), cultural exchanges between them reached culmination. Japan and Korea have introduced the Chinese calendar and time system. For the improvement and compilation of the calendar, they have preserved more time observations of eclipses than China. The records preserved in Japan from A.D. 840 to 1639 amount to 232 (including 2 from Korea) (Ohura Sinkishi 1916; Wanabe Toshio 1978). In the same period, there were only some 80 records from China (Li Zhisen et al. 1985). Therefore, Japanese records can not only

Table 2. Eclipse timing records left over by the ancient civilizations of the world

Region	Time interval	Duration	Number of records	Mean fr. per 10 years
Babylonia and Assyria	B.C. 721 – 382	349	11	0.3
Greek (Mediterranean)	B.C. 201 – A.D. 364	560	10	0.2
East Asia	B.C. 145 – A.D. 1630	1800	420	2.1
Islam	A.D. 829 – 1004	175	33	1.9
European	?			

Note: The eclipse records in the inscription on bones in China are not included.

bridge the gaps in the Chinese series and improve its continuity, but they also remedy its intrinsic defect. From Table 1 it is seen that the mean frequency of eclipse timing records in the combined series is greater than in the individual ones.

The time records of these three countries of East Asia may constitute one combined series, which extends over more than 2000 years and is rarely found elsewhere in the world. For comparison, the preserved time records of eclipses for several ancient civilizations of the world are listed in Table 2, in order to show the value and importance of the joint records series of East Asia.

As for the ancient time system of China, analyses have been made elsewhere (Zheng Tianjie 1977), so it is unnecessary to describe them here.

3 Importance of Central Eclipse

Work on the Earth's rotation based on astronomical records was initiated by J. Fotheringham at the beginning of this century (Fotheringham 1909, 1915, 1918, 1920). Although he used many kinds of records, the most effective results were achieved by using central eclipses. Today we know that timing records are valuable but difficult to obtain, and their history is shorter than that of central eclipses. Central eclipse records have a long history, their amount is large and, besides, their precision can be assured by rational selection.

Chinese records of central eclipses have attracted the attention of scientists all over the world. R. R. Newton (1967) and F. R. Stephenson (1978) are eminent in this field of study. Chinese records have played an important role in their work (see Table 3) (Fotheringham 1915, 1918; Newton 1967; Stephenson 1978).

It is unfortunate that the ancient records of Japan and Korea are not interesting for scientists in the study of the Earth's rotation. Their records refer to later years than those in China, but they are more abundant, and have been analyzed and collect-

Table 3. The importance of the Chinese records in the study of Earth's rotation

Authors	Sampling duration	Sampling number	Number of Chinese records
Newton, R.R.	before B.C. 2	25	9
Stephenson, F.R.	before B.C. 2	15	17
Stephenson, F.R.	A.D. 2 – 1567	44	7

Table 4. Comparison of the central eclipse records in ancient China, Japan and Korea (by the end of A.D. 862)

Content	China	Japan	Korea
Most ancient	B.C. 776.9.6	A.D. 628.4.10	B.C. 54.5.9
Total record	(Total)	(Total)	(Total)
Number of solar eclipses	415	134	67
Number of central eclipses	363	111	38
Number of useful eclipses	97	35	10

ed (Wanabe Toshio 1978). Chinese records have a long history, and are of considerable size. The Korean ones are fewer than those of Japan, but they date back to more ancient years. Particulars of the ancient records of the three countries are listed in Table 4.

It is necessary to point out, that both Japan and Korea are narrow countries lying in a south-north direction. This is an ideal condition for using astronomical records in studying the Earth's rotation. Therefore, the Japan-Korea records should be used as far as possible in improving and corroborating the Chinese record series of before A.D. 862.

4 Use of the East Asian Series of Ancient Records

In recent years, we have analyzed the records of East Asia (Li Zhisen et al. 1984; Li Zhisen 1987), and attempted to present related methods (Han Yanben et al. 1984; Li Zhisen et al. 1985). Our preliminary results seem to be satisfactory, and are plotted in Figure 1.

It is seen from Figure 1 that

1) The results from time records are obviously better than those from central eclipses.
2) Results from Japan-Korea records are obviously better than those from Chinese records alone of the same kind.
3) The combined series of records has good continuity.
4) All results reflect not only the tendency of deceleration of the Earth's rotation, but also provide a possibility to deduce the fluctuations.

The use of East Asian series has great potentiality. Our work has been only preliminary and remains to be improved upon through further efforts.

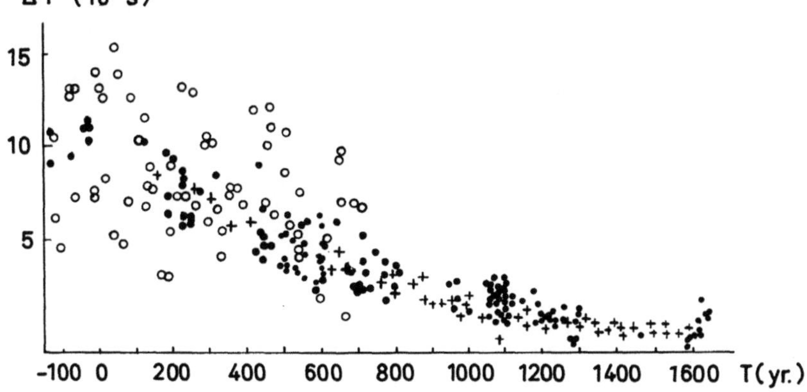

Fig. 1. ΔT (correction of universal time) as determined from East Asian series of ancient eclipses. ○ Results determined from central eclipse records in China; ● results determined from timing records in China; + results determined from timing records in Japan

References

Brosche P et al. (eds) (1978) Tidal friction and the Earth's rotation. Springer, Berlin Heidelberg New York, pp 5–27

Brosche P et al (eds) (1982) Tidal friction and the Earth's rotation II. Springer, Berlin Heidelberg New York Tokyo, pp 190–209

Fotheringham J (1909) The eclipse of Hipparchus. MNRAS 69:204–210

Fotheringham J (1915) The secular acceleration of the Moon's mean motion, as determined from occultations in almagest. MNRAS 75:377

Fotheringham J (1918) The secular acceleration of the Sun as determined from Hipparchus' equinox. MNRAS 78:406

Fotheringham J (1920) Secular acceleration of Sun and Moon as determined from ancient lunar and solar eclipses, occultation and equinox observations. MNRAS 80:578

Han Yanben et al (1984) The parameter of variation in the rotation speed of the Earth from ancient central eclipses records of China. Acta Astrophys Sin 4:107

Li Zhisen et al (1984) Contact time records of the eclipses in ancient Japan with their application in modern sciences. Proc Annu Meeting of the Astrometry of Chinese Astronomical Society, Beijing

Li Zhisen (1987) Joint use of the eclipse records in China, Japan and Korea in study of the Earth's rotation. Beijing Astron Observ 10:1–2

Li Zhisen et al (1985) Ancient eclipse records of China and Earth's rotation. Sci Sin Ser A 28:12

Li Zhisen (1987) The contact time records of eclipses in ancient China with their use. Selected works of historical naturology. Ocean Press, Beijing

Newton RR (1967) Secular accelerations of Earth and Moon. Science 166 (3907):825–831

Newton RR (1970) Ancient astronomical observations and the acceleration of the Earth and Moon. J Hopkins, Baltimore

Newton RR (1972) Medieval chronicles and the rotation of the Earth. J Hopkins, Baltimore

Okura Sinkishi (1916) Eclipse records of ancient Japan. Astron Herald 9:2–6

Stephenson FR (1978) Pretelescopic Astronomical Observations. In: Brosche P, Sundermann J (eds) Tidal friction and the Earth's rotation. Springer, Berlin Heidelberg New York Tokyo, pp 5–20

Stephenson FR (1978) The applications of early astronomical records. Hilger, Bristol

Wanabe Toshio (1978) Canon of eclipses in China, Japan and Korea. Yuuyama, Tokyo

Zheng Tianjie (1977) The story of the calendar. Sinica Culture, Taibei

Tidal Acceleration of the Moon

XX Newhall, J.G. Williams, and J.O. Dickey

Tidal dissipation causes a misalignment of the tidal bulge of the Earth with the Earth-Moon direction. The bulge exerts a secular torque which expands the lunar orbit while slowing Earth rotation. Because the lunar orbit is neither circular nor coplanar with the Earth's equator, the tide can be separated into several distinct frequency components with bands near 0, 1, and 2 cycles/day. The tidally induced secular acceleration $\dot{n}$ of the lunar mean longitude is particularly sensitive to the dominant semi-diurnal M_2 constituent, with smaller contributions from the diurnal O_1 and semi-diurnal N_2 components.

The precession of the lunar orbit about the ecliptic every 18.6 yr causes a $\pm 5.2°$ variation of the apparent inclination of the lunar orbit with respect to the Earth's equator. This 18.6-yr variation in latitude results in a periodic variation in $\dot{n}$ whose magnitude depends on the relative contribution of the semi-diurnal and diurnal tides to the secular portion of $\dot{n}$ (Williams et al. 1978). For this reason we have implemented a tidal force model in which we can specify three Love numbers k_{20}, k_{21}, k_{22} and time lags τ_{20}, τ_{21}, τ_{22} for the contribution of the long-period, diurnal, and semi-diurnal tide bands. The time delays can be solved for, but good separation has not resulted; therefore, in making the following solution we required that $k_{21}\tau_{21} = k_{22}\tau_{22}$ and $\tau_{20} = 0$.

The value of $\dot{n} = -24.9 \pm 1.0$ arcsec/century^2 results from a solution using data acquired from August 1969 through July 1988. The error is largely due to uncertainty in the relative proportions of τ_{21} and τ_{22}. The value of $\dot{n}$ corresponds to a linear increase of 3.7 ± 0.2 cm/yr in the mean Earth-Moon separation. There have been major improvements in data accuracy in the past 3 years. Current LLR data consist of ranges from each of three sites (McDonald Laser Ranging System, Texas; Maui, Hawaii; and CERGA, France) to four lunar reflectors, with accuracies of $3-6$ cm. This determination of $\dot{n}$ agrees with the value inferred from the analysis of satellite laser ranging data (-25.27 ± 0.61 arcsec/century^2) (Christodoulidis et al. 1988).

References

Christodoulidis DC, Smith DE, Williamson RG, Klosko SM (1988) Observed tidal braking in the Earth/Moon/Sun system. J Geophys Res 93 (B6):6216–6236

Williams JG, Sinclair WS, Yoder CF (1978) Tidal acceleration of the Moon. Geophys Res Lett 5:943–946

Jet Propulsion Laboratory, California Institute of Technology, Pasadena, CA 91109-8099, USA

The Variation in J_2 and in the Moments of Inertia: Satellite Results and Consequences for the Angular Momentum Budget of the Earth-Moon-Sun System

M. Burša

1 Introduction

The latest results of determination of the secular variation in the Moon's mean motion n purely of tidal origin on the basis of Lunar Laser Ranging (LLR) and Satellite Laser Ranging (SLR) techniques, as well as of the nontidal variation in the second zonal geopotential harmonic J_2 on the basis of LAGEOS orbit dynamics, make it possible to refine the residuum in the Earth-Moon-Sun angular momentum budget. The tidal and nontidal secular variation in the angular velocity ω of the Earth's rotation can be also refined.

2 Recent Values of Quantities Needed for the Angular Momentum Budget

2.1 Secular Variation in the Moon's Mean Motion Due to Tidal Dissipation

Recently, the techniques of LLR and SLR made it possible to derive the secular variation in the Moon's mean motion with an actual accuracy $\pm 1'' \, cy^2$. From LLR (Dickey et al., this Vol.)

$$\frac{dn}{dt} = -(24.9'' \pm 1.0'')\, cy^{-2} \; ; \tag{1}$$

from SLR (Christodoulidis et al. 1988):

$$\frac{dn}{dt} = -(25.27'' \pm 0.61'')\, cy^{-2} \; . \tag{2}$$

Values (1) and (2) differ by about 0.5 of the standard error of their determination which may adopted realistic. The weighted mean and its standard error is

$$\frac{dn}{dt} = -(25.17'' \pm 0.8'')\, cy^{-2} \; . \tag{3}$$

Astronomical Institute, Czechoslovak Academy of Sciences, Prague, ČSFR.

Value (3) will be used as given. The corresponding secular increase in the semi-major axis a of the lunar orbit comes out as

$$\frac{da}{dt} = \frac{2}{3} a^{-5/2} [G(M_\oplus + M_☾)]^{-1/2} \frac{dn}{dt} = (3.72 \pm 0.1) \text{ m cy}^{-1} ; \quad (4)$$

$a = 384\,400\,000$ m, $GM_\oplus = (398\,600.434 \pm 0.002) \times 10^9 \text{ m}^3\text{s}^{-2}$, $GM_☾ = (4902.799 \pm 0.003) \times 10^9 \text{ m}^3 \text{ s}^{-2}$. Note that if value (3) be assumed constant during the tidal evolution of the Earth-Moon-Sun system, value (4) is not constant, being dependent on increasing a.

2.2 Secular Variation in J_2 and Related Quantities

The most recent value of the variation in J_2 detected by LAGEOS orbital analysis has been given in the Report of IAG Special Study Group 5–100 by Chovitz (August 1987):

$$\frac{dJ_2}{dt} = -(2.8 \pm 0.3)\ 10^{-9} \text{ cy}^{-1} ; \quad (5)$$

$J_2 = [C - \frac{1}{2}(A+B)]\ (M_\oplus a_0^2)^{-1}$, $a_0 = 6\,378\,137$ m, $C > B > A$ are the principal moments of the Earth's inertia.

Assuming $\delta A + \delta B + \delta C = 0$,

$$\frac{dC}{dt} = \frac{2}{3} M_\oplus a_0^2 \frac{dJ_2}{dt} = -(4.536 \pm 0.48)\ 10^{29} \text{ kg m}^2 \text{ cy}^{-1} , \quad (6)$$

$$\frac{dA}{dt} + \frac{dB}{dt} = (4.536 \pm 0.48)\ 10^{29} \text{ kg m}^2 \text{ cy}^{-1} ; \quad (7)$$

$M_\oplus a_0^2 = 2.429989 \times 10^{38} \text{ kg m}^2$.

From (7) the nontidal positive variation in ω comes out as

$$\left(\frac{d\omega}{dt}\right)_{\text{nontidal}} = -\frac{\omega}{C}\frac{dC}{dt} = (1.304 \pm 0.138)\ 10^{-22} \text{ rad s}^{-2} . \quad (8)$$

The origin of (6) and/or (8) as suggested by Rubincam (1984); Wu and Peltier (1984) is the pleistocene deglaciation. However, the dynamics of the Earth's interior, e.g., at the core-mantle boundary, might also be responsible for this extremely important recent global geodynamic phenomenon (Burša 1987).

2.3 Secular Variations in the Orbital Elements of the Moon's and Earth's Orbit Due to Tidal Dissipation

There are two recent solutions for the Moon's orbit made by (α) (Kostelecký, this Vol.) and (β) (Christodoulidis et al. 1988):

Eccentricity:

$$(\alpha)\ \frac{de}{dt} = (2.1 \pm 0.3)\ 10^{-9}\ cy^{-1}\ ,$$

$$(\beta)\ \frac{de}{dt} = (1.83 \pm 1.10)\ 10^{-9}\ cy^{-1}\ ;$$

weighted mean

$$\frac{de}{dt} = (2.08 \pm 0.8)\ 10^{-9}\ cy^{-1}\ . \tag{9}$$

Inclination:

$$(\alpha)\ \frac{di}{dt} = -(4.4 \pm 0.7)\ 10^{-8}\ deg\ cy^{-1}\ ,$$

$$(\beta)\ \frac{di}{dt} = -(6.65 \pm 0.53)\ 10^{-8}\ deg\ cy^{-1}\ ;$$

weighted mean

$$\frac{di}{dt} = -(5.83 \pm 0.6)\ 10^{-8}\ deg\ cy^{-1}\ . \tag{10}$$

A recent solution for the orbital elements of the Earth-Moon barycenter (semi-major axis $a_{\oplus}$, eccentricity $e_{\oplus}$, inclination $i_{\oplus}$) by Christodoulidis et al. (1988) provides:

$$\frac{da_{\oplus}}{dt} = (1.43 \pm 0.25)\ 10^{-4}\ m\ cy^{-1}\ , \tag{11}$$

$$\frac{de_{\oplus}}{dt} = (1.3 \pm 9.4)\ 10^{-17}\ cy^{-1}\ , \tag{12}$$

$$\frac{di_{\oplus}}{dt} = (-5 \pm 51)\ 10^{-15}\ deg\ cy^{-1}\ ; \tag{13}$$

(12) and (13) are only estimates of the orders of magnitudes. However, the contribution of these variations to the angular momentum budget is negligible as demonstrated earlier by Kaula (1964).

3 Secular Tidal Variation in the Orbital Angular Momentum of the Earth-Moon-Sun System

We shall deal with the component J of the orbital angular momentum vector of the Earth-Moon-Sun system directed along the vector ω of the Earth's rotation. Its total tidal variation dJ/dt is composed from two parts:

$$\frac{dJ}{dt} = \left(\frac{dJ}{dt}\right)_{\oplus ☾} + \left(\frac{dJ}{dt}\right)_{B\odot}\ . \tag{14}$$

The first ist due to the tidal variations in the geocentric orbital elements a, e, i of the Moon, the second in the heliocentric orbital elements $a_\oplus$, $e_\oplus$, $i_\oplus$ of barycenter B of the Earth-Moon system:

$$\left(\frac{dJ}{dt}\right)_{\oplus ☾} = \frac{M_\oplus M_☾}{M_\oplus + M_☾} [G(M_\oplus + M_☾) a (1-e^2)]^{1/2} \cos i \times \left[\frac{1}{2a}\frac{da}{dt} - \frac{e}{1-e^2}\frac{de}{dt} - \tan i \frac{di}{dt}\right] , \tag{15}$$

$$\left(\frac{dJ}{dt}\right)_{B\odot} = \frac{M_\odot (M_\oplus + M_☾)}{M_\odot + M_\oplus + M_☾} \times [G(M_\odot + M_\oplus + M_☾) a_\oplus (1-e_\oplus^2)]^{1/2} \cos \varepsilon_0 \frac{1}{2a_\oplus}\frac{da_\oplus}{dt} ; \tag{16}$$

$GM_\odot = 13\,271\,244.0 \times 10^{13}\, m^3\, s^{-2}$, $a_\oplus = 1.4959787 \times 10^{11}$ m, $\varepsilon_0 = 23.4412°$, $i = \varepsilon_0 \pm 5°09'$.

The numerical values of the various partial contributions are:

da/dt:	$(4.00 \pm 0.11)\, 10^{35}\, kg\, m^2\, cy^{-2}$,
de/dt:	$-(0.09 \pm 0.04)\, 10^{35}\, kg\, m^2\, cy^{-2}$,
di/dt:	$(0.42 \pm 0.03)\, 10^{35}\, kg\, m^2\, cy^{-2}$,
$da_\oplus$/dt:	$(0.37 \pm 0.06)\, 10^{35}\, kg\, m^2\, cy^{-2}$.

The numerical values of (15) and (16) are:

$$\left(\frac{dJ}{dt}\right)_{\oplus ☾} = (4.33 \pm 0.12)\, 10^{35}\, kg\, m^2\, cy^{-2} , \tag{17}$$

$$\left(\frac{dJ}{dt}\right)_{B\odot} = (0.37 \pm 0.06)\, 10^{35}\, kg\, m^2\, cy^{-2} . \tag{18}$$

Total variation (14):

$$\frac{dJ}{dt} = (4.70 \pm 0.13)\, 10^{35}\, kg\, m^2\, cy^{-2} . \tag{19}$$

4 Secular Tidal Variations in the Angular Velocity of the Earth's Rotation

The conservation of the total angular momentum prescribes (17) to be compensated by the secular tidal decrease in ω due to the Moon

$$\left(\frac{d\omega}{dt}\right)_{tidal ☾} = -\frac{1}{C_m}\left(\frac{dJ}{dt}\right)_{\oplus ☾} = -(6.10 \pm 0.17)\, 10^{-22}\, rad\, s^{-2} ; \tag{20}$$

analogously

$$\left(\frac{d\omega}{dt}\right)_{tidal\odot} = -\frac{1}{C_m}\left(\frac{dJ}{dt}\right)_{B\odot} = -(0.52\pm0.09)\,10^{-22}\,\text{rad s}^{-2}\;; \tag{21}$$

$C_m = 7.123\times10^{37}$ kg m^2 is the principal moment of inertia of the Earth's mantle plus crust. The total secular tidal variation in ω comes out as

$$\left(\frac{d\omega}{dt}\right)_{tidal} = -(6.62\pm0.20)\,10^{-22}\,\text{rad s}^{-2}\;. \tag{22}$$

The contribution by the secular decrease in the angular velocity of the Moon's rotation was not taken into account; however, it is quite negligible.

Note that values (20)–(22) are dependent on C_m. If the total value $C = 8.036 \times10^{37}$ kg m^2 is used instead of C_m, the absolute values of (20)–(22) are about 7/8 of the given values.

Variations (17) and (18) should be equal to the corresponding tidal torques which enables to estimate the product of Love number k_2 and phase lag angle ε:

$$(k_2\varepsilon)_{Moon} \doteq \frac{1}{3}\,\frac{a}{GM_{☾}^2}\left(\frac{a}{a_0}\right)^5\left(\frac{dJ}{dt}\right)_{\oplus ☾} = (1.23\pm0.04)\,10^{-2}\;, \tag{23}$$

$$(k_2\varepsilon)_{Sun} = \frac{1}{3}\,\frac{a_\oplus}{GM_\odot^2}\left(\frac{a_\oplus}{a_0}\right)^5\left(\frac{dJ}{dt}\right)_{B\odot} = (0.50\pm0.08)\,10^{-2}\;. \tag{24}$$

Assuming $k_2 = 0.295$ we get $\varepsilon_{Moon} = 2.39°\pm0.08°$, $\varepsilon_{Sun} = 0.97°\pm0.16°$.

5 Residuum in the Total Angular Momentum Budget

No directly observed total secular variation in ω was used in the above calculations. It can be precomputed as the sum of (22) and (8):

$$\frac{d\omega}{dt} = \left(\frac{d\omega}{dt}\right)_{tidal} + \left(\frac{d\omega}{dt}\right)_{nontidal} = -(5.32\pm0.25)\,10^{-22}\,\text{rad s}^{-2}\;. \tag{25}$$

The deviation of (25) from the total secular decrease in ω directly observed is the actual residuum in the angular momentum budget in the Earth-Moon-Sun system. On the basis of the recent estimates, there is no discrepancy exceeding the standard error in the direct determination of the secular term in $d\omega/dt$.

6 Conclusions

1. The observed values of dn/dt, dJ_2/dt, $d\omega/dt$ fit well to the angular momentum budget in the Earth-Moon-Sun system.

2. The tidal decrease in ω
   a) due to the Moon is about $-(6.1 \pm 0.2)\ 10^{-22}$ rad s^{-2},
   b) due to the Sun $-(0.5 \pm 0.1)\ 10^{-22}$ rad s^{-2}.
3. The total decrease in ω corresponding to the zero-residuum in the angular momentum budget in the Earth-Moon-Sun system is $-(5.3 \pm 0.3)\ 10^{-22}$ rad s^{-2}.

References

Burša M (1987) Secular deceleration of the Moon and of the Earth's rotation and variation in the zonal geopotential harmonic. Bull Astron Inst Czech 38:309–313

Chovitz BH (1987) Parameters of common relevance of astronomy, geodesy, and geodynamics. XIX Gen Assoc IUGG, Vancouver, Canada

Christodoulidis DC, Smith DE, Williamson RG, Klosko SM (1988) Observed tidal braking in the Earth/Moon/Sun system. J Geophys Res 93 (B6):6216

Kaula WM (1964) Tidal dissipation by solid friction and the resulting orbital evolution. Rev Geophys 2 (4):661

Rubincam DP (1984) Postglacial rebound observed by Lageos and the effective viscosity of the lower mantle. J Geophys Res 89 (B2):1077–1087

Wu PW, Peltier WR (1984) Pleistocene deglaciation and the Earth's rotation: a new analysis. Geophys J R Astron Soc 76:753–791

Celestial Mechanics of Present Tidal Friction

J. Kostelecký

1 Introduction

The purpose of this paper is to outline the main features of celestial-mechanical aspects of the influence of tidal friction in the Earth-Moon-Sun system.

We will especially concentrate on the tidal evolution of the Moon's orbit, because the contemporary observational technique enables us in this case to confront theoretical estimations with the results of observations. The solar effects will hence be somewhat behind. This subject was already discussed in detail by many authors, in some cases in a quite exhaustive way (see, e.g., Lambeck 1977; Brosche and Sündermann 1982).

2 Equations of Motion

One of the possible and from the practical point of view very suitable attempt is to leave aside the classical theories of the motion (be it older or newer ones) and to use the tool of astrodynamics.

From the viewpoint of astrodynamics the motions of the Moon and/or the Sun (or of the Earth round the Sun) may be investigated in the same way (that means by the same mathematical tool) as the motion of an artificial satellite. The investigation of the whole system stems from Newton's law of gravitation which in a simple form is given by well-known Newton's equations in the Cartesian coordinates system.

In our case it is more suitable to change from rapid time varying Cartesian coordinates to Keplerian elements – the semi-major axis a, eccentricity e, argument of perigee ω, node Ω, inclination I and mean anomaly M, which define the elliptic motion. The advantage of Kepler's elements is the fact that in the case of so-called non-perturbed motion (solution of the problem of the motion of two spherically symmetrical bodies) they are constant, except the mean anomaly which varies regularly with time. In the case of perturbed motion – this is our case– their time changes are small.

The transformation of Newton's equations of motion into Kepler's elements is performed by means of Lagrange's planetary equations (LPE) (see, e.g., Kaula 1966). From the system of six Lagrange's planetary equations we give the first three – for the change of the semi-major axis a, of the eccentricity e, and of the orbit inclination I:

Geodetic Observatory Pecný, 251 65 Ondřejov 244, ČSFR.

$$\frac{da}{dt} = \frac{2}{na}\frac{\partial V}{\partial M},$$

$$\frac{de}{dt} = \frac{1}{na^2 e}\left[(1-e^2)\frac{\partial V}{\partial M} - (1-e^2)^{1/2}\frac{\partial V}{\partial \omega}\right] \tag{1}$$

$$\frac{dI}{dt} = \frac{1}{na^2(1-e^2)^{1/2}\sin I}\left[\frac{\partial V}{\partial \omega}\cos I - \frac{\partial V}{\partial \Omega}\right].$$

The quantity V is called the perturbing potential and it can include all the effects which cannot be represented as an effect of a central symmetrical body.

Further, we will be interested in the effects of non-homogeneous Earth gravitational field and of tidal forces on the motion of artificial satellites and a natural satellite – the Moon.

3 Effect of Gravitational Field

The effect of the external gravitational field on the motion of a satellite may be properly described by a system of spherical harmonic function and its parameters – the so-called Stokes' parameters. In a system of Keplerian elements it can be written as

$$V = \mathrm{Re}\sum_{l,m} \bar{\mathrm{C}}_{l,m}\left(\frac{a_e}{a}\right)^l f(a,e,I)\exp i[g(\omega,\Omega,M,\theta_G)] \tag{2}$$

where

$$\bar{\mathrm{C}}_{lm} = \bar{C}_{lm} - i\bar{S}_{lm}\,. \tag{3}$$

$\bar{C}_{lm}, \bar{S}_{lm}$ are fully normalized Stokes' parameters (harmonic coefficients) and θ_G is the Greenwich sidereal time. Function g is a linear function of the elements ω, Ω, M and θ_G; a_e is the radius of the Earth.

The crucial factor for the magnitude of disturbances is the ratio $(a_e/a)^l$. In the case of low satellites $a_e/a \sim 1$ and, accordingly, the effect of higher order terms is considerable. In the case of the Moon it is $(a_e/a) \sim (6/384) \ll 1$ and consequently the Moon's motion is essentially affected only by the coefficient $\bar{\mathrm{C}}_{2,0}$ which expresses the dynamical flattening of the Earth.

Putting (2) into (1) we obtain

$$\frac{da}{dt} = \mathrm{Re}\left\{\sum \bar{\mathrm{C}}_{lm}\left(\frac{a_e}{a}\right)^l f(a,e,I)\, i^{(l+m+1)}\exp i[\psi - m\theta_G]\right\} \tag{4}$$

and equivalent relations for $\frac{de}{dt}, \frac{dI}{dt}$. For the function ψ is valid:

$$\psi = (.)\omega + (..)M + (...)\Omega$$

where (.), (..), (...) are constant numbers – different for different frequencies.

4 Earth's Body Tides

As a result of the Earth's elasticity, the tidal forces of the Moon and the Sun cause the deformation of the Earth. This deformation gives rise to the change of the external gravitational field and, consequently, also to the change of Stokes' parameters.

Let us now outline the derivation of the equations of motion of a satellite due to the Moon's tidal effect (for the effect of the Sun we proceed identically).

As we are interested in the effect of an "additional" potential ΔV_T caused by the deformation, let us recall that it holds (from the definition)

$$\Delta V_T(P) = \sum_{l=2}^{\infty} k_l V_l(P) \tag{5}$$

where k_l are the Love numbers – elastic parameters.

In order to obtain the value of the additional potential at an external point, we have to solve Dirichlet's problem for the exterior of the sphere. Substituting the real $V_l(P)$ into (5) and comparing the result with (2) (see, e.g., Kostelecký 1988) we obtain for the "increment" of the Stokes' parameters $\Delta\bar{C}_{lm}$ the expression:

$$\Delta\bar{C}_{l,m} = \mathrm{Re}\{\sum k_l f(a_☾, e_☾, i_☾, \varepsilon) \cdot \exp i [g(\omega_☾, \Omega_☾, M_☾, \theta_G, \varepsilon_☾)\}$$

where ε is obliquity and $\varepsilon_☾$ phase shift. In case of solar tides we get an analogous result, only we write everywhere "$\odot$" instead of "☾" and it holds identically $i_\odot = 0$.

5 Ocean Tides

If the oceans were spread over the entire Earth, it would be easy to express their effect through the potential of a spherical layer. Even if this condition is not actually fulfilled, it is possible to take advantage of this attempt.

The elevation ξ of the tidally deformed sea surface above mean sea level at a point with coordinates φ, λ at a moment t may be expressed – (see Lambeck 1977) – as

$$\xi(\varphi, \lambda, t) = \sum_{lm} \sum_{+}^{-} D_{lm}^{\pm} \sin(\sigma t \pm m\lambda + \varepsilon_{lm}^{\pm}) \bar{P}_{lm}(\sin\varphi) \tag{6}$$

where $D_{lm}^{\pm}$ and $\varepsilon_{lm}^{\pm}$ denote the amplitude and the phase of corresponding ocean tidal waves, σ is the corresponding frequency of these waves, t is the mean sidereal time and $\bar{P}_{lm}$ is Legendre associated function.

The additional potential ΔV corresponding to the ξ from Eq. (6) may be written in the infinite sum of spherical harmonic expansion

$$\Delta V = \sum_{l} \Delta V_l \ . \tag{7}$$

This additional potential represents, however, only the effect of water masses. It must be completed by an effect of the potential change due to the deformation of the subsoil – the ocean bottom. This effect may be expressed by means of "loading numbers" k_l'. The values of k_l' for the Earth have been derived by Farell 1971). The resulting additional potential then reads

$$\Delta V = \sum_l (1+k_l')\Delta V_l \ . \tag{8}$$

After substitution of decomposition ΔV_l into the spherical harmonics and comparison with (2) we find that the effect of additional potential of that type may be included into the Stokes' parameters.

It holds

$$\Delta \bar{C}_{l,m} = \Delta \bar{C}_{l,m} - i\Delta \bar{S}_{l,m} = \sum_s \sum_{\substack{-\\+}} (\bar{C}^{\pm}_{slm} \mp i S^{\pm}_{slm}) F_{lm} \exp(\pm i\theta_s) \tag{9}$$

where

$$F_{lm} = \frac{4\pi\varrho_w G}{g}\left[\frac{(l+m)!}{(l-m)!\,(2l+1)(2-\delta_0^m)}\right]\frac{1+k_e'}{2l+1} \ ,$$

G is universal gravity constant, ϱ_w is water density, g is the mean gravity value on the Earth's surface, θ_s is a linear combination of "angular" elements of the Sun and the Moon. Corresponding linear combinations then form well-known "classical" tidal frequencies with respective denotations – for definition see, e.g., Kostelecký (1988).

6 Effect of Tides on the Motion of Artificial Satellites, Solution of the Inverse Problem: Determination of Tidal Parameters

Summarizing the results of preceding sections we can state that the effect of tidal perturbations on satellite orbits may generally be characterized by the expression

$$\frac{dE}{dt} = \mathrm{Re}\{A_E \exp i[\psi - \psi_{☾(\odot)} + \varepsilon_{☾(\odot)}]\} \tag{10}$$

where E denotes an arbitrary element from the six Keplerian elements, A_E is the amplitude corresponding to this element according to LPE [it depends on the semi-major axis a, eccentricity e and inclination I of a satellite and on $(a,e,i)_{☾}$ or $(a,e)_{\odot}$, and possibly on the inclination of the ecliptic ε], ψ is a linear combination of the angular elements of a satellite ω, Ω and M, and $\psi_{☾(\odot)}$ is a linear combination of the angular elements $\omega_{☾}, \Omega_{☾}, M_{☾}$ or those of the Sun, $\varepsilon_{☾(\odot)}$ being the phase shift.

Dividing the types of perturbations as expressed by (10) into the parts a) secular (non-periodic perturbations), b) long-periodic (periodic perturbations with a period longer than 1 day) and c) short-periodic, we may draw the following conclusions.

In consequence of the fact that "almost everywhere" $\psi \neq \psi_{☾(\odot)}$ and that for the elements a, e, I a linear combination l, m, p, ... (which forms secular perturbations) cannot be found, the tidal variations of the elements a, e, I are only periodical. (With the rest of elements it is possible to select suitable combinations of coefficients l, m, ... forming also secular perturbations.)

Accordingly, the elastic parameters k_l and the coefficients in the expansion of ocean tides have been determined by spectral analysis of periodic variations of the orbital elements of a satellite as a function of time (after elimination of other perturbing effects). The analysis of secular effects did not lead to exact results since in the modeling of secular variations a significant part is due to non-gravitational effects, the characteristics of which are not reliable known.

At present one does not proceed in this way yet and the tidal parameters are being determined by another method. Before we outline the presently used method we have to pay attention to one difficulty inherent in determination of the tidal parameters from the analysis of satellite orbits.

The expressions for the effect of solid earth (6) and ocean tides (9) contain the same terms describing the motion of the Moon (or Sun). Hence it follows that two effects cannot be separated and from the analysis of satellite observations we may directly determine only one of them, whereas the other must be theoretically modeled in advance.

In most satellite solutions the solution for solid tides $k_2 = 0.3, \varepsilon_2 = 0$! is taken as known and the parameters of ocean tides are determined (Christodoulides et al. 1988; Gendt and Dietrich 1988). Periodic variations of in advance (from observations) determined elements of satellite orbits are not analyzed but one uses directly the observation equation

$$\varrho_c - \varrho_{obs} = \sum \left(\frac{\partial f(p)}{\partial p} \Delta p + p_0 \right) \tag{11}$$

where ϱ denotes either the measured distance station-satellite or the component of the direction toward the satellite or the difference of distances for two time instants etc., p is the parameter to be determined (e.g., Stokes's dynamical parameters of the Earth's gravitational field, coordinates of a terrestrial station, coordinates of the Earth's rotation pole or tidal parameters). The function f(p) is a function of these parameters and p_0 is a corresponding approximation as the equations are non-linear and must be linearized before using the least squares adjustment.

7 Effect of Tides on the Moon's Motion

As the Moon is a natural satellite of the Earth, we may, of course, apply LPE for the motion of artificial satellites also to the motion of the Moon.

In the following we will deal only with the character of perturbations of the semi-major axis a, eccentricity e and inclination of the orbit $i_☾$. We already know that in the case of artificial satellites there are only periodic perturbations of these elements. However, as the Moon is a tide-generating body on the one hand and a satellite on the other hand, it is possible to find such coefficients of linear combinations in ψ so that it might hold

$$\left.\frac{dE}{dt}\right|_0 = \mathrm{Re}\{A_E \exp i[\psi_☾ - \psi_☾ + \varepsilon_☾]\} \rightarrow \frac{dE}{dt} = \text{const.} \tag{12}$$

in case of the ocean tides and

$$\left.\frac{dE}{dt}\right|_b = A_E \sin \varepsilon_{☾} \doteq (.) k_2 \varepsilon_2 / _{☾} \ . \tag{13}$$

in case of the body tides.

Thus we can see that, contrary to satellites, the *secular* terms exist in the motion of the Moon.

8 Parameters of Ocean Tides – Satellite Solutions

In Table 1 the parameters of ocean tides are given for the main tidal waves which exert a decisive influence on the variation of the Moon's orbital elements.

It is a case of amplitudes and phases in form

$$(C^{\pm}_{nm,f} - i S^{\pm}_{nm,f}) = -i \hat{C}^{\pm}_{nm,f} \exp i(\varepsilon^{\pm} + \chi_f) \ . \tag{14}$$

Most of these parameters have been derived by way of spectral analysis of the elements of satellite orbits and only few of them have been determined by a common processing of parameters of gravitational field and ocean tides.

As a result of inseparability of body and ocean tides most parameters of ocean tides were obtained on the assumption of zero phase shift $\varepsilon_{☾}$ for body tides.

Goad and Douglas (1978) demonstrate the dependence of amplitudes on the magnitudes of the phase shifts ε_2. This dependence amounts for $\varepsilon_2 = 1°$ up to 30%, which is not a negligible value (practically 1 cm in amplitude).

Hitherto the theory of tidal friction has presupposed the phase shift ε_2 to be about 0.5° (Schwiderski, person. commun.), even if Melchior indicates (in Melchior 1986) that tidal measurements corrected for an indirect oceanic effect give almost zero phase.

Table 1 is further completed by the data obtained with the help of cotidal maps (Schwiderski 1980; Seiler 1988).

It is possible to say that the values of the expansion of ocean tides into the spherical harmonics that are necessary for the determination of the tidal friction are presently known with an accuracy of about 20%.

9 Parameters of Body Tides

A nominal value of the elasticity parameter k_2 is at present supposed to be equal to 0.3 taking account of corresponding corrections for different frequencies as derived by Wahr (Project Merit Standards 1983). As far as the phase shift ε_2 is concerned, the present analyses indicate that

$$\varepsilon_2 < 0.5° \ . \tag{15}$$

(Zschau and Dehant, suppose that ε_2 can maximally reach a few tenths of a degree – see Melchior 1986).

Table 1. Amplitudes and phases of ocean tides ($\hat{C}^{+}_{nm,f}, \varepsilon^{+}_{nm,f}$)

References	M_2		N_2		K_1		O_1	
	cm	o	cm	o	cm	o	cm	o
Lambeck (1977)	3.64 ± 0.3 2.6 ± 0.6	338*	0.69 ± 0.2	350*	2.00 ± 0.3	304*	1.17 ± 0.3	310*
Felsentreger et al. (1978)	3.1 ± 0.7	319						
Goad and Douglas (1977)	3.8 ± 0.6	331						
Goad and Douglas (1978)	1.9 ÷ 3.2	317 ÷ 331						
Christodoulidis et al. (1985)	3.46	319	0.59	346			2.84	327
Williamson and Marsh (1985)	2.57	319	0.76	323			2.85	344
Moore (1987)	2.70	342	0.46	8.5			3.61	19
Marsh et al. (1988)	3.26	321	0.70	334			2.69	318
Gendt and Dietrich (1988)	3.26 ± 0.02	320 ± 0.4	0.62 ± 0.03	321 ± 2			3.06 ± 0.05	312 ± 1.2
Christodoulidis et al. (1988)	3.26 ± 0.05	321 ± 0.9	0.70 ± 0.07	334 ± 5	2.61 ± 0.2	321 ± 5	2.69 ± 0.17	310 ± 3.6
Goad by cotidal charts of Schwiderski (1980)	2.95	311	0.65	321	2.8	315	2.42	314
Seiler (1988)	3.26	332*	0.79	339*	2.71	302*	2.96	308*

* after recomputation.

10 Secular Tidal Variation of Parameters of the Moon's Orbit

a) Effect of Oceans

On the basis of (12) can be written

$$\frac{da_{☾}}{dt}\left[\frac{m}{y}\right] = 0.007\,S^{+}_{20,M_m} + 0.010\,S^{+}_{20,M_f} - 0.071\,C^{+}_{21,Q_1} - 0.245\,C^{+}_{21,O_1} + 0.340\,S^{+}_{22,N_2} + 1.181\,S^{+}_{22,M_2} + \ldots,$$

$$\frac{de_{☾}}{dt}\left[\frac{1}{10^9 y}\right] = 0.159\,S^{+}_{20,M_m} - 0.555\,C^{+}_{21,Q_1} + 0.009\,C^{+}_{21,O_1} + 2.674\,S^{+}_{22,N_2} - 0.042\,S^{+}_{22,M_2} + \ldots, \tag{16}$$

$$\frac{di_{☾}}{dt}\left[\frac{deg}{10^9 y}\right] = -0.161\,(-0.035)\,S^{+}_{20,M_f} + 0.729\,(0.158)\,C^{+}_{21,Q_1} + 3.793\,(0.821)\,C^{+}_{21,O_1} - 3.516\,(-0.761)\,S^{+}_{22,N_2} - 18.281\,(-3.959)\,S^{+}_{22,M_2} + \ldots$$

(the data in brackets hold for the change of inclination defined with respect to the ecliptic).

The quantities C^+, S^+ are connected with the amplitudes and phases from Table 1 by the following relation

$$(C^{\pm}_{nm,f} - i S^{\pm}_{nm,f}) \cdot 10^2 = -i\hat{C}^{\pm}_{nm,f} \exp i(\varepsilon^{\pm}_{nm,f} + \chi_f)\,,$$

$$\chi_f = \begin{cases} 0 & \text{semidiurnal, longperiodic} \\ \pi/2 & K_1 \\ -\pi/2 & O_1, P_1, Q_1 \end{cases} \tag{17}$$

In Table 2 the results of secular variation of the eccentricity and inclination of the Moon's orbit are presented. The values by Lambeck (1977) are based on older results from the analysis of satellite orbits, the values by Kostelecký (1988) have been calculated on the basis of results derived from Schwiderski's cotidal maps. The Seiler (1988) result is a new solution from oceanography. (The values in brackets are deter-

Table 2

Influence of ocean	$de_{☾}/dt$ 10^{-9} yr	$di_{☾}/dt$ o/10^9 yr	$da_{☾}/dt$ cm/yr
Lambeck (1977)	0.018 ± 0.006	−0.754 ± 0.08	4.54 ± 0.5
Kostelecký (1988)	0.014	−0.433 (−0.086)	2.87
Seiler (1988)	0.018	−0.620 (−0.134)	4.09

Table 3

Influence of body tides	$de_{☾}/dt$ 10^{-9} yr	$di_{☾}/dt$ o/10^9 yr
Kaula	0.012	−0.54
Gravimetry without ocean corr.	0.007	−0.35
Satellite sol.	0.0026	−0.12

mined for $i_{☾}$ defined with respect to ecliptic.) Lambeck's solution gives too large a value of the change of the semi-major axis of the Moon's orbit (4.5±0.5 cm/yr) as compared to the observed one 3.7 cm/yr – see Burša (1987). By contrast, the solution Kostelecký (1988) made on the basis of Schwiderski's (1980) results leaves about 1 cm/yr for the effect of the body tides.

b) Effect of Body Tides

On the basis of (13) can be written for $de_{☾}/dt, di_{☾}/dt$

$$\frac{de_{☾}}{dt}\left[\frac{1}{10^9 y}\right] = 0.53\, k_{2,N_2} \sin \varepsilon_{2201} + \dots$$
$$\frac{di_{☾}}{dt}\left[\frac{|\deg|}{10^9 y}\right] = -19.421\, k_{2,M_2} \sin \varepsilon_{2200} + \dots \qquad (18)$$

where

$$\varepsilon_{☾} \equiv \varepsilon_{lmpq} = -(l-2p+q)n_{☾}\Delta t + m\theta\Delta t$$

$n_{☾}$ being the mean diurnal motion of the Moon and Δt the lime-lag of the tidal response.

In Table 3 the results are given for secular variations of the eccentricity and inclination of the Moon's orbit for different values of the $\varepsilon_{☾}$ factor, taken from Kostelecký (1985). The original Kaula's solution seems to be exaggerated from the point of view of present geophysical results, the results of gravimetry are a bit more realistic. The tabulated values, however, make use of $\varepsilon_{☾}$ factors which are not corrected for an indirect effect of ocean tides. The satellite solution by Lambeck et al. (1974) corresponds to the present reality.

c) Cumulative Effect

In Table 4 the values given are of the total effect of secular tidal variations of the eccentricity and inclination of the Moon's orbit, which may be used for the balance equation. The values I. are calculated on the basis of oceanic effects from Kostelecký (1988) and of body tides from gravimetry (not corrected for the indirect ocean effect). The values II. are based on satellite estimation of body tides. The results show good agreement for the secular variation of the eccentricity and inclination with the solu-

Table 4

Total effect	$de_{☾}/dt$ 10^{-9} yr	$di_{☾}/dt$ o/10^9 yr	$da_{☾}/dt$ cm/yr
I	0.021 ± 0.003	-0.78 ± 0.07 (-0.44)	
II	0.017 ± 0.003	-0.55 ± 0.07 (-0.21)	
Christodoulidis et al. (1988)	0.018 ± 0.01	-0.66 ± 0.05	3.73 ± 0.09

tion of Christodoulidis et al. (1988). Another question is whether the values from Christodoulidis et al. express a total effect. On the other hand, it is true that the results by Christodoulidis et al. have been obtained from processing of satellite observations supposing identially $\varepsilon_{☾} = 0$ – that is zero effect of body tides.

11 Conclusion

Thanks to precise satellite observations – first of all to laser ranging to satellites *Lageos* and *Starlette* (see Christodoulidis et al. 1988; and Gendt and Dietrich 1988) it has recently become possible to obtain the characteristics of ocean tides with a fairly high accuracy. The results agree with the values obtained on the basis of cotidal maps.

Also, in the case of determination of the tidal friction due to body tides it becomes evident – in accordance with geophysical models – that the value of the phase lag is less than originally assumed. If the ocean tides cause at least 70% or 80% of the secular regression of the Moon (compare the results in Tables 2 and 4 and Seiler's oceanographic results) it holds $\varepsilon_2 \ll 0.5°$ for body tides.

The dominating role of the ocean tides is thus presently confirmed also by the results of precise satellite observations.

References

Brosche P, Sündermann J (1978 and 1982) Tidal friction and Earth's Rotation I and II. Springer, Berlin Heidelberg New York Tokyo

Burša M (1987) Tidal friction and the Earth's rotation. Final report of IAG. Special study group 5-99 (Submitted to XIX General Assembly of IUGG, Vancouver)

Christodoulidis DC, Smith DE, Klosko SM, Torrence MH, Dunn PJ (1985) A GSFC alternative to the SLR MERIT constants. In: Mueller II (ed) Proc Int Conf on Earth Rot and Terr Ref Frame. Dep Geod Sci Ohio State Univ, Columbis, Ohio, 675 p

Christodoulidis DC, Klosko SM (1988) Observed tidal braking in the Earth/Moon/Sun system. Geoph Res 93 (B6):6216

Farell WE (1971) Deformation of the Earth by surface loads. Rev Geoph Space Phys 10:761

Felsentreger TL, Marsh JG, Agreen RW (1976) Analyses of the solid earth and ocean tidal perturbations on the orbits of the Geos 1 and Geos 2 satellites. Geoph Res 81, No 14:2557

Gendt G, Dietrich R (1988) On the determination of tidal parameters using LAGEOS laser ranging data. 6th Int Sym Geodesy and Physics of the Earth, Potsdam

Goad CC, Douglas BC (1978) Lunar tidal acceleration obtained from satellite-derived ocean tide parameters. JGR 83 (35):2306

Kaula WM (1966) Introduction to satellite geodesy. Bleisdel

Kostelecký J (1985) Tidal friction. Secular variations of the inclination and eccentricity of the Moon's orbit. Studia Geoph Geod 29:319

Kostelecký J (1986) A few remarks about rotational deformation of the Earth and motion of the Moon. Proc Int Symp. Figure and Dynamics of the Earth, Moon and Planets. Prague, 581 p

Kostelecký J (1988) Tidal friction. The effect of oceans on the secular variation of the inclination and eccentricity of the Moon's orbit. Studia Geoph Geod 32:9

Lambeck K (1977) Tidal dissipation in the oceans: astronomical geophysical and oceanographic consequences. Phil Trans Soc Lond 287 (1347):545

Lambeck K, Cazenave A, Balmino G (1974) Solid Earth and ocean tides estimated from satellite orbit analyses. Rev Geoph Space Phys 12:421

Marsh JG, Lerch FJ, Putney BH et al. (1988) A new gravitational model for the Earth from satellite tracking data: GEM-T1. Geoph Res 93 (B6):6169

Melchior P (1986) Earth rotation and solid tides. In: Cazenave A (ed) Earth Rotation: Solved and Unsolved Problems. Reidel, Dordrecht, 79 p

Moore P (1987) Ocean tidal parameters from Starlette data. Bull Geod 61:223

Project MERIT Standards (1983) USNO Circular No 167, US Naval Observatory, Washington DC

Schwiderski EW (1980) On charting global ocean tides. Rev Geoph Space Phys 18:243

Seiler U (1988) Eine Untersuchung der Gezeiten des Weltozeans und ihres instantanen Drehimpulshaushalts. Dissertation, Univ Hamburg

Williamson RG, Marsh JG (1985) Starlette geodynamics: The Earth's tidal response. Geoph Res 90 (B11):9346

The Consideration of Solid Earth Effects in Ocean Tide Modeling

W. Zahel

1 Introduction

Computations performed with realistic ocean tide models and with a hemispherical analytical model including loading and self-attraction effects show (Zahel 1988) that the latter model represents a proper tool for estimating how tidal oscillation systems and quantities being relevant to variations of the Earth's rotation depend on ocean and solid Earth parameters.

Besides the analytical approach providing among other things the possibility to comprise a wide range of, e.g., forcing frequencies and parameter values at acceptable expense, the free oscillation properties of the oceanic area can be calculated with the same degree of precision for exactly the same internal forces acting as in the forced oscillation model. Thus, in particular, consequences of the different ways of taking into account the gravitational attraction varying due to solid Earth and sea-surface deformations can concurrently be computed for free and forced oscillations, and the results obtained can be related to each other. This procedure promises a better understanding of dynamical effects a proper inclusion of which might be decisive for the reliability of realistic ocean tide models and for the quantities which are derived from the computed tidal fields of pressure and motion. In fact, it must be regarded as rather unknown how the oceanic response to the complete astronomical tidal potential depends on the way loading and self-attraction are included in different recent realistic ocean tide models. Further, the expansion of the hemispherical model variables in terms of spherical harmonics in a straightforward fashion permits to assimilate prescribed sea-surface values and to investigate their contribution to changes of the oscillation systems dependent on which defect of the dynamics included in the model is to be adjusted by the data. In view of the expected availability of uniformly distributed open ocean tidal data from satellite altimetry, indeed models are needed which simulate the data assimilation scenario and which have the power to provide an idea of the efficiency of introducing data into models.

2 The Computation of Free and Forced Oscillations

Starting from the long wave equations taken as the basis of the realistic ocean tide model applied by Zahel (1980), one obtains when in particular introducing the special

---

Institut für Meereskunde, Universität Hamburg, Troplowitzstraße 7, D-2000 Hamburg 54, FRG.

topography of a hemispherical ocean with constant depth h the linear algebraic equations (Zahel 1988)

$$-\sigma^2 p_r - iR\sigma p_r - 2i\sigma\omega\mu_r^{-1} \sum_{s=-\infty}^{\infty} \beta_{r,s} p_s + gh\mu_r (1-\alpha'_r/2) p_r$$

$$-\frac{1}{2} gh \sum_{\substack{r'=1 \\ r' \neq r}}^{\infty} \mu_{r'} q(r,r') p_{r'} = g\gamma_2 \bar{\zeta}_r \tag{1}$$

$$-\sigma^2 p_{-r} - iR\sigma p_{-r} - 2i\sigma\omega\nu_r^{-1} \sum_{s=-\infty}^{\infty} \beta_{-r,s} p_s = 0 \tag{2}$$

for the expansion coefficients p_r, p_{-r} where

$$q(r,r') = R_e^4 a_{\bar{n}}^{\bar{m}} a_{n'}^{m'} \sum_n \alpha'_n \sum_m (2m a_n^m)^2 I\binom{m\,\bar{m}}{n\,\bar{n}} I\binom{m\,m'}{n\,n'} / ((m^2-\bar{m}^2)(m^2-m'^2)) \ ,$$

$$I\binom{m\,m'}{n\,n'} = \int_{-1}^{+1} P_n^m(\mu) P_{n'}^{m'}(\mu)\, d\mu \quad \text{and} \quad \nu_r = \mu_r = n(n+1)/R_e^2 \ , \quad \bar{\zeta}_r = \iint_B \Phi_r \bar{\zeta}\, dA \ ,$$

$$\alpha'_n = (1+k'_n-h'_n)\alpha_n \ , \quad \gamma_2 = 1+k_2-h_2 \quad \text{with Love numbers } k'_n,\ h'_n \text{ and } k_2,\ h_2 \ .$$

The horizontal current velocity vector $\mathbf{v}_h$ and the sea-surface elevation ζ relative to the sea bottom are given by

$$\mathbf{v}_h = \frac{\partial}{\partial t}(\nabla\Phi + \nabla\Psi \times \mathbf{z}) \quad \text{and} \quad \zeta = h \sum_{r=1}^{\infty} p_r \mu_r \Phi_r$$

where $\Phi = \sum_{r=1}^{\infty} p_r \Phi_r e^{-i\sigma t}$, $\Psi = \sum_{r=1}^{\infty} p_{-r} \Psi_r e^{-i\sigma t}$, $\Phi_r = a_n^m P_n^m(\sin\phi)\cos(m\lambda)$, $\Psi_r = a_n^m P_n^m(\sin\phi)\sin(m\lambda)$ and λ, ϕ denote geographical longitude and latitude, respectively. While σ denotes the frequency of an oscillation in the ocean, ω stands for the Earth's angular speed of rotation about its axis. R is the coefficient of linear bottom friction. The suffixes, r, $-$r are associated with pairs of suffixes (n, m). The factors a_n^m are chosen such that each of the systems $\{\Phi_r\}$ and $\{\Psi_r\}$ is orthonormal in the hemispherical domain denoted by B. The validity of $\nabla\Phi_r \cdot \mathbf{n} = 0$ and $\Psi_r = 0$ in $\lambda = 0$, π guarantees exact fulfillment of $\mathbf{v}_h \cdot \mathbf{n} = 0$ there, i.e., the normal current velocity component is zero at the meridional boundaries. The $\beta_{r,s}$ are the gyroscopic coefficients (Webb 1980) and $\bar{\zeta}$ denotes the oceanic equilibrium tide on a rigid Earth with radius R_e.

As introduced by Accad and Pekeris (1978) and used, e.g., by Schwiderski (1981) and Seiler (this volume), the loading and self-attraction effect (abbr. LSA) can be parameterized by a term being proportional to the local sea-surface elevation. When using this parameterization in the above model the coefficients $\alpha'_r/2$ in (1) are replaced by a degree-independent constant ε' for which Accad and Pekeris (1978) give the value 0.085. In case of introducing this simplification the Love-number dependent off-diagonal terms in (1) disappear. In the following the abbreviations CLSA and

PLSA designate the property of a model to take into account LSA by incorporating the full LOVE-number approach and by parameterizing LSA using a proper degree-independent coefficient, respectively.

Free oscillations are computed by solving the eigenvalue problem

$$\bar{\sigma}\mu_r p_r = (1-\alpha_r'/2)\mu_r^2 R_e^2 \tau p_r - \frac{1}{2} R_e^2 \tau \mu_r \sum_{\substack{r'=1 \\ r' \neq r}}^{\infty} \mu_{r'} q(r,r') p_{r'}$$
$$-iR/(2\omega)\mu_r p_r - i \sum_{s=-\infty}^{\infty} \beta_{r,s} p_s \qquad (3)$$

$$\bar{\sigma}\nu_r p_{-r} = -iR/(2\omega)\nu_r p_{-r} - i \sum_{s=-\infty}^{\infty} \beta_{-r,s} p_s \qquad (4)$$

with the notations $\bar{\sigma} = \sigma/(2\omega)$, $\tau = 1/(\varepsilon\bar{\sigma})$, $\varepsilon = 4\omega^2 R_e^2/(gh)$. Equations (3), (4) result from (1), (2) as the corresponding homogeneous equations. The solution is obtained by giving τ as a complex number and varying it in the frame of an iteration procedure such that $1/(\tau\bar{\sigma}(\tau))$ takes the prescribed value $\varepsilon = 4\omega^2 R_e^2/(gh)$.

Finally, a data assimilation scenario can properly be defined and evaluated in this case of the tides in a schematic ocean basin. The role of the data to be assimilated into the model described by equations (1), (2) is played by values of the tidal sea-surface topography computed by a corresponding model which differs from the former one solely by the set of parameters used or the way LSA is taken into account. In case of this spectral model where the sea-surface elevation is explicitly expressed in terms of the expansion coefficients, a least squares procedure can be formulated in a straightforward way. This procedure provides an adjustment between the fulfillment of the equations of motion and the adoption of prescribed sea-surface values and is a finite dimensional analog to the procedure developed by Bennett and McIntosh (1982). To introduce dimensional consistency and to make the terms of the data equations take values of the same order of magnitude as the terms of (1) and (2), the data equations are multiplied by $R_e\pi/g$ thus writing

$$(kR_e\pi/g)\zeta_j = (kR_e\pi/g)h \sum_{r=1}^{\infty} p_r\mu_r \Phi_r(\phi_j,\lambda_j) \quad j=1,2,\ldots \qquad (5)$$

where ζ_j denotes a prescribed sea-surface value at the position (ϕ_j,λ_j) and k is a weighting coefficient. The least squares solution to the system (1), (2), (5) is then obtained by standard methods.

3 Free and Forced Hemispherical Oscillations

When evaluating (1), (2) and (3), (4), respectively, constituents up to degree and order 23 are considered, and the time interval regarded comprises the semi-diurnal and diurnal tidal bands and, as the largest period, that of the fastest rotational free oscillation.

Table 1. Computed real eigenperiods T and imaginary parts σ_I of eigenfrequencies of oscillations in a hemispherical ocean

CLSA		PLSA		LSA neglected	
T [s]	$\sigma_I [10^{-6} s^{-1}]$	T [s]	$\sigma_I [10^{-6} s^{-1}]$	T [s]	$\sigma_I [10^{-6} s^{-1}]$
Symmetric eigenoscillations					
Fastest rotational eigenoscillation					
384348	−7.9628	385983	−7.9281	380894	−7.9605
Slowest gravitational eigenoscillations					
162728	−4.2875	161521	−4.2313	153806	−4.2231
84273	−4.4669	84809	−4.4320	80960	−4.4390
59452	−5.1319	60016	−5.0084	57644	−5.1284
52989	−6.1891	53016	−6.2622	51476	−6.1098
48870	−5.5802	49277	−5.5187	47541	−5.5263
42762	−5.0242	43360	−5.0670	41635	−5.0057
40753	−5.2533	41298	−5.2485	39733	−5.2205
Antisymmetric eigenoscillations					
Fastest rotational eigenoscillation					
172478	−7.9612	172954	−7.9429	170780	−7.9901
Slowest gravitational eigenoscillations					
110937	−6.2428	110410	−6.1736	107145	−6.1237
74672	−5.3368	74793	−5.3149	71992	−5.2776
54628	−5.0022	55224	−4.9914	52999	−4.9707
43821	−5.6154	44339	−5.4281	42778	−5.6286
41553	−5.8475	41897	−6.0244	40561	−5.7437

The ocean depth and the friction used are h = 4420 m and $R = 9.26 \cdot 10^{-6} s^{-1}$, respectively.

The real eigenperiods, as given in Table 1, demonstrate that their prolongation due to CLSA corresponds to a comparable one when applying PLSA with $\varepsilon' = 0.085$, where the remaining differences are relatively smaller for the slower gravitational oscillations than for the faster ones. However, the latter type of model appears to yield no systematic change in the absolute magnitudes of the imaginary parts of eigenfrequencies which corresponds to the general increase due to CLSA. Of course, with $1/\sqrt{\varepsilon} = 0.2242$ being valid for the computations performed (h = 4420 m), this gravitational effect is of minor importance for the rotational oscillations leading to a slight prolongation of the eigenperiods, just as was obtained for the free oscillations on a globe covered with water (Zahel 1986). As Figures 1–4 show, the property of a free oscillation to be tidally effective is generally not changed when neglecting LSA, applying PLSA or applying CLSA, and the same is true of the main spatial features of the free oscillation systems. E.g., in the case of the sixth, fifth and fourth symmetric oscillations determining semi-diurnal resonances, the amphidromes are preserved with their senses of rotation. The sixth and the fourth oscillations show a northward propagation of phase at the eastern boundary north of the equator, while the fifth shows a southward propagation in the southern part of this section. This behavior clearly corresponds to that of the realistic free oscillations in this period range in the

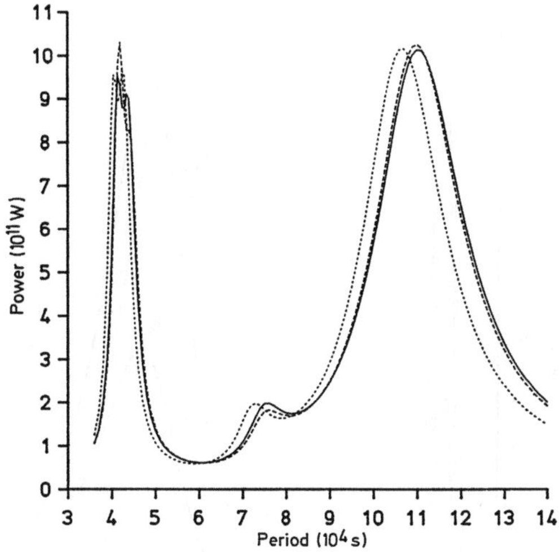

Fig. 1. Rate of work done by tidal forces and moving sea bottom. Model ocean with h = 4420 m and $R = 9.26 \cdot 10^{-6}\,s^{-1}$ driven by Y_2^{-1}-tidal potential with K_1-amplitude and varying periods. *Solid, dashed* and *dotted lines* refer to computations with CLSA, PLSA and LSA neglected, respectively

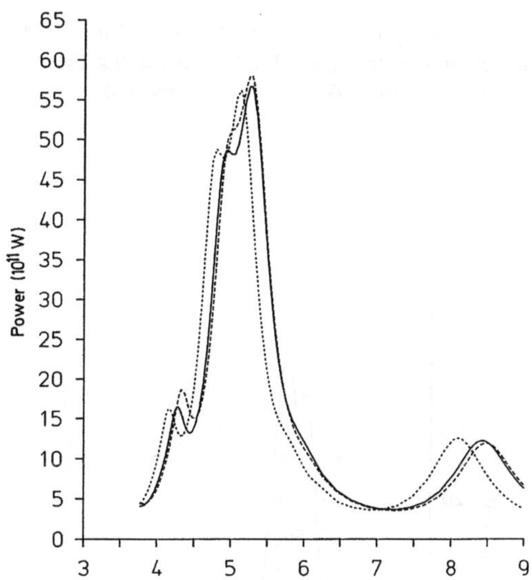

Fig. 2. Same as Fig. 1 except for the model ocean driven by Y_2^{-2}-tidal potential with M_2-amplitude

Pacific (Platzman 1984). Nevertheless, in a few cases tidally effective changes in the spatial structure of the free oscillations can be recognized when inspecting the tidal power curves. Applying PLSA, e.g., makes the fourth antisymmetric free oscillation considerably less effective and the sixth symmetric free oscillation more effective as compared with their behavior when LSA is neglected or CLSA is applied.

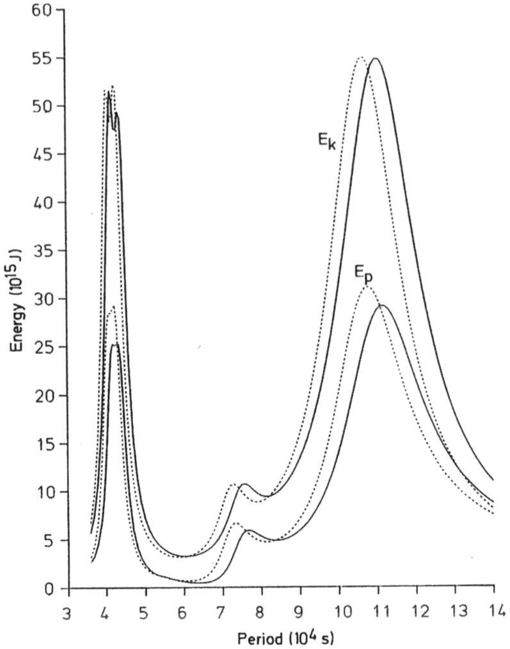

Fig. 3. Kinetic and potential energy contents. Model ocean with $h = 4420\,m$ and $R = 9.26 \cdot 10^{-6}$ s^{-1}, driven by Y_2^{-1}-tidal potential with K_1-amplitude and varying periods. *Solid* and *dotted lines* refer to computations with CLSA and LSA neglected, respectively

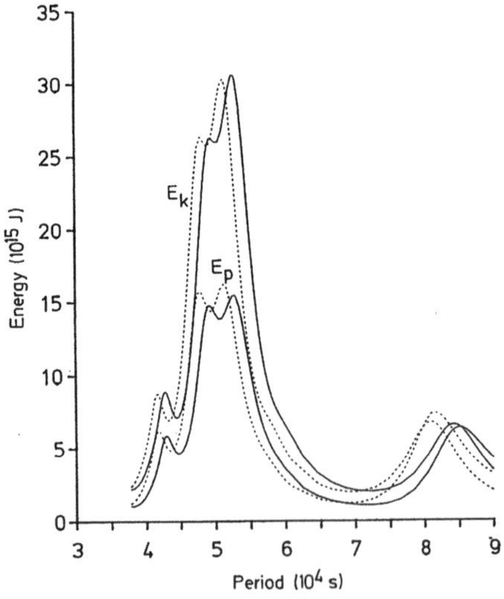

Fig. 4. Same as Fig. 3 except for the model ocean driven by Y_2^{-2}-tidal potential with M_2-amplitude

The oceanic response to forcing by Y_2^{-1}- and Y_2^{-2}-potentials shows a considerable dependence upon whether and by which way, i.e., applying CLSA or PLSA, LSA is taken into account. This dependence becomes obvious in the distribution and partially in the shape of resonance peaks of the work done by external forces and the moving sea bottom and of the peaks of the time-averaged potential energy contents (see Figs. 1–4). The kinetic energy contents are proportional to the rate of work done in case of this hemispherical model. These integrated properties of the forced oscillations clearly reflect the properties of the free oscillations described above. Hence differences which result from applying PLSA instead of CLSA appear predominantly in the semi-diurnal period range. Particularly close to the period of the sixth symmetric free oscillation (K_{2m}, L_2, M_2) the rates of tidal work and energy contents are overestimated by PLSA to up to 28% and 41%, respectively.

In view of the experiences gained with respect to changes in frictional decay time and mean depth in the hemispherical model (Zahel 1988), it is expected that changes in the topography will lead to individual tidal constituents changing roles concerning closeness to or distance from resonance. In that case the tidal constituents would also change roles concerning sensitivity to the way LSA is taken into account, i.e. to whether PLSA or CLSA is applied.

4 Near-Resonance Semi-Diurnal Oscillation Systems

As pointed out in the previous section, in near-resonance situations quantities which describe the overall behavior of the ocean tide, as for example the tidal pressure work and equivalently the Moon's torque on the Earth, suffer alterations in their rates which amount to up to more than 100% when LSA is neglected and still up to 28% when PLSA with $\varepsilon' = 0.085$ is applied, as compared with the rates obtained by applying the model with CLSA. Assigning smaller or larger values than 0.085 to the

Table 2. Computed rates of work $\tilde{W}$ done by tidal forces and moving sea bottom, and of potential energy contents E_p in a hemispherical ocean when applying CLSA, PLSA and when LSA is neglected (LSAn)

Period [s]	$\tilde{W}$ [10^{12} W]			E_p [10^{16} J]		
	CLSA	PLSA	LSAn	CLSA	PLSA	LSAn
42316.28 (η_2)	1.560	1.448	1.462	5.226	4.980	5.807
43082.05 (K_{2m})	1.614	1.829	1.280	5.862	6.908	4.960
43889.83 (L_2)	1.378	1.764	1.352	5.066	7.160	4.915
44714.16 (M_2)	1.330	1.516	1.696	4.634	6.151	5.861
45570.05 (N_2)	1.555	1.539	2.393	5.092	5.910	8.058
47384.04 ($3N_2$)	3.064	2.667	4.670	9.523	9.348	15.381
86164.06 (K_1)	0.197	0.197	0.218	0.500	0.500	0.541
92949.62 (O_1)	0.309	0.314	0.380	0.704	0.764	0.936

The model is driven by Y_2^{-2}- and Y_2^{-1}-equilibrium tides with M_2-amplitude and K_1-amplitude, respectively. The ocean depth is $h = 4420$ m, the frictional decay used is 30 h, i.e., $R = 1/(30 \cdot 3600)\ s^{-1}$.

parameterization coefficient ε' leads to corresponding deviations which are larger than 28%.

It applies also to the real ocean that the semi-diurnal tides are near to resonance, and the rates of tidal power and energy contents given in Table 2 compare well with the rates obtained by realistic global models when regarding the different areas (Schwiderski 1985). As far as the main features are concerned, also the semi-diurnal

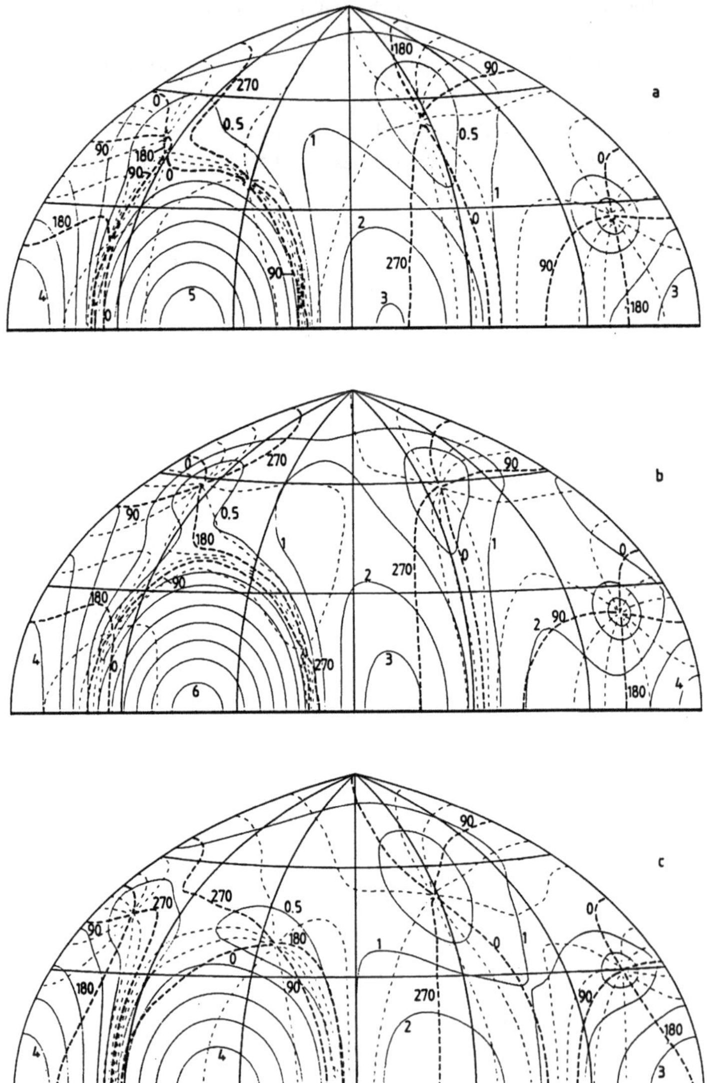

Fig. 5a–c. L_2-tide as obtained by a hemispherical ocean model with $h = 4420$ m and $R = 9.26 \cdot 10^{-6}\,s^{-1}$ and with CLSA (**a**), PLSA (**b**) and LSA neglected (**c**), respectively. Cotidal lines are *dashed*, with phases in degrees (in steps of 30°) referred to meridian passage at the western boundary. Coamplitude lines are *solid*, with amplitudes in units (0.5, 1, 2, 3, 4, 5, 6) of the maximum equilibrium tide amplitude

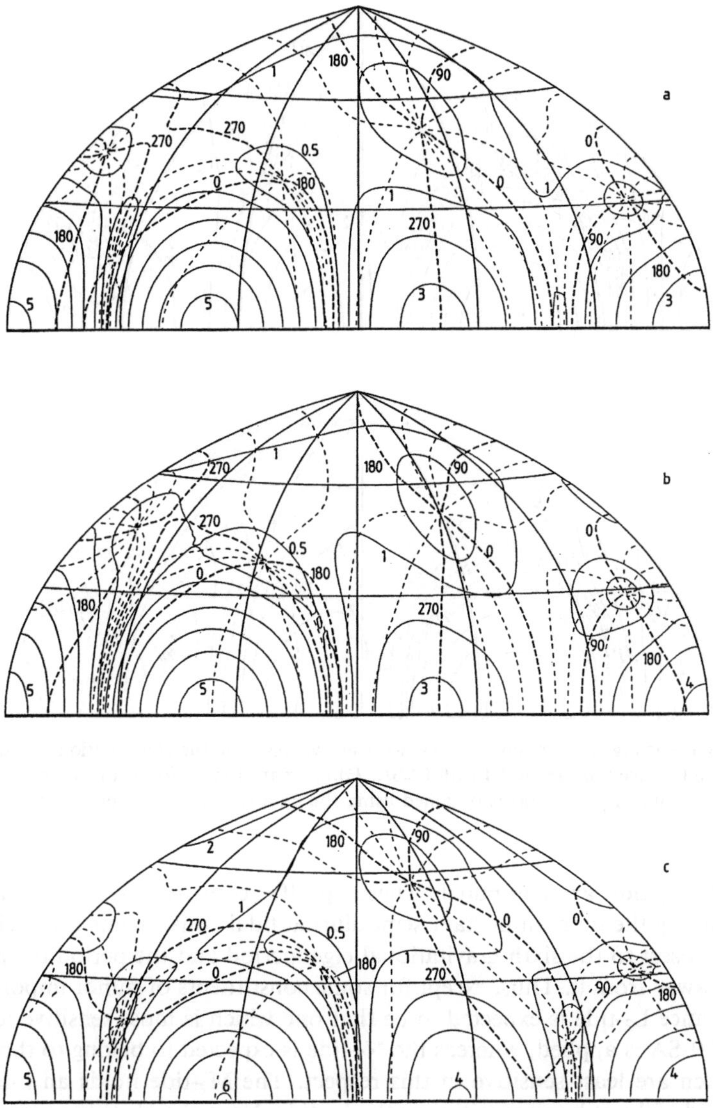

Fig. 6a–c. Same as Fig. 5 except for the computational results referring to the N_2-tide

tidal oscillation systems compare with the realistic ones in the Pacific basically, as can be recognized, e.g., in the case of the N_2-tide from Figure 6 and Schwiderski (1981). In Figures 5–7 only the northern half of the hemispherical domain is depicted because the semi-diurnal tidal oscillation systems obtained by the hemispherical model are symmetric about the equator.

In order to give an idea of how time- and space-dependent properties of near-resonance semi-diurnal tidal oscillation systems may depend on whether models with CLSA or with PLSA are applied, the different hemispherical systems of the L_2-tide

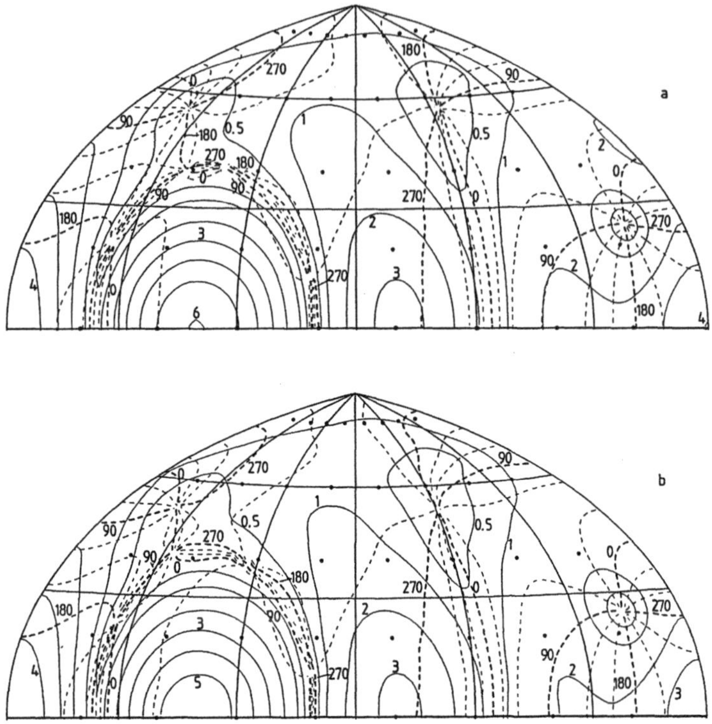

Fig. 7a, b. Same as Fig. 5 (**b**) except for 40 sea-surface values from the computational results depicted in Fig. 5 (**a**) introduced in the model with PLSA. *Points* mark the positions of the sea-surface values introduced. The upper system (**a**) is obtained using the data weight k = 1, the lower (**b**) using k = 3

and of the N_2-tide are computed. These specific constituents have been selected because among the more important semi-diurnal tidal constituents the period of the L_2-tide is closest to the sixth gravitational eigenperiod and the period of the N_2-tide is farthest away from it. Thus, except for some constituents of minor importance, the near-resonance L_2-tide is expected to be that one which is most sensitive to whether CLSA or PLSA is applied, whereas the N_2-tide is expected to belong to those constituents which are least sensitive in this respect. The M_2-tide holds an intermediate position and will not be as sensitive as the L_2-tide. Now, the N_2-tide system obtained by the model neglecting LSA is indeed considerably improved toward agreement with the system obtained by the model with CLSA when that one with PLSA is applied (Fig. 6). The three L_2-tide systems depicted in Figure 5 show that typical LSA dynamic effects, as the more pronounced equatorial anti-nodes, the displacements of eastern amphidromes and the phase delay at the eastern boundary (concerning these effects as given by a realistic ocean tide model, see Zahel 1980) are overestimated when applying the model with PLSA. In the western part of the schematic ocean area even the existence of characteristic features can depend on whether and in which way, i.e., by CLSA or PLSA, LSA is taken into account.

Regarding sea-surface values taken from the L_2-tide obtained by the model with CLSA as realistic ones, the actual problem of introducing measured values in realistic

Table 3. Tidal power $\bar{W}$, potential energy contents E_p and residuals when introducing 40 uniformly distributed values of sea-surface elevation into the hemispherical L_2-tide model with PLSA

L_2-tide model	$\bar{W}$ [10^{11} W]	E_p [10^{16} J]	r.m.s. residuals (relative and absolute, respectively)	
			Equations of motion (per cent)	Data (equilibrium units)
CLSA	13.784	5.066	3.13	0.0
PLSA without data introduced	17.636	7.160	0.0	0.349
PLSA: k = 1/3	17.288	6.907	0.58	0.310
PLSA: k = 1	16.161	6.166	2.59	0.187
PLSA: k = 3	15.192	5.693	5.26	0.078
PLSA: k = 6	14.839	5.619	6.98	0.034

Model parameters as in Table 1, sea-surface values taken from the L_2-tide computed with CLSA. PLSA: k = 1, e.g., denotes the results obtained by the model with PLSA ($\varepsilon' = 0.085$) and introducing data with the weight k = 1. Residuals of equations of motion referred to PLSA: k = 0.

models, which are always based some way or other on dynamical equations not exactly describing reality, can be simulated by solving the system (1), (2), (5). Figure 7 and Table 3 show results for different weighting coefficients k demonstrating that by this procedure the rates of tidal power and energy can significantly be improved by incorporating data into the model with PLSA. Also, the oscillation system regarded as realistic (Fig. 5a) is approximated rather well in the eastern area, whereas in the western part the improvement remains restricted in case of introducing 40 uniformly distributed values as the residuals of the equations of motion indicate (Table 3).

5 Conclusions

The hemispherical model presented has proven to be adequate to give an idea of free and forced oscillation properties of extended realistic ocean areas on a non-rigid Earth. The results obtained for the schematic ocean area suggest that the free oscillation behavior of the open ocean is significantly influenced by loading and self-attraction and that the simplified consideration of the latter effect cannot be expected to be satisfactory for all relevant modes and period ranges. In particular, in near-resonance situations, as they are typical of the range of semi-diurnal tidal constituents, the forced oscillation systems are expected also in case of a more realistic topography to be clearly dependent on the way LSA is taken into consideration.

Only precise and well-distributed data information contributes to a significant improvement of the results obtained by applying models which are simplified with respect to the dynamics included. The rates of tidal pressure work undergo alterations that demonstrate the difficulties which might appear when an important effect as that referred to is to be taken into consideration by parameterization and data assimilation.

References

Accad Y, Pekeris CL (1978) Solution of the tidal equations for the M_2 and S_2 tides in the World Oceans from the knowledge of the tidal potential alone. Philos Trans R Soc London A290:235–266

Bennett AF, McIntosh PC (1982) Open ocean modeling as an inverse problem: tidal theory. J Phys Oceanogr 12:1004–1018

Platzmann GW (1984) Normal modes of the World Ocean. Part IV: Synthesis of diurnal and semidiurnal tides. J Phys Oceanogr 14:1532–1550

Schwiderski EW (1981) Global ocean tides. Part VI: The semidiurnal elliptical lunar tide (N_2), atlas of tidal charts and maps. Naval Surface Weapons Center, NSWC TR 81-218, Dahlgren

Schwiderski EW (1985) On tidal friction and the decelerations of the Earth's rotation and Moon's revolution. Marine Geodesy 9:399–450

Webb DJ (1980) Tides and tidal friction in a hemispherical ocean centered at the equator. Geophys J R Astr Soc 61:573–600

Zahel W (1980) Mathematical modelling of global interaction between ocean tides and earth tides. Phys Earth Planet Inter 21:202–217

Zahel W (1986) Astronomical tides. In: Sündermann J (ed) Oceanography. Landolt-Börnstein Vc, Springer, Berlin Heidelberg New York, pp 83–134

Zahel W (1988) The influence of ocean and solid earth parameters on oceanic eigenoscillations, tides and tidal dissipation. In: McCarthy D (ed) Variations in earth rotation. Am Geophys Union (in press)

Variations of the Angular Momentum Budget for Tides of the Present Ocean

U. Seiler

1 Introduction

Modern observation techniques like VLBI provide information about the Earth's rotation with an accuracy of centimeters (Schuh, this Vol.). Consequently, the need will arise to interpret small signals, which in turn requires more reliable theoretical estimates of fluctuations in the Earth's rotation.

As a possible source of small, periodic changes in the angular velocity of the Earth ocean tides have to be considered, but these have as yet only seldom been investigated in this respect. Estimates given by Yoder et al. (1981) for the major semi-diurnal and diurnal tides are inaccurate, due to their method of analysis and the fact that only variations in the moments of inertia were considered. As has been shown for the M_2 (Brosche 1982; Baader et al. 1983), fluctuations have to be expected from periodic changes in the content of relative angular momentum rather than from mass redistributions. The 0.025 ms amplitude of the UT1-variation given by the authors is near the present VLBI-precision, a systematic investigation of partial tides from all tidal bands seems thus promising.

In this study, four long-period tides in addition to the major semi-diurnal and diurnal constituents are considered. Tidal elevations and currents were calculated by a recently developed numerical model, which will be briefly presented in the following section. The angular momentum budgets and the corresponding theoretical fluctuations in Universal Time and axis orientation will be discussed in the subsequent sections. Further informations on any of these topics are given by Seiler (1988).

2 Simulation of Ocean Tides

The tidal model comprises prognostic equations for the vertical mean velocity components (the hydrodynamic equations of motion) and the sea surface elevation (the continuity equation). The formulation is essentially the same as in Zahel's (1978) model and is not given here. Apart from a few differences, which are insignificant for the instantaneous angular momentum budgets, the most important one is the way in which the effects of "loading and self attraction" are considered here:

$$\tilde{\Phi} = g\varepsilon\zeta$$

Institut für Meereskunde der Universität Hamburg, Troplowitzstr. 7, D-2000 Hamburg 54, FRG.

where $\tilde{\Phi}$ is the potential arising from the ocean – solid earth interaction, g the mean gravitational acceleration, ζ the sea surface elevation and ε a constant. In this formulation, which was introduced by Accad and Pekeris (1978), the force associated with loading and self attraction is assumed to have a fraction ε of the pressure gradient force *everywhere* and to act opposite to the latter. With $\varepsilon = 0.1$ this means in terms of angular momentum that the topographic coupling between hydro- and lithosphere is reduced by 10%. The complete (and numerically lengthy) treatment (Zahel 1978) yields a similar representation of $\tilde{\Phi}$ with ε depending on latitude and longitude. Thus, a considerably better agreement of the predicted elevations with observations can be expected in the global mean.

The integration is performed with a semi-implicit scheme (Backhaus 1983), which was extended for computations on a global, spherical grid (1°-resolution). As in such a model the timestep is not limited by the mesh size, a special treatment of the polar latitudes is not necessary, as it is in other models for the sake of computational efficiency. The ocean bathymetry was taken from Schwiderski (1978).

The reason for developing a new model, instead of using results from existing models, is that the latter, however successful they may be in the classic subject of tidal prediction, not necessarily meet the requirements of subtle balance computations, the main subject here. Two items will elucidate this further:

This model does not constrain the solution by observed elevations, as for example in Schwiderski's (1980) model. Prescribed elevations are inconsistent with the tidal equations, i.e., differ from those that would be predicted from the continuity equation. This "continuity gap" problem (Schwiderski 1980), which manifests itself in more or less severe distortions of the velocity field, can be overcome only partly and only at the expense of volume conservation. Imbalances in the angular momentum equations, resulting from such inconsistencies are certainly intolerable, if the physics is to be understood. Furthermore, relative angular momentum is predominant in the budgets of the semi-diurnal and – to a less extent – also in those of the diurnal tides, stressing the significance of realistic tidal currents. Verification of calculated tidal currents is not possible, but their reliability can be established from a good reproduction of the elevations, provided both are consistent with the tidal equations.

As the model equations are non-linear, strictly harmonic behavior of prognostic variables cannot be assumed. The integration has to start from some initial state and

Table 1. Characteristics of simulated partial tides (see text)

	Amplitude	$\sigma\,[s^{-1}]$
M_2	0.90812 (100%)	$1.4052 \cdot 10^{-4}$
S_2	0.42286 (46.6%)	$1.4544 \cdot 10^{-4}$
N_2	0.17387 (19.2%)	$1.3788 \cdot 10^{-4}$
K_1	0.53050 (100%)	$7.2921 \cdot 10^{-5}$
O_1	0.37689 (71.0%)	$6.7597 \cdot 10^{-5}$
P_1	0.17544 (33.1%)	$7.2523 \cdot 10^{-5}$
Mf	0.15642 (100%)	$5.3234 \cdot 10^{-6}$
Mf'	0.06481 (41.4%)	$5.3341 \cdot 10^{-6}$
Mm	0.08254 (52.8%)	$2.6392 \cdot 10^{-6}$
Ssa	0.07287 (46.6%)	$0.3982 \cdot 10^{-6}$

has to be performed, until the ocean had adjusted to the external force, i.e., amplitudes and phases in a harmonic expansion with respect to time are sufficiently constant. It turned out, that "sufficient" for the reproduction of the tidal oscillation system is by no means sufficient, when angular momentum budgets are considered (see also Baader 1982). The latter need about five times as much simulation time as the previous one, until the budgets exhibit three constant decimal digits in their instantaneous values and vary by not more than about 10% in the temporal mean. This criterion was chosen to exclude the possibility that uncertainties due to incomplete adjustment are comparable or even larger in magnitude than those inherent in the physical and numerical treatment of the problem (Hövel 1982).

Table 1 lists all simulated tides as characterized by the frequency σ and amplitude of the tidal potential; the amplitudes relative to the major constituent in the same tidal band are given in brackets.

Due to their frequency, the S_2 and O_1 stand out against the other two tides of the same band, which will be evident in the angular momentum budgets. The M_2 and N_2 as well as the K_1 and P_1 have nearly the same period, but are excited by potentials of quite different amplitude. Their angular momentum budgets allow conclusions on the linearity of the response, in which case the amplitudes in the tidal frequency should exhibit the relative values and the corresponding phases should be the same. In general, such behavior cannot be expected. In the long-period band, the response will approach the equilibrium with increasing period and decreasing amplitude of the forcing, as the ocean then adjusts immediately.

Examples of the calculated oscillation systems are shown in Figures 5–7. Generally, the distributions for the semi-diurnal and diurnal tides agree very well with observations and other tidal models. Errors are somewhat larger in those areas, where an individual tide finds good resonance conditions. All tides of the same band have essentially the same oscillation system, but those of the S_2 and the O_1 exhibit marked differences in some regions. Generally, all short-period tides can be explained by the eigenoscillations of the world ocean. Among the long-period tides, only the Ssa is in global equilibrium (which will also be valid for any tide of longer period). The fortnightly and monthly tides more or less depart from equilibrium. The distribution of the Mf′ is exactly the same as that of the Mf, with correspondingly lower amplitudes. As this also holds for the angular momentum budget, this tide will not be discussed in the following sections: any instantaneous quantity can be extrapolated by means of the relative value in Table 1.

3 The Angular Momentum Budgets

In what follows, quantities are represented in a body-fixed frame of reference

$$x = \alpha \cos\varphi \cos\lambda$$

$$y = \alpha \cos\varphi \sin\lambda$$

$$z = \alpha \sin\varphi$$

where α denotes the mean radius of the Earth. The contributions to the variable part of oceanic angular momentum $\vec{J}$ are denoted by $\vec{J}^{\theta}$ (rotational, i.e., related to the

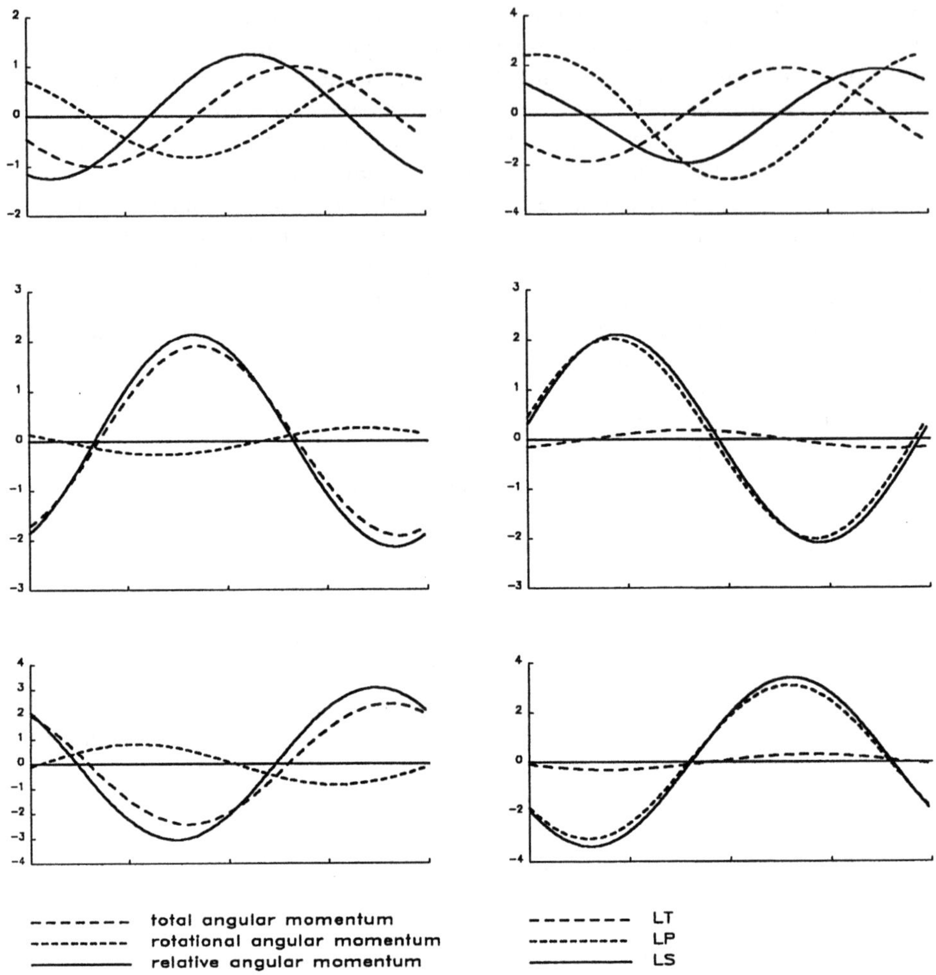

Fig. 1. Angular momentum (*left*) and balance of angular momentum (*right*) of the M_2 for one tidal period in x, y, z (*top to bottom*). Ordinates: 10^{25} kg m^2/s (*left*), $10^{20}(x)$, 10^{21} kg m^2/s^2 (y, z) (*right*)

moments of inertia) and J^{rel} (relative, i.e., due to motions relative to the frame of reference). Quantities having the dimension of a torque are denoted by L and a suffix that indicates their origin. The formal balance of oceanic angular momentum, as derived from the hydrodynamic equations of motion is then for any component

$$LR + LC = LP + LT + LF$$

where LR denotes the time-derivative of relative angular momentum, LC the torque associated with the Coriolis force (in the z-component this is $\dot{J}_z^{\theta}$) and LS the sum of both terms. On the right hand side, LP denotes the pressure torque, reduced by a factor $1-\varepsilon$ for loading and self attraction. LT is the external torque and LF the sum of frictional torques. The latter are negligible in the *instantaneous* budgets (2–3 orders of magnitude less than the other terms) and will therefore not be discussed.

As an example for the semi-diurnal tides, the time-dependent angular momentum budget of the M_2 is shown in Figure 1. As can be seen, relative angular momentum is predominant in each component, where the largest amplitude is in J_z, the lowest in J_x. These features are common to all semi-diurnal tides, but the anisotropy is less pronounced in the N_2 and least in the S_2, where J_x has the same, J_z half the value as in the M_2. Furthermore, corresponding phases differ by 30°–50°, reflecting the different response of these tides. On the whole, the N_2 resembles more the M_2, the phases being almost the same and the amplitudes amounting to 13% (x)–22% (z) of the M_2-values. The balance is essentially constituted by LS and LP, the contribution from LT is approximately 10%. In the x-component, the balance is different in that the amplitudes are extremely low and LT is more important. This also is less pronounced in N_2, and the S_2 has the same kind of balance in all components.

Figure 2 shows the angular momentum budget of the O_1 as an example for the diurnal tides. In contrast to the semi-diurnal tides, relative angular momentum is not predominant in the equatorial components. This fundamental difference can be ex-

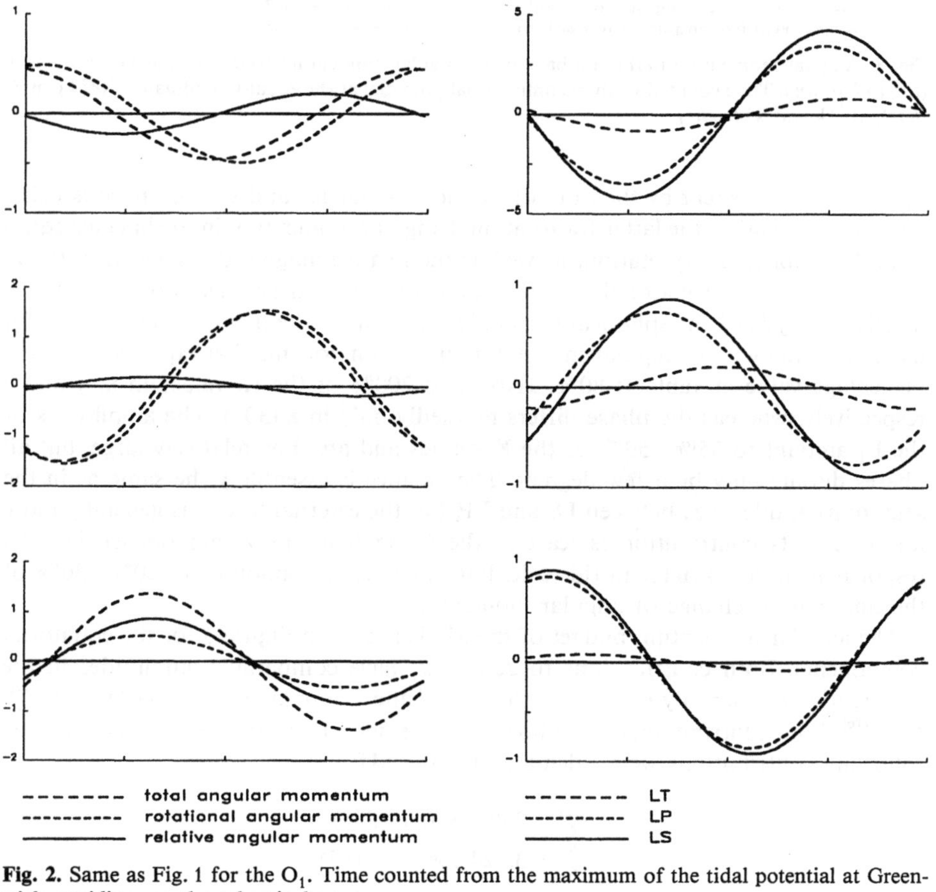

Fig. 2. Same as Fig. 1 for the O_1. Time counted from the maximum of the tidal potential at Greenwich meridian, northern hemisphere

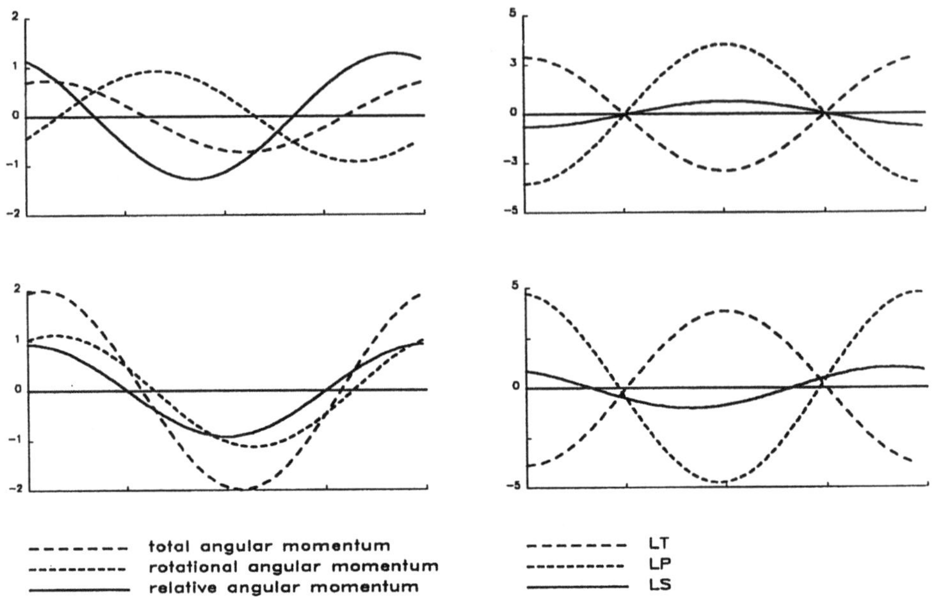

Fig. 3. Angular momentum (*left*) and balance of angular momentum (*right*) of the Mf in x and y (*top to bottom*). Time counted from maximum tidal potential at the equator. Ordinates: 10^{25} kg m^2/s (*left*), 10^{19} kg m^2/s^2 (*right*)

plained to some extent by the ratio of the tidal frequency and Ω, the previous being a weight for relative, the latter for rotational angular momentum in the balance equation. A complete interpretation, as well as the relative magnitude of terms in the z-component, where rotational angular momentum is still smaller, would require a detailed analysis of the spherical harmonic spectrum of the mass transports. The O_1 has considerably more angular momentum in all components that expected from the percentage value in Table 1: 80%, 100% and 200% of the K_1-values in x, y and z respectively, whereas the phase differs markedly only in z (30°). The amplitudes of the P_1 amount to 35%−50% of the K_1-values and are thus relatively large, but all phases deviate only by a few degrees. The balance is essentially the same as in the semi-diurnal tides, i.e., between LS and LP, but the external torque is generally more important. Its contribution is least in the O_1 (8% in the z-component), i.e., the response is more resonant in this tide, but otherwise LT amounts to 20%−30% of the time rate of change of angular momentum.

The angular momentum budget of the Mf is shown in Figure 3 for the equatorial components. Both contributions to J_x and J_y have comparable amplitudes in the fortnightly and monthly periods, in the Ssa J_z^θ and J_y^θ are about twice as large at J_x^{rel} and J_y^{rel}, both contributions being however extremely small. In the z-component, the following contributions were calculated for the Mf:

$$J_z^{\mathrm{rel}} = 2.86 \cos(\sigma t - 98°)$$
$$J_z^\theta = 33.24 \cos(\sigma t - 16°)$$
$$J_z = 33.76 \cos(\sigma t - 21°)$$

where the amplitudes are given in 10^{23} kg m^2/s. J_z^θ prevails by an order of magnitude, but J_z^{rel} is still larger than in the equatorial components. With increasing period, equatorial angular momentum as well as J_z^{rel} decrease somewhat more than would be expected from the percentages in Table 1, whereas J_z^θ decreases less. This has to be attributed to the approach of equilibrium in the longer periods: in the Ssa any spectral coefficient of the elevation in other wavelengths than that of the tidal potential is entirely due to the presence of continents.

In the z-component, only LP can balance LS, i.e., the pressure torque generates a corresponding time rate of change of (rotational) angular momentum. Thus, the balance is essentially the same as in the semi-diurnal tides. But in the equatorial components LT and LP balance each other to a large amount, the more, the longer the period, and LS, resulting from the incomplete balance of the previous terms, is very small.

4 Variations of the Earth's Rotation

The equation for changes in Universal Time (UT1), derived from the z-component of the Liouville equations (Lambeck 1980), is

$$\Delta UT1 = \int \psi_3 dt \ .$$

The excitation function is given by

$$\psi_3 = -\frac{J_z}{C\Omega} \ .$$

The system of equations for the equatorial components $\Omega \mathrm{m}_1$, $\Omega \mathrm{m}_2$ of the angular velocity is

$$\dot{\mathrm{m}} - i\sigma_0 \mathrm{m} = -i\sigma_0 \psi$$

where

$$\psi = \frac{1}{\sigma_0 A \omega}(\Omega J - i\dot{J}) \ ,$$

σ_0 is the frequency of the Chandler Wobble and the conventional complex notations $\mathrm{m} = \mathrm{m}_1 + i\mathrm{m}_2$, $\psi = \psi_1 + i\psi_2$ and $J = J_x + iJ_y$ have been used. For periodic excitation

$$\psi_l = \hat{\psi}_l \cos(\sigma t - \chi_l) \ , \quad l = 1,2$$

with arbitrary amplitudes and phases, the forced part of the solution to the above equation is given by

$$\mathrm{m} = -\frac{\sigma_0}{\sigma^2 - \sigma_0^2}(\sigma_0 \psi - i\dot{\psi})$$

which is an elliptically polarized process, superimposed on the free nutation.

Table 2. Amplitudes [μs] and phases [°] of variations in UT1 due to the excitation constituents of the headline

	J_z^{rel}	J_z^{θ}	J_z		J_z^{rel}	J_z^{θ}	J_z		J_z^{rel}	J_z^{θ}	J_z
M_2	37	10	30	K_1	15	2	17	Mf	9	107	108
	42	187	53		233	215	231		106	188	111
S_2	22	9	14	O_1	22	13	35	Mm	5	126	127
	68	220	86		204	197	201		101	181	103
N_2	7	<1	7	P_1	6	1	7	Ssa	4	793	794
	50	158	52		235	273	241		97	179	92

Contributions by external torques have been omitted from the excitation functions, i.e., only variations due to exchange of angular momentum between the oceans and the solid earth are considered here. Though external torques are important in the oceanic angular momentum budgets, they will not affect the Earth's rotation. As the Earth's crust is essentially in isostatic equilibrium, tidal torques exerted upon the oceans are compensated by the contributions upon the lithosphere.

From the above equations, both in UT1 and in axis orientation the largest signals have to be expected in the long-period tides considered here. Whereas in UT1 this is due to cumulation of decreases or increases of the length of day over longer time-scales, in the forced wobble this result from the vicinity of the tidal frequency to that of the free wobble, i.e., from resonant response.

Amplitudes and phases of variations in UT1 are given in Table 2. As for the M_2, the amplitudes of the angular momentum contributions agree very well with those calculated by Baader et al. (1983), whereas the phases differ typically by 30°. This can be attributed to the consideration of loading and self attraction in the present study. As can be seen, the variations in the semi-diurnal tides are of the same magnitude, the major part being due to relative angular momentum. The diurnal tides differ from the semi-diurnals in that both angular momentum contributions are almost in phase. The UT1 variations in the long-period tides clearly reveal the "amplitude-period" dependence: whereas the variations in the forthnightly and monthly tides amount to ≈ 0.1 ms, 0.8 ms are found for the Ssa, corresponding to 37 cm for a point on the equator.

Figure 4 shows the calculated wobble for the short-period tides. In the semi-diurnal tides (left diagram) the wobble is highly elliptical, which is extremely pronounced in the M_2 (Fig. 5), where the path of the pole has almost degenerated to a straight line. This is the result of an amplitude ratio of 1:2 and a 90° phaseshift in the equatorial angular momentum components. The eccentricity is somewhat less in the N_2 and least in the S_2. The major (r) and minor (s) semi-axes are:

$$r = \begin{cases} 0.492 \\ 0.090 \\ 0.266 \end{cases} \qquad s = \begin{cases} -0.002 \\ 0.023 \\ -0.134 \end{cases} \qquad \beta = \begin{cases} 127^\circ & \text{for } M_2 \\ 110^\circ & \text{for } N_2 \\ 45^\circ & \text{for } S_2 \end{cases}$$

and β and $\beta + \pi$ denote the phases when the tips of the major axis are passed. The negative sign of s indicates retrograde wobble, but in view of the generally large eccen-

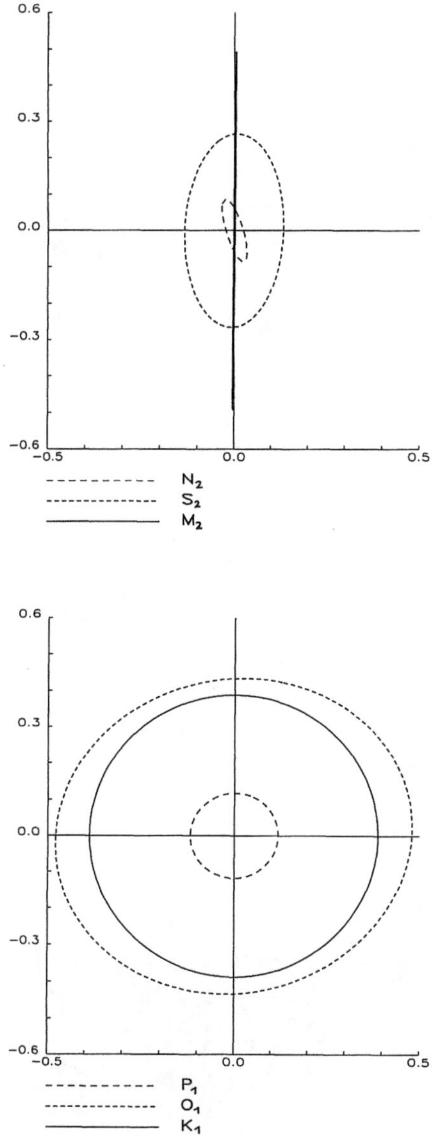

Fig. 4. Wobble [masec] due to the semi-diurnal (*upper panel*) and diurnal tides (*lower panel*)

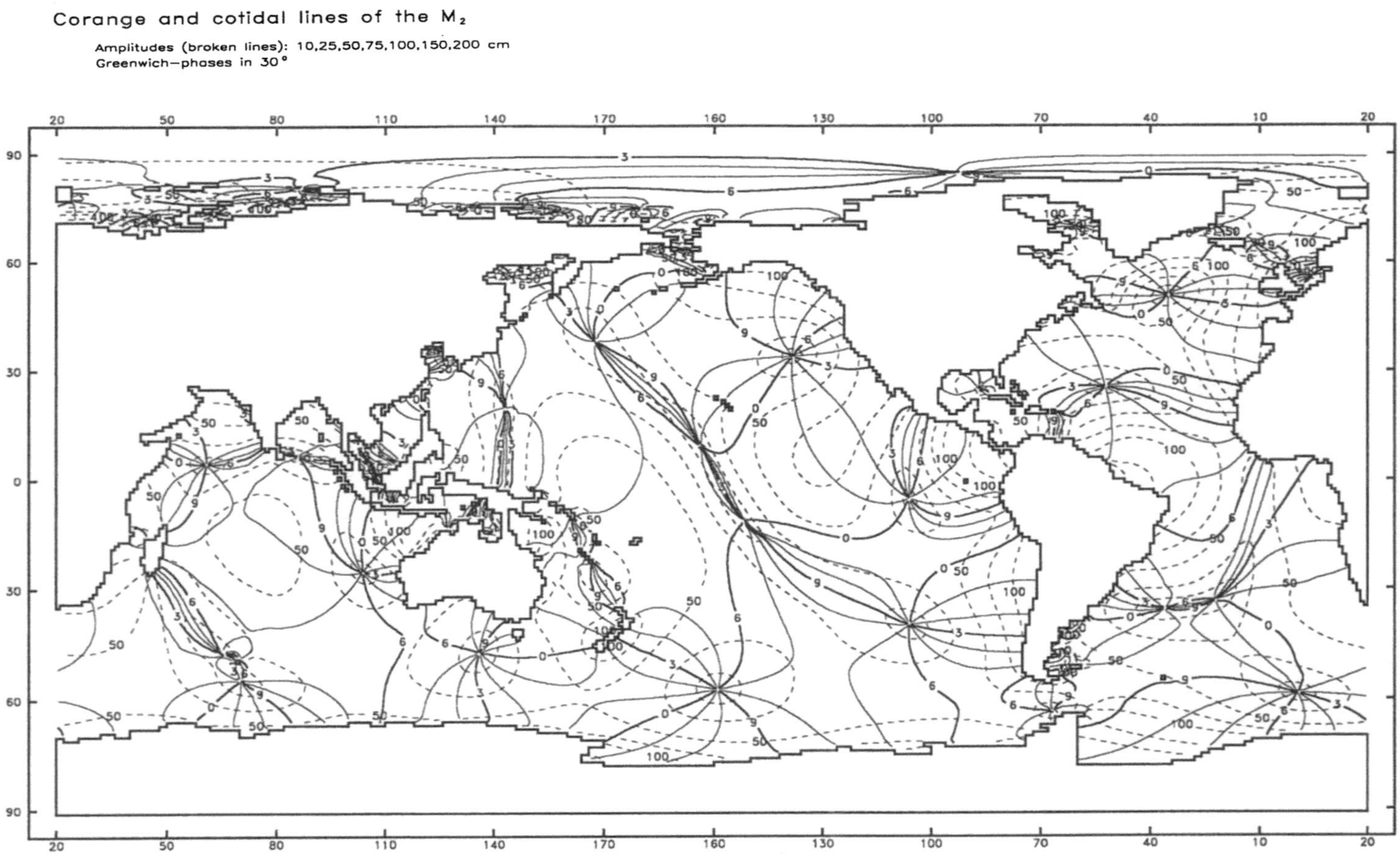

Fig. 5. M_2-Tide. Greenwich phases (*full lines*) in 30°, amplitudes (*broken lines*): 10, 25, 50, 75, 100, 150, 200 cm

tricities, the sense of propagation is not too significant, as it can be opposite if the excitation functions are slightly changed. This would be the case in the M_2 if the small (fictitious) contributions by external torques were considered. It can thus be stated that the forced semi-diurnal wobble is highly elliptical and can be retrograde or prograde.

The diurnal wobble (see Fig. 4, right diagram) differs fundamentally from the semi-diurnal wobble. From the above definitions of the excitation function it follows that for $\sigma = \Omega$ the wobble is circularly polarized and always prograde. This holds for the K_1 and nearly for the P_1, whereas in the frequency of the O_1 (Fig. 6), ellipticity is noticeable:

$$r = \begin{cases} 0.389 \\ 0.119 \\ 0.483 \end{cases} \qquad s = \begin{cases} 0.389 \\ 0.117 \\ 0.432 \end{cases} \qquad \beta = \begin{cases} & \text{for } K_1 \\ 74^\circ & \text{for } P_1 \\ 55^\circ & \text{for } O_1 \end{cases}$$

Finally, semi-axes and phases for the long-period tides are

$$r = \begin{cases} 0.103 \\ 0.111 \end{cases} \qquad s = \begin{cases} 0.008 \\ -0.007 \end{cases} \qquad \beta = \begin{cases} 28^\circ & \text{for Mf} \\ 322^\circ & \text{for Mm} \end{cases}$$

Due to the very small amount of equatorial angular momentum exchange between the oceans and the solid Earth, the wobble is considerably smaller than in the short-period tides. As in the semi-diurnal tides, the wobble is highly elliptical and can thus be prograde (Mf; Fig. 7) or retrograde (Mm). In the semi-annual tide, the exchange of the equatorial angular momentum is extremely small, as the torques of the tidal and pressure forces balance each other by more than 99%. Due to the vicinity of the tidal frequency and σ_0, a wobble comparable in magnitude to those of the Mf and Mm results, its reliability – in view of the smallness of the excitation function – being however doubtful.

5 Summary and Conclusions

Ocean tides with periods from half a day to half a year have been investigated with respect to their angular momentum budget and associated periodic variations in the Earth's rotation. The angular momentum balances clearly reveal the physical nature of the individual tides. In the short-period tides, resonant amplification manifests itself in a relatively small influence of external torques on the angular momentum budgets, the major part of the response being thus due to interactions between the oceans and the solid Earth. In the long-period tides, equilibrium response is represented by the mutual balance of the tidal and the pressure torques in the equatorial components and that of the pressure torque and the time rate of change of rotational angular momentum in the z-component. Oceanic variations in the Earth's rotation are therefore caused by a periodic exchange of angular momentum between the oceans and the solid Earth, mediated by that part of the pressure torque that does not balance external torques.

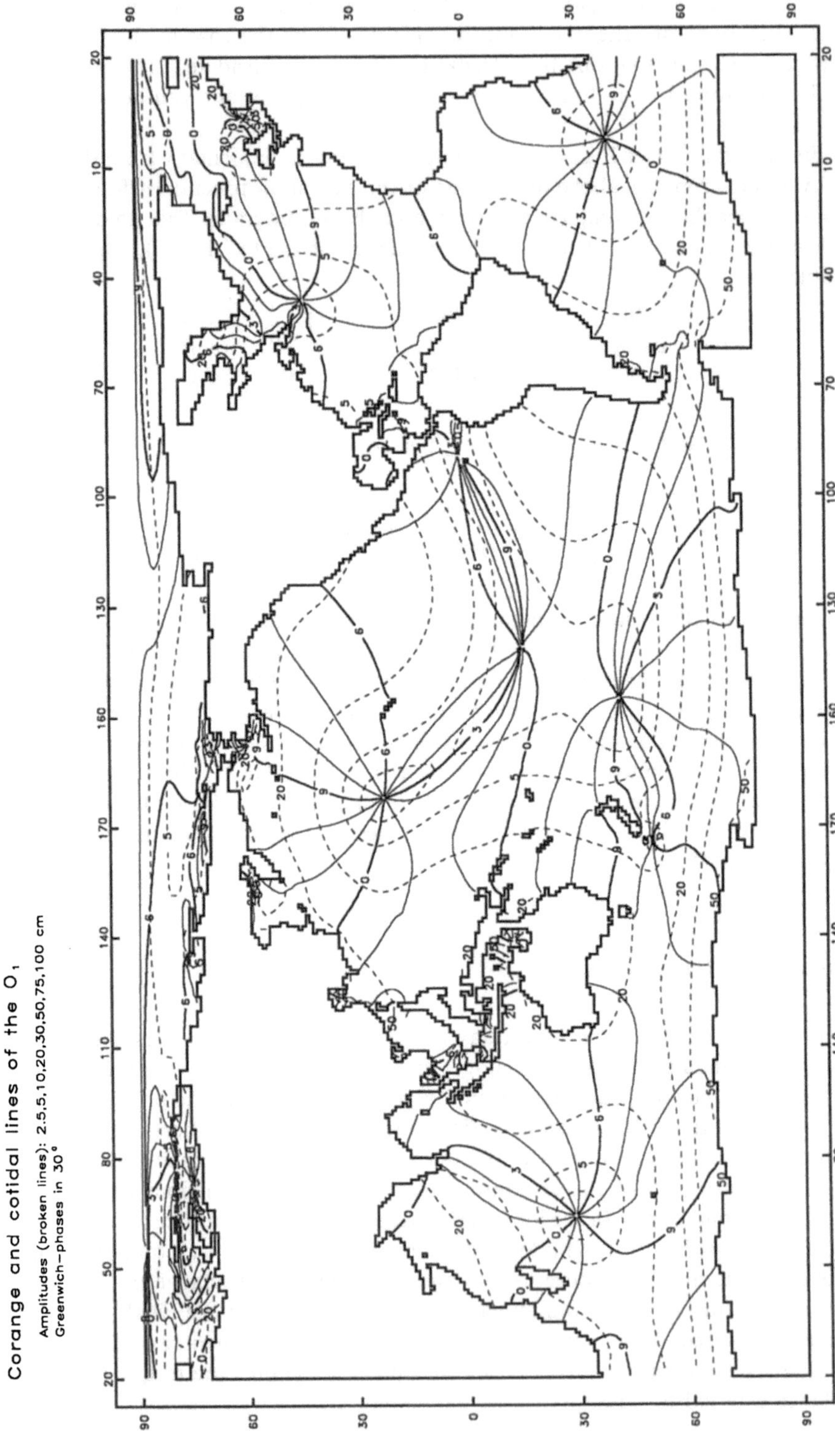

Fig. 6. O_1-Tide. Greenwich phases in 30°, amplitudes: 2.5, 5, 10, 20, 30, 50, 75, 100 cm

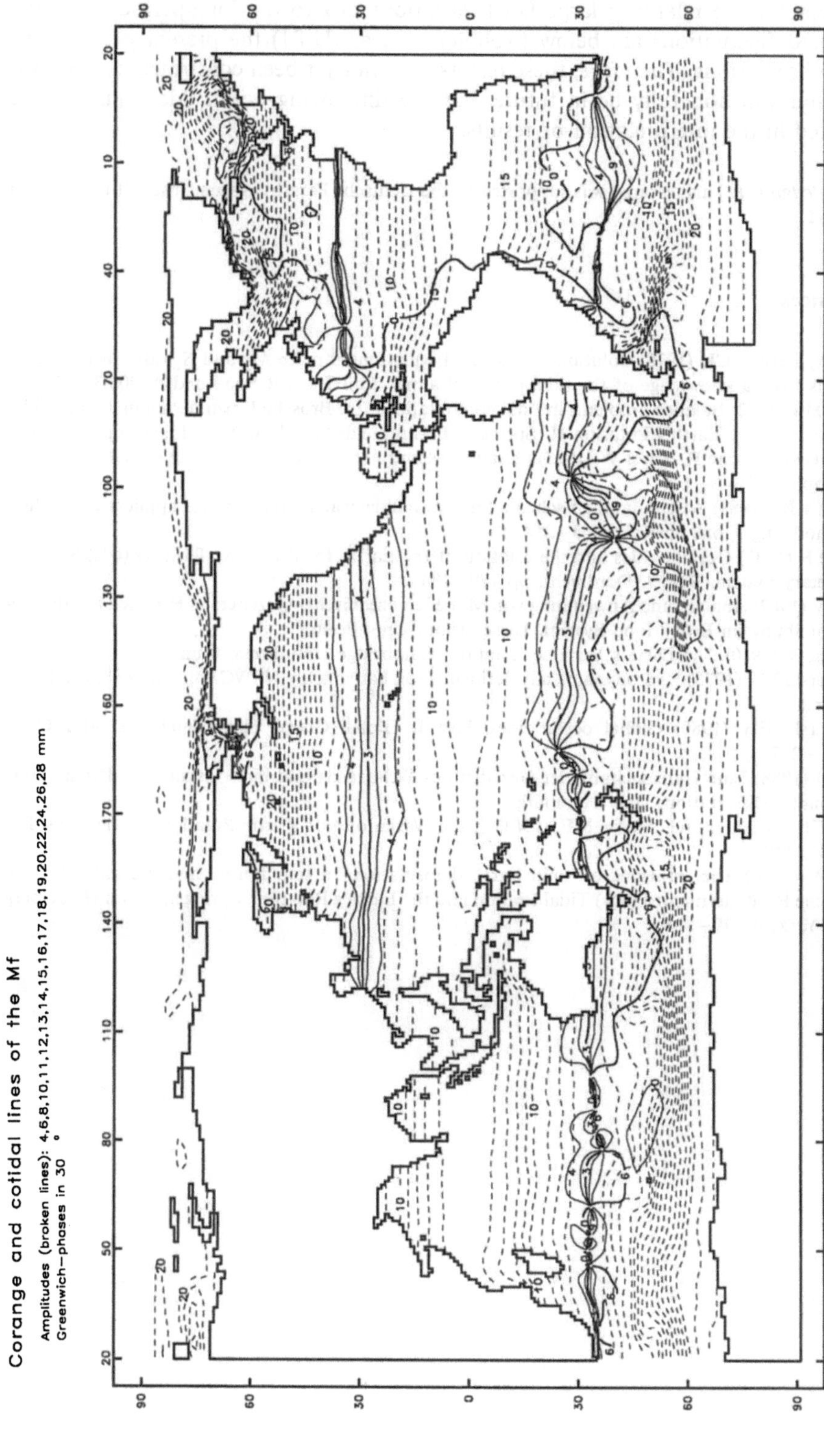

Fig. 7. Mf-Tide. Phases in 30°, amplitudes: 4, 6, 8, 10, 11, 12, 13, 14, 15, 16, 17, 18, 19, 20, 22, 24, 26, 28 mm

Except for the relatively large UT1 variations due to the long-period tides, the calculated fluctuations are below (wobble) or near (UT1) the present precision of VLBI-observations. Therefore, these results have not yet been confirmed by observations, but will hopefully be in future. For the time being ocean tides must not be neglected in the interpretation of results.

Acknowledgment. This research was supported by the Deutsche Forschungsgemeinschaft under grant Su 69/9-1.

References

Accad Y, Pekeris CL (1978) Solution of the tidal equations for the M_2 and S_2 tides in the world oceans from a knowledge of the tidal potential alone. Phil Trans R Soc Lond A290:235–266

Baader HR (1982) Balance problems in tidal computations. In: Brosche P, Sündermann J (eds) Tidal friction and the Earth's Rotation II. Springer, Berlin Heidelberg New York Tokyo, pp 122–164

Baader HR, Brosche P, Hövel W (1983) Ocean tides and periodic variations of the Earth's rotation. J Geophys 52:140–142

Backhaus JO (1983) A semi-implicit scheme for the shallow water equations for application to shelf sea modeling. Cont Sh Res 2:243–254

Brosche P (1982) Oceanic tides and the rotation of the Earth. In: Fricke W, Teleki G (eds) Sun and Planetary System. Reidel, Dordrecht, pp 179–184

Hövel W (1982) Analyse und Anwendung von Modellen ozeanischer Gezeiten im Hinblick auf die Gezeitenreibung im Erde-Mond-System. Dissertation, Univ Bonn

Lambeck K (1980) The Earth's Variable Rotation. Cambridge Univ Press, Cambridge

Schwiderski EW (1978) Hydrodynamically defined ocean bathymetry. NSWC/DL TR 3888, Dahlgren, Virginia

Schwiderski EW (1980) Global ocean tides, Part I: Global Ocean Tidal Equations. Mar Geod 3:161–217

Seiler U (1988) Eine Untersuchung der Gezeiten des Weltozeans und ihres instantanen Drehimpulshaushalts. Dissertation, Univ Hamburg

Yoder CF, Williams JG, Parke ME (1981) Tidal Variations of Earth Rotation. J Geophys Res 86:881–891

Zahel W (1978) The influence of solid Earth deformation on semidiurnal and diurnal tides. In: Brosche P, Sündermann J (eds) Tidal friction and the Earth's Rotation. Springer, Berlin Heidelberg New York, pp 98–124

The Pole Tide and the Damping of the Earth's Free Nutation

H. Lenhardt

1 Introduction

Since the discovery of the Earth's free nutation at the end of the last century the problem of its damping has not yet been solved satisfactorily. Tidal friction in the oceans was suspected to be a possible cause, and there has been a long discussion on this topic.

In order to distinguish clearly from other tidal phenomena, the subject of the present treatment is the 14 month free gyroscopic nutation, sometimes called "Chandler wobble", and its associated tide, referred to as "pole tide". The variation of the rotation axis inside the Earth's body induces changes in the centrifugal potential; and the ocean is also assumed to experience a varying centrifugal potential due to this free nutation. It ensues that the pole tide is a curious tidal phenomenon because lunisolar tides are engendered by extraterrestrial objects, and the pole tide is not.

An equilibrium pole tide (in the sense that the ocean surface remains an equipotential surface) does not contribute to the damping of the free nutation. Tidal friction, however, may cause departures from equilibrium and dampen gradually the Chandler motion. The present theory can exactly determine the implications for the Chandler motion, as will be shown in Sect. 4.

In the next section two former contributions to the pole tide problem will be sketched briefly. The subsequent sections are devoted to an equilibrium tide model and its comparison with actual observations. Proceeding from this tide model some results of the recent theory will be discussed in the last part.

2 A Glance at the History

The subsequent treatment is based on the discussions in Klein and Sommerfeld (Reprint 1965), Bondi and Gold (1955), and Munk and MacDonald (1960). It will be shown that the mechanical concepts were already known at the beginning of this century; but eight decades of discussion have not solved the problem, and the question is still open.

Institut für Physikalische Geodäsie, Technische Hochschule Darmstadt, Petersenstraße 13, D-6100 Darmstadt, FRG.

2.1 At the Beginning: Klein and Sommerfeld (1910)

First of all, I shall follow the discussion in Klein and Sommerfeld (Reprint 1965, p. 727). The position of the Earth's rotation axis with respect to a geographic reference system may be described by the vector $(\omega_1, \omega_2, \omega_3)$ at some initial epoch. ω_3 is assumed to differ only slightly from the Earth's mean angular velocity, and consequently ω_1 and ω_2 are small quantities. According to our preceding discussion, a tidal bulge is associated with this particular location of the Earth's rotation axis. In a subsequent moment, the position of the rotation axis may be described by $(\omega_1+\dot{\omega}_1 \mathrm{dt}, \omega_2+\dot{\omega}_2 \mathrm{dt}, \omega_3+\dot{\omega}_3 \mathrm{dt})$, and the tidal bulge will change accordingly. The tidal bulge can be imagined to be transported from its initial position by rotating around an axis *perpendicular* to both the initial rotation axis and the changed one; the frictional torque may be assumed to act around the same axis, resisting any variation of ocean mass. Presupposing that the frictional torque is proportional to the velocity of rotation around this axis, the components of tidal friction may be assumed to be proportional to the vectorial product $(\omega_1, \omega_2, \omega_3)\times(\dot{\omega}_1, \dot{\omega}_2, \dot{\omega}_3)$ or, with i, j, k denoting unit vectors:

$$\begin{matrix} \mathrm{i} & \mathrm{j} & \mathrm{k} \\ \omega_1 & \omega_2 & \omega_3 \\ \dot{\omega}_1 & \dot{\omega}_2 & \dot{\omega}_3 \end{matrix} = \mathrm{i}(\omega_2\dot{\omega}_3-\omega_3\dot{\omega}_2)+\mathrm{j}(\omega_3\dot{\omega}_1-\omega_1\dot{\omega}_3)+\mathrm{k}(\omega_1\dot{\omega}_2-\omega_2\dot{\omega}_1) \ . \tag{1}$$

All vector components except ω_3 ($\omega_3 \approx \Omega$, Earth's mean rotation rate) are small, and their products can be neglected. Hence we obtain for Eq. (1) the vector $(-\Omega\dot{\omega}_2, \Omega\dot{\omega}_1, 0)$ and the components of tidal friction can be written proportionally to these quantities. A torque is obtained by multiplication of this vector with $\lambda \mathrm{A}/\Omega$, with λ being a dimensionless number and A being the equatorial moment of inertia: $(-\lambda \mathrm{A}\dot{\omega}_2, \lambda \mathrm{A}\dot{\omega}_1, 0)$. Euler's equations can now be written, in complex notation and with the polar moment of inertia C:

$$\mathrm{A}(\dot{\omega}_1+\mathrm{i}\dot{\omega}_2) = (\mathrm{C}-\mathrm{A})\mathrm{i}\Omega(\omega_1+\mathrm{i}\omega_2)+\mathrm{i}\lambda \mathrm{A}(\dot{\omega}_1+\mathrm{i}\dot{\omega}_2) \ . \tag{2}$$

With $\sigma = (\mathrm{C}-\mathrm{A})/\mathrm{A}\Omega$ and by regarding λ as small quantity, Eq. (2) can be rearranged if higher-order products are neglected:

$$(\dot{\omega}_1+\mathrm{i}\dot{\omega}_2) = \mathrm{i}\sigma(\omega_1+\mathrm{i}\omega_2)(1+\mathrm{i}\lambda) \ . \tag{3}$$

Integration leads to:

$$\omega_1+\mathrm{i}\omega_2 = \mathrm{const.}\ \mathrm{e}^{\mathrm{i}\sigma t-\sigma\lambda t} \tag{4}$$

with σ being the Euler frequency. It can be seen that the term containing λ disturbs the otherwise purely harmonic oscillation. It causes a gradual decrease of the oscillation which is no more a circular motion but an inward-directed spiral. The amount of this damping depends on the magnitude of λ, but λ was unknown at the beginning of this century and Klein and Sommerfeld could not give a quantitative description. However, it can be seen in Eq. (4) that friction does not influence the frequency of the damped oscillation, if higher-order products are disregarded.

2.2 A Milestone: Bondi and Gold (1955)

Bondi and Gold (1955) published an investigation which aimed at an estimate of the maximum degree of damping. Even if their approach to the problem was originally designed to investigate the influence of the Earth's core, it can also be applied to the ocean. For this purpose, let us imagine an ocean of uniform depth completely enclosing the Earth's mantle. We understand $\omega = (\omega_1, \omega_2, \omega_3)$ as the angular velocity vector of the Earth's mantle, and $\chi = (\chi_1, \chi_2, \chi_3)$ will denote the angular velocity vector of the ocean. By doing this the ocean is imputed to behave as some kind of "solid body"; this is desirable in a maximum estimation because the highest friction between ocean and mantle will occur when the ocean has maximum inertia. That is the case when the ocean behaves like a solid body. The equations of motion can be found straightforwardly from conservation of angular momentum:

$$\begin{aligned} A\dot{\omega}_1+(C-A)\omega_2\omega_3 &= J(\chi_1-\omega_1) = -D(\dot{\chi}_1+\omega_2\chi_3-\omega_3\chi_2) \\ A\dot{\omega}_2-(C-A)\omega_3\omega_1 &= J(\chi_2-\omega_2) = -D(\dot{\chi}_2+\omega_3\chi_1-\omega_1\chi_3) \\ C\dot{\omega}_3 &= J(\chi_3-\omega_3) = -D(\dot{\chi}_3+\omega_1\chi_2-\omega_2\chi_1) \end{aligned} \quad (5)$$

Here, A and C denote the equatorial and polar moments of inertia of the Earth's mantle, and D represents the moments of inertia of the ocean. It is implicitly assumed that the frictional couple is J times the differential angular velocity between the Earth's mantle and ocean. Note that J represents an angular momentum.

These equations will be simplified in the usual manner by assuming that the motion differs only slightly from rigid-body rotation about the third axis of coordinates with angular velocity Ω. Then the third equation is independent from the other ones because small products can be neglected. It will have no significance for the further discussion and only the first two equations will be required. Complex notation is again convenient, with $\omega = \omega_1+i\omega_2$, $\chi = \chi_1+i\chi_2$. Introducing the quantities ψ, β and α by $J = \Omega D\psi$, $C = A(1+\beta)$, $D = A\alpha$, leads to

$$1/\Omega\,\dot{\omega}-i\beta\omega = \alpha\psi(\chi-\omega) = -\alpha[(1/\Omega)\dot{\chi}-i(\omega-\chi)] \ . \quad (6)$$

The normal modes can be found by assuming that ω and χ are proportional to $\exp(iq\Omega t)$, and $\chi = \kappa\omega$, with q and κ being complex numbers to be determined. We obtain

$$i(q-\beta) = \alpha\psi(\kappa-1) = -i\alpha(\kappa q-1+\kappa) \ . \quad (7)$$

It can be seen immediately that

$$\kappa = 1+i(\alpha\psi)^{-1}(q-\beta) \quad \text{and} \quad q^2+q[1-\beta-i\Psi(1+\alpha)]-\beta(1-i\omega) = 0 \ . \quad (8)$$

For small β one root is

$$q = \beta(1-i\psi)[1-i\psi(1+\alpha)]^{-1} \ . \quad (9)$$

The imaginary part of q describes the damping of the oscillation. The damping will reach its maximum value for a maximal imaginary part, and it can be computed after decomposing Eq. (9) in its real and imaginary part.

Differentiation shows that the maximum is reached for $\psi = (1+\alpha)^{-1}$, and inserting into Eq. (9) yields:

$$q = \frac{2+\alpha+i\alpha}{2(1+\alpha)}\beta . \tag{10}$$

The amplitude will be reduced by a factor 1/e after the relaxation time has passed which can be computed by inserting Eq. (10) in $\exp(iq\Omega t)$. From Eq. (9) follows that for $\psi = 0$ and for $\psi \rightarrow \infty$ the quantity q is purely real, i.e., there is no damping. At first sight it may be astonishing that infinite friction results in zero damping. However, infinite friction means that the two components move together without slip, and this corresponds to the motion of a single solid body. For a rigid Earth model with the mantle's moments of inertia being $A = 7.1005 \cdot 10^{37}\,m^2\,kg$ and $C = 7.1242 \cdot 10^{37}\,m^2\,kg$ (Smith and Dahlen 1981), $\beta = 0.00334$. An appropriate value for the moments of inertia of the oceans is $A_0 = B_0 = C_0 = D = 0.37 \cdot 10^{35}\,m^2\,kg$. The values have been obtained for an ocean of uniform depth (2630 m) entirely covering the globe. A_0, B_0 and C_0 would not be equal for a more realistic Earth model, due to ellipticity and the irregular distribution of the continents. However, ellipticity will be only significant if one figure more in the value of D is considered; the irregular ocean – continent distribution will be more important by one magnitude but does also not alter the forthcoming result. By using the value of A as given above we obtain $\alpha = 0.0005$. This amounts to a relaxation time of about $\tau = 520$ years, and the quality factor is found to be roughly $Q = 2000$. Hence we can proceed presumably from the assumption that the oceanic contribution to the damping of the Earth's free nutation is negligibly small.

Let us now look to Bondi and Gold's (1955) approach in its entirety: It does not provide an estimate of the real damping but instead the maximum effect is obtained. No particular friction mechanism has to be specified, and one might suppose that the result is fully general and, moreover, reliable. Which circumstances give rise to doubts? A discomfort was formulated by Munk and MacDonald (1960): They criticized that the application of Euler's equations as in Eq. (5) implies that the frictional torque acts around the Earth's instantaneous rotation axis; but this is not the only choice. Rather, it seems to be more appropriate to choose an axis similar to the one discussed by Klein and Sommerfeld; see also Sect. 2.1.

The rigid body eigenfrequency is modified by the oceanic surface layer via the factor of β in Eq. (10). Its difference to one depends on the magnitude of α, or equivalently on the magnitude of the ocean's moments of inertia.

Consequently, the essence of Bondi and Gold's (1955) treatment may be seen in the fact that the ocean's moment of inertia is far to small to contribute perceptibly to damping. Elasticity lengthens the relaxation time further.

3 A Spherical Harmonic Model of the Pole Tide

The subsequent treatment is based on a static equilibrium tide model being presented by Dahlen (1976). The algorithm is quite general, and in principle every tidal potential can be introduced if the condition of static equilibrium is fulfilled. The pole tide as well as other long-period tides may be assumed to follow equilibrium; see also Dickman (1988). The algorithm serves to compute the oceanic equilibrium tide for one particular position of the rotation axis; it is assumed implicitly that the superposition of spatial dependence and time dependence holds in reality.

3.1 The Algorithm

Naturally, every attempt to get a realistic, global feature of the pole tide entails that a model of the land – ocean distribution has to be provided. A theoretically very convenient but nevertheless numerically laborious way is a spherical harmonic expansion of the ocean function. The ocean function $C(\varphi,\lambda)$ is simply defined by the condition $C(\varphi,\lambda) = 1$ where there is water and $C(\varphi,\lambda) = 0$ where there is land. Hence, it follows

$$C(\varphi,\lambda) = \sum_{l=0}^{\infty} \sum_{m=-l}^{l} C_l^m Y_l^m \tag{11}$$

for the surface spherical harmonic expansion of the ocean function. Subsequently, we shall follow the normalization convention of the Y_l^m as given by Eq. (2.5.29) of Edmonds (1964); see also Dahlen [1976, Eq. (64)].

The polar motion-induced tidal potential can be represented by means of complex spherical harmonics of degree 2 and order 1:

$$V(r,\theta,\lambda) = -m_1 \Omega^2 (2\pi/15)^{0.5} r^2 [Y_2^1 - Y_2^{-1}] - m_2 \Omega^2 (2\pi/15)^{0.5} r^2 i^{-1} [Y_2^1 + Y_2^{-1}] \ . \tag{12}$$

Here, m_1 and m_2 represent direction cosines describing the pole position, "r" is the mean radius of the Earth, and Ω stands as before for the Earth's rotation rate.

It is appropriate to introduce the Newtonian equilibrium tidal height η, which can be expressed in terms of the tidal potential V and gravity acceleration g, by

$$\eta(\theta,\lambda) = -(1/g) V(r,\theta,\lambda) \ . \tag{13}$$

By confining it to this direct effect (the ocean's response to a variation of the centrifugal potential) the problem would be easily solvable in the spherical harmonic domain by considering only spherical harmonics of degree 2.

Unfortunately, things become intricate by accounting for the "indirect effects". This term comprises three different physical effects: first there will be an additional potential caused by the self-attraction of the tidal surface layer, second there will be effects owing to loading effects of the tide, and the procedure must also consider that ocean mass has to be conserved. The complicated dependence of the ocean tide on a disturbing potential is described by Eq. (78) of Dahlen (1976):

$$\xi(\theta,\lambda) = C(\theta,\lambda)\left[\sum_{l=2}^{\infty}\sum_{m=-l}^{l}(1+k_l-h_l)\eta_l^m Y_l^m + \sum_{l=0}^{\infty}\sum_{m=-l}^{l}\frac{3}{2l+1}\frac{\varrho_w}{\varrho_m}(1+k_l'-h_l')\xi_l^m Y_l^m + \frac{\Delta\Phi}{g}\right] . \quad (14)$$

k_l and h_l are the well-known Love numbers, whereas k_l' and h_l' are the so-called load Love numbers describing the effect of a surface load on the elastic Earth (Munk and MacDonald 1960, Sect. 5.8). $\varrho_w = 1.025\ \mathrm{g/cm^3}$ denotes the density of sea water, and $\varrho_m = 5.517\ \mathrm{g/cm^3}$ denotes the mean density of the Earth model. In this equation both the irregular distribution of the continents and the effects of surface loading are evident. Mass conservation in the ocean implies that the ocean surface may have changed its potential from the initial value Φ to $\Phi+\Delta\Phi$, and is taken into account by $\Delta\Phi/g$.

It will be useful to write explicitly a direct consequence of the normalization according to Edmonds (1964, Eq. 2.5.4)

$$\int_S Y_l^m Y_{l'}^{m'} dS = \delta_{ll'}\delta_{mm'} . \quad (15)$$

δ_{ll} and δ_{mm} stand, as usual, for the Kronecker symbol. Also, a formal simplification will be achieved by introducing

$$A_{lrp}^{msq} = \int_S Y_l^{*m} Y_r^s Y_p^q dS . \quad (16)$$

By using both relations and Eq. (11), Eq. (14) leads to an infinite system of complex linear algebraic equations, if both sides are multiplied consecutively with Y_l^{*m}, $l \geq 1$, $-l \leq m \leq l$. After integration over the surface of the sphere (above denoted by "S"), we finally obtain:

$$\begin{pmatrix} B_{1\,1}^{-1\,-1} & B_{1\,1}^{-1\,0} & B_{1\,1}^{-1\,1} & B_{1\,2}^{-1\,-2} & \cdot \\ B_{1\,1}^{0\,-1} & B_{1\,1}^{0\,0} & B_{1\,1}^{0\,1} & B_{1\,2}^{0\,-2} & \cdot \\ B_{1\,1}^{1\,-1} & B_{1\,1}^{1\,0} & B_{1\,1}^{1\,1} & B_{1\,2}^{1\,-2} & \cdot \\ B_{2\,1}^{-2\,-1} & B_{2\,1}^{-2\,0} & B_{2\,1}^{-2\,1} & B_{2\,2}^{-2\,-2} & \cdot \\ \vdots & \vdots & \vdots & \vdots & \end{pmatrix} \begin{pmatrix} \xi_1^{-1} \\ \xi_1^{0} \\ \xi_1^{1} \\ \xi_2^{-2} \\ \vdots \end{pmatrix} = \begin{pmatrix} b_1^{-1} \\ b_1^{0} \\ b_1^{1} \\ b_2^{-2} \\ \vdots \end{pmatrix} - \left(\frac{\Delta\Phi}{g}\right) \begin{pmatrix} C_1^{-1} \\ C_1^{0} \\ C_1^{1} \\ C_2^{-2} \\ \vdots \end{pmatrix} \quad (17)$$

where the symbols B_{lp}^{mq} and b_l^m are represented by

$$B_{lp}^{mq} = \frac{3}{2p+1}\frac{\varrho_w}{\varrho_m}(1+k_p'-h_p')\left[\sum_{r=0}^{\infty}\sum_{s=-r}^{r} A_{lrp}^{msq} C_r^s\right] - \delta_{lp}\delta_{mq} \quad (18)$$

and

$$b_l^m = -\sum_{p=2}^{\infty}\sum_{q=-p}^{p}\left[\sum_{r=0}^{\infty}\sum_{s=-r}^{r} A_{lrp}^{msq} C_r^s\right](1+k_p-h_p)\eta_p^q . \quad (19)$$

The symbols A_{lrp}^{msq} describe triple-product integrals of surface spherical harmonics, as can be seen in Eq. (16). Procedures for their numerical computation can be found in Edmonds (1964). Equation (17) is not consistent because the pole tide model was subjected to the condition of mass conservation in the oceans. The latter is responsible for the additional unknown quantity $\Delta\Phi/g$. Therefore, Eq. (17) has to be solved jointly with

$$\left(B_{0\;1}^{0\;-1} \quad B_{0\;1}^{0\;0} \quad B_{0\;1}^{0\;1} \quad B_{0\;-2}^{0\;-2} \;\ldots \right) \begin{pmatrix} \xi_1^{-1} \\ \xi_1^{0} \\ \xi_1^{1} \\ \xi_2^{-2} \\ \vdots \end{pmatrix} = b_0^0 - \left(\frac{\Delta\Phi}{g} \right) C_0^0 \tag{20}$$

which has been obtained similarly by multiplying Eq. (14) with Y_0^{*0}. By using matrices and vectors, Eq. (17) and Eq. (20) may be formally described by

$$B\xi = b - (\Delta\Phi/g)C \tag{21}$$

and

$$B_0^0 \xi = b_0^0 - (\Delta\Phi/g) C_0^0 \;, \tag{22}$$

respectively. A solution of Eq. (21) and Eq. (22) for the desired pole tide ξ follows immediately:

$$\frac{\Delta\Phi}{g} = \frac{b_0^0 - B_0^0 B^{-1} b}{C_0^0 - B_0^0 B^{-1} C} \tag{23}$$

and

$$\xi = B^{-1} b - (\Delta\Phi/g) B^{-1} C \;. \tag{24}$$

In principle, each desired accuracy can be achieved by truncating the spherical harmonic expansion at a sufficiently high degree. Here, the ocean function has been developed up to degree and order 24. The resolution is displayed in Figure 1.

These linear complex equations will be used, as they stand, for computing the harmonic coefficients ξ_l^m of the static equilibrium pole tide. Only the symbols B_{lp}^{mq} and b_l^m will be computed in the abbreviated manner described in Dahlen (1976).

Finally, it should be mentioned that the ocean function behaves as "step-function" at the continental boundaries by jumping from 1 to 0. It is well known that truncated spherical harmonics reproduce imperfectly these jumps. Hence, the present algorithm is not particularly well suited for computing tide values along continents. We try to diminish these disturbances of the ocean function by inserting its theoretical value 1 in Eq. (14). However, it remains valid that the main disadvantage of Dahlen's (1976) algorithm is associated with the truncation of spherical harmonics, and we refer to Dahlen (1976) or Dickman and Steinberg (1986) for further details.

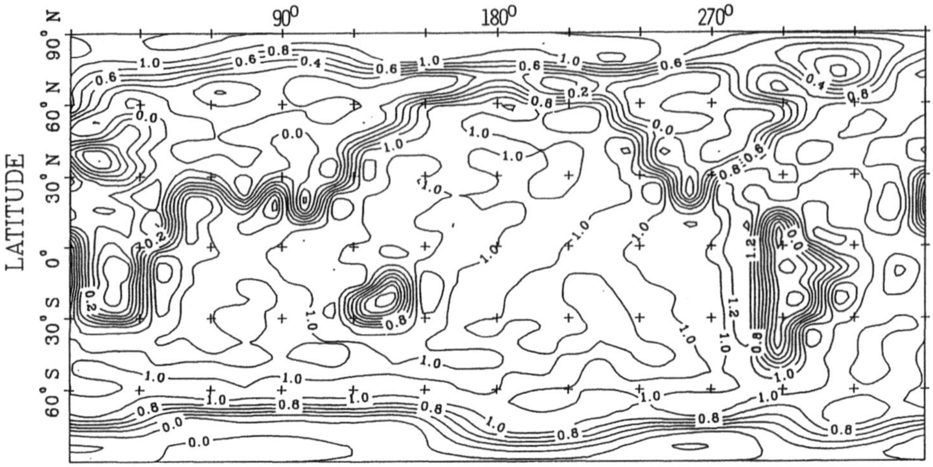

Fig. 1. Spherical harmonic development of the ocean function (should be 1 for sea, 0 for land) truncated at degree and order 24

3.2 The Comparison with Observed Values

Unfortunately, it must be realized that pole tide observations are sparse, irregularly distributed over the globe, and, moreover, not very reliable. This statement refers particularly to the observation of oceanic phase lags. The observed amplitudes are hardly above the noise level in most observation sites. Nevertheless, the global amplitude values as published by Miller and Wunsch (1973) will be used.

It is a well-known fact that a spectral investigation of ILS polar motion values reveals two distinct Chandler peaks at about 436 [solar] days and 428 days, see, e.g., Lenhardt and Groten (1985). Miller and Wunsch (1973) assigned pole tide amplitudes to spectral peaks with corresponding Chandler periods ranging from 419 days to 441 days. Because no physical reason for this frequency splitting is obvious, we try to obviate all pitfalls associated with this effect by confining to pole tide amplitudes with corresponding periods from 426 days to 436 days. If in Table 1 of Miller and Wunsch (1973) several values in this range are given, the particular one closest to 433 days will be used. This choice may be only one of several possibilities but provides unique values. The theoretically computed pole tide values are gained as follows: an amplitude of 0″.16 is assumed to represent an average Chandler amplitude for this century; a verification can be performed by a simple spectral investigation of ILS polar motion data. It is now widely recognized that the amplitude of the Earth's free nutation has not been stable throughout this century. However, we proceed in this manner because other ways appear to be even more slippery.

By inserting this value in Eq. (12) and by an appropriate variation of the phase angle, thus simulating one Chandler cycle, a different harmonic set of pole tide coefficients for each particular phase angle is obtained. Consecutive pole tide values for each desired site on the globe are gained via the spherical harmonic expansion. A subsequent amplitude estimation for each observation site finally provides the theoretically predicted amplitude which can be compared with the observed one.

Figure 2 displays the amplitude of the equilibrium pole tide in a global scale. The following table will be clear after these preliminaries:

φ [degree]	λ	Equi. Pole Tide [cm]	Observ. Pole Tide [cm]	Δ [cm]	Geographic description
Atlantic Ocean:					
28.483	343.767	0.59	0.77	−0.18	Santa Cruz (Canary Islands)
46.233	296.417	0.66	0.64	0.01	Charlottetown (Prince Edward Is.)
45.267	293.933	0.65	1.32	−0.67	St. John (Fundy Bay)
43.667	289.750	0.64	0.53	−0.11	Portland (Maine)
42.350	288.950	0.64	0.66	−0.02	Boston (Mass.)
39.267	283.417	0.62	0.35	0.27	Baltimore (Del.)
29.317	265.200	0.52	0.48	0.04	Galveston (Texas)
Pacific Ocean:					
55.333	228.367	0.59	0.98	−0.40	Ketchikan (Alaska, Alexander Arch.)
47.600	237.667	0.62	0.30	0.32	Seattle (Washington)
37.800	237.533	0.62	0.61	0.00	San Francisco (California)
8.967	280.433	0.19	0.60	−0.41	Balboa (Panama Canal Zone)
21.300	202.133	0.46	0.56	−0.11	Honolulu (Hawaii)
−33.850	151.233	0.67	0.10	0.55	Sydney (Australia)
Other Oceans:					
43.300	5.350	0.68	0.70	−0.03	Marseille (Mediterranean)
58.783	265.800	0.52	1.29	−0.77	Churchill (Hudson Bay)
44.100	39.067	0.64	0.72	−0.07	Port Tuapse (Black Sea)
13.450	100.600	0.26	0.24	0.07	Bangkok Bar (Gulf of Thailand)
11.800	99.817	0.28	0.98	−0.70	Phrachuapkirikhan (Gulf of Thailand)

It is not easy, and perhaps impossible, to derive a general feature of the global behavior of the pole tide from these few tide comparisons. It can be seen that several observation sites with unexpectedly large differences between theoretically computed values and observed ones are situated in gulf zones or fissured coastal areas. This statement refers particularly to St. John, Ketchikan, Seattle, Balboa and Churchill which are observation sites with large deviations from equilibrium. The European counterpart may be found in the anomalously large non-equilibrium pole tides of the North- and Baltic Sea which have been investigated in detail by Wunsch (1986). Also, the large difference of the two observed values in the Gulf of Thailand may be interpreted in terms of inconsistent observations, rather than a departure from equilibrium. With some doubts it may be assumed that the pole tide on the northern hemisphere follows equilibrium.

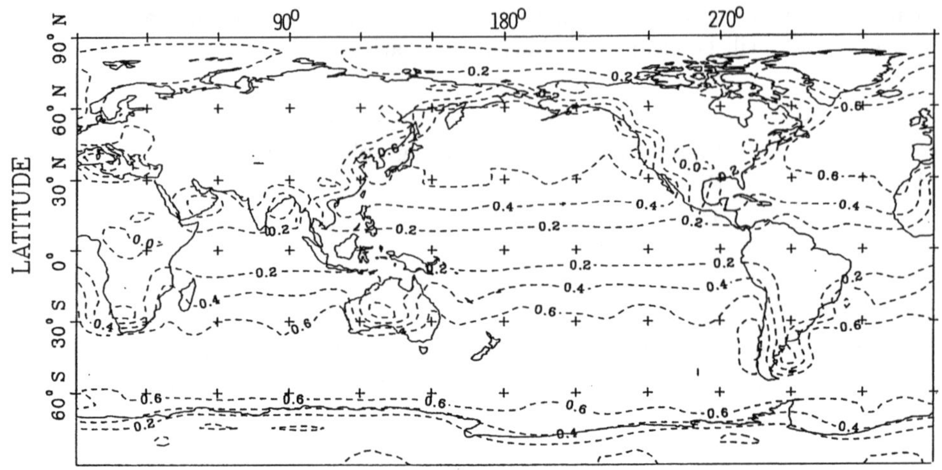

Fig. 2. Amplitudes of the equilibrium pole tide based on the ocean function given in Fig. 1. Unit is centimeter. Assumed Chandler amplitude 0″.16

4 Some Aspects of the Modern Theory

An equilibrium pole tide model as described in the preceding section enables us to impose further constraints on the contribution of the pole tide to the damping of the Chandler motion. This can be done by computing the effects of the oceans to the lengthening of the Chandler period. Other constituents, like the Earth's core, also influence this period; however, a description of the core's effect would be beyond the frame of the present treatment. The relevant values are taken from Smith and Dahlen (1981); but the forthcoming results might be different if these parameters prove to be inaccurate.

As far as the influence of the pole tide is concerned, the algorithm as described above offers a convenient procedure to compute the elements D_{ij} of the rotational response tensor, as given by Dahlen (1976). This tensor provides the relation between the (sign-reversed) products of inertia c_{ij} and the dimensionless pole coordinates m_j: $c_{i3} = D_{ij} m_j$. The second-order potential can be expressed in terms of the pole tide coefficients ξ_2^1, ξ_2^{-1} and ξ_2^0, and also via MacCullagh's formula in terms of the c_{ij}. The desired relation can then be found straightforwardly by comparison:

$$\begin{aligned} D_{11} &= [k_2 - 0.3\,(\varrho_w/\varrho_m)(1+k_2')(\xi_2^1 - \xi_2^{-1})]\,a^5\,\Omega^2/(3\,G) \\ D_{22} &= [k_2 + 0.3\,(\varrho_w/\varrho_m)(1+k_2')(\xi_2^1 + \xi_2^{-1})\,i^{-1}]\,a^5\,\Omega^2/(3G)\ . \end{aligned} \quad (25)$$

These are in the present case the elements of particular interest. "a" denotes the radius of the outer surface of the Earth model, and "G" denotes Newton's constant of gravitation. The application of these tensor elements in the eigenvalue equation provides the most accurate determination of the theoretical eigenfrequency which is presently available, subject only to the accuracy with which some basic parameters of the Earth are known. A broad discussion is given by Smith and Dahlen (1981), and their approximate solution of the eigenvalue equation is given below, see Eq. (26). The normalized tensor elements (divided by $a^5\Omega^2/3G$) are:

$$d_{11} = 0.35340 \qquad d_{22} = 0.34194 \ .$$

The values of these tensor elements differ slightly from those given by Smith and Dahlen (1981), because of the independent computation of the ocean function. The eigenfrequency can now be determined by using the tensor elements d_{11} and d_{22}. After introduction of $k_e = k_w + \Delta k$, with $\Delta k = 0.5\cdot(d_{11}+d_{22})-k_2$:

$$\sigma_e = \frac{C-A-k_e a^5\Omega^2/(3G)}{A_m+\varepsilon_b A_c+k_e a^5\Omega^2/(3G)}\,\Omega \ . \tag{26}$$

k_e is a new wobble-effective Love number, as indicated by the index. $k_w = 0.30158$ denotes a dynamical wobble Love number for an oceanless Earth. The indices $_m$ and $_c$ denote mantle and core, respectively, and ε_b is a measure for the core's ellipticity. We confine to the ocean's effect which is obviously expressed by k_e. Using $\varepsilon_b = 1/393$, $k_w = 0.30158$, $k_2 = 0.30088$ and $(C-A)/A = 304.4$, a period of 427.1 (solar) days is obtained which differs only slightly from the observed value of about 433 days. As far as the pole tide is suspected to produce this discrepancy, there are several possibilities which should be discussed: The modification of the frequency due to a phase lag ε, an amplification δ of the amplitude, or a combination of these effects.

The preceding algorithm proves to be an efficient tool also for this purpose. Evidently, the modification of the local pole tide amplitude ξ_0 can be written, if both effects are taken into account: $\xi = \delta\cdot\xi_0\cdot\exp(-i\varepsilon)$. It follows from the spherical harmonic development that each harmonic coefficient is modified accordingly, and, by the definition of the tensor elements in Eq. (25), also Δk as introduced above. A global oceanic phase lag, or amplitude amplification, or combination of both effects enter only via a new "wobble-effective" Love number k_e. The latter becomes complex for a phase lag ε, and consequently a complex eigenfrequency σ_e is obtained in Eq. (26). Now σ_e can be solved for its real and imaginary part, thus providing us with a slightly modified real frequency and the Q-factor. Vice versa, the eigenfrequency of an oceanless Earth model can be computed by inserting $\Delta k = 0$.

Some numerical values should be added: For the last-mentioned case ($\Delta k = 0$), an eigenperiod of 395.9 [solar] days is obtained. The equilibrium pole tide lengthens this value to 427.1 days, as already noted above. A global phase lag of $\varepsilon = 5°$ reduces the period to 427.0 days, and yields $Q = 73$. $\varepsilon = 20°$ results in a period of 425.1 days, and $Q = 19$. Hence it can be seen that an oceanic phase lag can easily be responsible for damping, if we assume that the Q-value is in the range between 50 and 200, as suggested from astronomical observations. On the other hand, the period is only slightly affected.

An amplification of the amplitude by a factor δ has a quite different consequence: $\delta = 1.2$, for instance, will lengthen the eigenperiod to 433.9 (solar) days. Hence it may entirely account for the discrepancy between an Earth model with equilibrium pole tide and the observed Chandler frequency. There is no dissipation for $\varepsilon = 0$, and Q is of course infinite. $\delta = 1.2$ and $\varepsilon = 10°$ reduces the period to 433.3 days, and $Q = 30$.

The eigenperiod for various combinations of δ and ε can be computed in this convenient manner. The preceding numerical values obey the general rule: A phase lag ε of less than 90° is responsible for the lengthening of the eigenperiod in comparison with an oceanless Earth model, but reduces the eigenperiod with respect to an Earth model having an equilibrium pole tide. An amplification of the tidal amplitude can considerably reduce the eigenfrequency. Some mechanical aspects may give additional insight: An elastic Earth model has a lower eigenfrequency in comparison with a rigid Earth model, because the "restoring force" of the equatorial bulge is reduced by elasticity. An equilibrium pole tide analogously reduces the "restoring force", because more movable constituents of the Earth have adapted to the changed centrifugal potential. A phase lag makes this adaptation partially incomplete, and leads consequently to a higher eigenfrequency as in the equilibrium case. On the other hand, an enhanced oceanic tidal bulge makes even more fluid parts of the Earth partaking in the diminution of "restoring force", and reduces the eigenfrequency.

5 Conclusions

The preceding discussion shows that a detailed theory of the equilibrium pole tide exists. A comprehensive extension to a dynamic pole tide theory has also been accomplished (Dickman 1988). From this investigation, it seems that the pole tide dissipates wobble energy at a very weak rate. It is worth mentioning that Dickman's (1988) values for the relaxation time are not too far from those obtained with Bondi and Gold's (1955) maximum estimation.

It has been shown that even a comparatively small departure from equilibrium may entirely account for the damping of the Earth's free nutation. However, the accuracy of the available tidal data does not permit to verify a small global phase lag, and theoretical considerations cannot replace tidal observations. This situation will not change perceptibly within the next decade. The comparison of the static equilibrium tide model with observations is ambiguous, as are the (ILS-) polar motion observations themselves; it suffers also from the fact that too few tidal observations are known. A global phase lag between oceanic tidal bulge and centrifugal potential may account for the observed Q-value, but phase determinations exist only for some particular regions and they cannot be generalized. Furthermore, the Q-value as determined from polar motion is also far from being clear and not generally accepted. After all Bondi and Gold's (1955) maximum estimation looks still attractive.

Acknowledgment. This research was supported by the Deutsche Forschungsgemeinschaft, Project GR-361/24.

References

Bondi H, Gold T (1955) On the Damping of the Free Nutation of the Earth. Astron Soc 115(1) :41–46

Dahlen FA (1976) The Passive Influence of the Oceans upon the Rotation of the Earth. Geophys J R Astron Soc 46:363–406

Dickman SR (1988) The self-consistent dynamic pole tide in non-global oceans. Geophys J R Astron Soc 94:519–543

Dickman SR, Steinberg DJ (1986) New aspects of the equilibrium pole tide. Geophys J R Astron Soc 86:1986

Edmonds AR (1964) Drehimpulse in der Quantenmechanik. Bibliogr Inst Mannheim. Hochschultaschenbücher-Verlag

Klein F, Sommerfeld A (1965) Über die Theorie des Kreisels. Teubner, Stuttgar. Nachdruck der Erstauflagen von 1897/1898/1903/1910

Lenhardt H, Groten E (1985) Chandler wobble parameters from BIH and ILS data. Manuscripta Geodaetica 10:296–305

Miller SP, Wunsch C (1973) The Pole Tide. Nat Phys Sci 246:98–102

Munk WH, MacDonald GJF (1960) The Rotation of the Earth. Univ Press, Cambridge

Smith ML, Dahlen FA (1981) The period and Q of the Chandler wobble. Geophys J R Astron Soc 64:223–281

Wunsch C (1986) Dynamics of the North Sea pole tide reconsidered. Geophys J R Astron Soc 87:869–884

The Seasonal Angular Momentum of the Thermohaline Ocean Circulation

A. Frische and J. Sündermann

1 Introduction

The variations in the Earth's rotation take place in different time scales reaching from geological periods down to days. Modern observation techniques as satellite measurements and Very Long Baseline Interferometry (VLBI) measure with increasing accuracy and resolution short-term variations in the range of seasonal cycles to tidal periods. Essential for the seasonal variations is the solar radiation and its cycle.

This mechanism cannot produce (as the tidal interaction) a transfer of angular momentum between two celestial bodies. But it causes a redistribution of angular momentum on the globe and an exchange between atmosphere, ocean and solid Earth. These processes are connected with significant short-period changes in the Earth's rotation parameters.

The main part of the seasonal and subseasonal variations obviously is produced by the dynamics of the atmosphere and its direct influence on the solid Earth (mountain torque, Hide 1984). On the other hand, a considerable part of the atmospheric angular momentum is transmitted to the ocean itself acting via friction and bottom torque on the solid Earth. It turned out from an estimation for the Antarctic Circumpolar Current (ACC) that its wind-driven variability alone corresponds to a semiannual deviation in Universal Time (ΔUT) of 9 ms (Brosche and Sündermann 1985).

Another part of the atmospheric forcing is due to the heat and salt fluxes at the sea surface showing also seasonal cycles. Besides the variable wind forcing they cause variations in the distribution of ocean water masses resulting in variable thermohaline currents. This means a changing seasonal angular momentum content of the world ocean connected with corresponding changes in Earth rotation. Its amount is presumably small compared to the wind-caused variations, nevertheless a complete balance requires also this estimate.

In the following, the seasonal variation of angular momentum contained in the world ocean is calculated from hydrographic data. The investigation is based on the formula expressing the axial component J (t) of the relative angular momentum, from now on referred to as axial angular momentum (Gill 1982):

$$J(t) = R^3 \int_{-\pi}^{+\pi} \int_{-\pi/2}^{+\pi/2} \int_{h}^{\zeta} \varrho(\lambda, \varphi, z, t)\, u(\lambda, \varphi, z, t) \cos^2 \varphi \, d\lambda \, d\varphi \, dz \; . \qquad (1)$$

Institut für Meereskunde, Universität Hamburg, Troplowitzstraße 7, D-2000 Hamburg 54, FRG.

The following notations have been used (the dimensions are given in brackets):

J axial component of relative angular momentum ($kg\,m^2\,s^{-1}$)
ϱ density of the water ($kg\,m^{-3}$)
u zonal component of the velocity (ms^{-1})
R radius of the Earth (m)
h water depth (m)
ζ water elevation (m)
λ, φ geographical longitude and latitude
z vertical coordinate (m), positive upwards
t time (s)

The origin of the coordinate system is at the undisturbed sea level ($z = 0$).

The quantities ϱ and u are calculated from three-dimensional fields of observed and interpolated sea temperatures and salinities for the four seasons:

winter – February, March, April
spring – May, June, July
summer – August, September, October
autumn – November, December, January

2 Calculation of Thermohaline Currents

2.1 Basic Equations

The general equation governing the hydrodynamic fields in the ocean is deduced from the fundamental law of conservation of momentum. It reads (Pedlosky 1979):

$$\frac{d\mathbf{v}}{dt} + 2\boldsymbol{\omega}\times\mathbf{v} - \nabla\cdot(A\nabla\mathbf{v}) = -\frac{1}{\varrho}\nabla P - \nabla\phi \ . \tag{2}$$

It is completed by the equation of state

$$\varrho = \varrho(S, T, p) \ . \tag{3}$$

The notations are:

$\mathbf{v}$ = (u, v, w) velocity of water (ms^{-1})
ω angular velocity of the Earth (s^{-1})
p pressure (Nm^{-2})
ϕ potential of external forces ($m^2\,s^{-2}$)
A eddy viscosity ($m^2\,s^{-1}$)
T temperature (K)
S salinity

Today there exist big numerical models integrating the three-dimensional non-linear prognostic system for given initial and boundary conditions (Bryan and Cox 1968; Maier-Reimer 1985). In the present case, however, only a rough estimate of the thermohaline currents and its fluctuations was required. For this purpose a simplified diagnostic system of equations seems to be sufficient.

2.2 Thermal Wind Equations

The following simplifications are introduced (Pedlosky 1979):

Motions and their variations are so slow that they can be considered as quasistationary

and inertia-free: $\frac{dv}{dt} = 0$

Motions are frictionless: $A = 0$

The only external force is gravity: $\nabla \phi = (0, 0, -g)$

g is the acceleration of gravity (ms^{-2})

Further, from scale analysis (using $w \ll u, v$) follows a separation of horizontal and vertical equilibria and Eq. (2) reads

$$\varrho f v = \frac{1}{R \cos\varphi} \frac{\partial p}{\partial \lambda} \tag{4a}$$

$$\varrho f u = -\frac{1}{R} \frac{\partial p}{\partial \varphi} \tag{4b}$$

$$0 = -\frac{1}{\varrho} \frac{\partial p}{\partial z} - g \tag{4c}$$

with $f = 2|\omega| \sin\varphi$ denoting the Coriolis parameter (s^{-1}). Equations (4a, 4b) are the so-called thermal wind equations. (4c) is the hydrostatic equation. Differentiating (4a) with respect to z and using (4c) gives

$$\varrho f \frac{\partial v}{\partial z} + f u \frac{\partial \varrho}{\partial z} = -\frac{\varrho}{R \cos\varphi} \frac{\partial \varrho}{\partial \lambda} .$$

For real oceanic conditions the second term on the left side is small compared with the term on the right side, resulting in:

$$\varrho f \frac{\partial v}{\partial z} = -\frac{g}{R \cos\varphi} \frac{\partial \varrho}{\partial \lambda}$$

and integration over a water layer (z, z_0) yields:

$$v(z) - v(z_0) = -\frac{g}{\varrho f R \cos\varphi} \int_{z_0}^{z} \frac{\partial \varrho}{\partial \lambda} dz . \tag{5}$$

This equation expresses that by means of the thermal wind equations only relative current velocities can be determined from a given density field $\varrho(\lambda, \varphi, z)$. But if a depth z_0 of no motion is known:

$$v(z_0) = 0 \ ,$$

from Eq. (5) follows

$$v(z) = -\frac{g}{\varrho f R \cos\varphi} \int_{z_0}^{z} \frac{\partial \varrho}{\partial \lambda} dz \tag{6a}$$

and correspondingly

$$u(z) = \frac{g}{\varrho f R} \int_{z_0}^{z} \frac{\partial \varrho}{\partial \varphi} dz \ . \tag{6b}$$

Equations (6a, b) together with the equation of state (3) are used in the following to determine the thermohaline currents in the ocean from observed temperature and salinity fields.

Estimations for the real sea show that the geostrophic equilibrium described by the system (6a, b) is fulfilled for almost the whole entire ocean. It is not valid for friction layers near the boundaries and in the vicinity of the equator ($\approx \pm 3°$).

The problem remains of finding a proper layer of no motion $z = z_0$ (reference layer). There are numerous approaches leading to different results, see Section 3.1 (Fomin 1964).

2.3 The Hydrographic Data

The following calculations are based on the Climatological Atlas of the World Ocean (Levitus 1982) presenting temperatures and salinities for a horizontal grid of $1° \times 1°$ and 33 vertical layers from 0 to 5500 m depth for four seasons (below 1500 m only annual means). It is the best global data set available today.

It should be mentioned, however, that there are still some severe defects in the data:

1. In the raw data are considerable lacks especially on the southern hemisphere.
2. The atlas has been completed therefore on the basis of interpolations. These do no take into account in all cases sufficiently the specific topographical and hence dynamical conditions. Obvious errors have been corrected.

3 Oceanographic Results

3.1 The Global Circulation

The main problem in the beginning is the proper choice of a reference level $z = z_0$. In the oceanographic literature, as a constant datum, depths from 1000 m to 3000 m are used. Defant (1941) defined a variable layer of no motion ranging from 500 m depth at subtropical latitudes up to more than 2000 m in polar waters.

In order to check the sensitivity of oceanographic and geodynamic quantities with respect to the specific choice of the reference depth, calculations have been carried out with constant values of 1500 m, 2000 m and 2500 m and a schematically varying level as well. Close to Defants approach, the latter has been chosen as a simple polygon depending on latitude only (see Fig. 1).

Areas of smaller depth than the reference layer are not considered. The equation of state (3) has been used according to Millero and Poisson (1981).

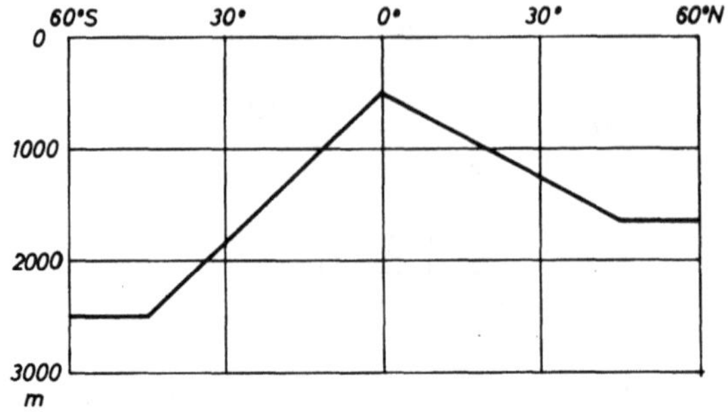

Fig. 1. Depth of the variable reference layer

As an example of calculated currents, Figure 2 shows the velocities at 500 m depth for a reference level of 2000 m. The large subtropical gyres are clearly seen, especially on the northern hemisphere. The strongest global current is the Antarctic Circumpolar Current. This picture agrees qualitatively with the known observed circulation pattern. A quantitative verification is not possible due to the insufficient data base of observed velocities. It will be done by means of integrated quantities (transports) in Sect. 3.3.

A better idea of the calculated general circulation is provided by isolines of the geopotential anomaly. They are derived from Eq. (4c) introducing a potential:

$$d\phi = \alpha\,dp = -g\,dz \ . \qquad (7)$$

Here $\alpha = \frac{1}{\varrho}$ is the specific volume which can be expressed as a sum of a reference value and a deviation:

$$\alpha(S,T,p) = \alpha(35,0,p) + \Delta\alpha \ .$$

The expression

$$\int_{z_0}^{z} \Delta\alpha\,dp$$

denotes the geopotential anomaly; it expresses the deviation from the ideal geopotential caused by the baroclinic pressure gradients.

Figure 3 shows the annual mean of the geopotential anomaly again for 500 m depth related to a reference depth of 2000 m. The isolines can be interpreted as streamlines, their density is proportional to the mass transport.

Figures 4 and 5 give for the same case the geopotential anomalies at the surface and at a depth of 1500 m. The surface picture shows the known circulation pattern (Gulf Stream, Kuroshio, equatorial currents, ACC). It cannot, however, be directly compared with observations because the upper ocean layer is mainly driven by the wind and its influence is not considered in the calculations. The only global structure still visible in Figure 5 is the ACC. It can be assumed that the major part of the axial angular momentum is stored in this water body.

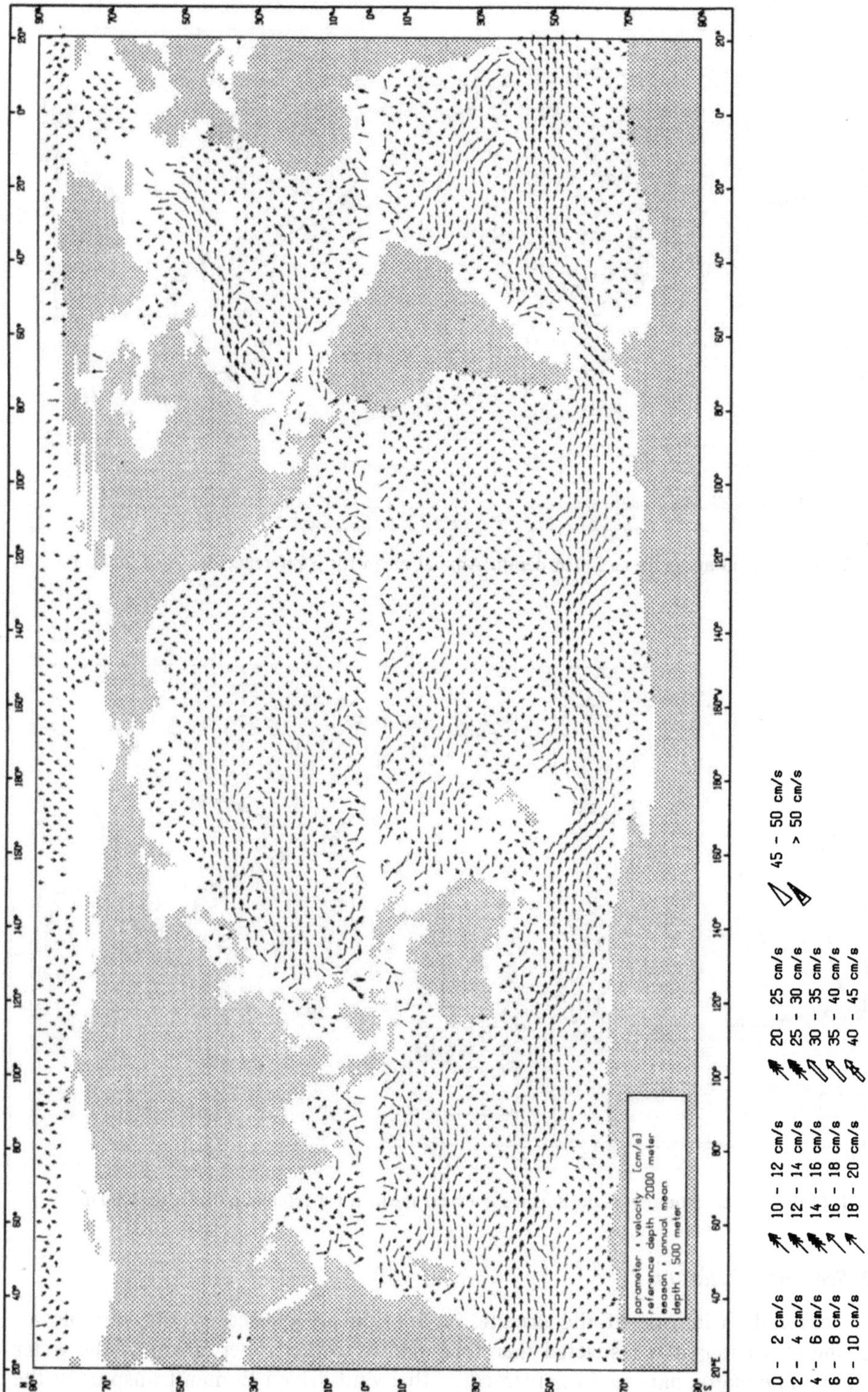

Fig. 2. Calculated geostrophic currents at 500 m depth. Annual mean for a reference depth of 2000 m. Only every fourth arrow is drawn

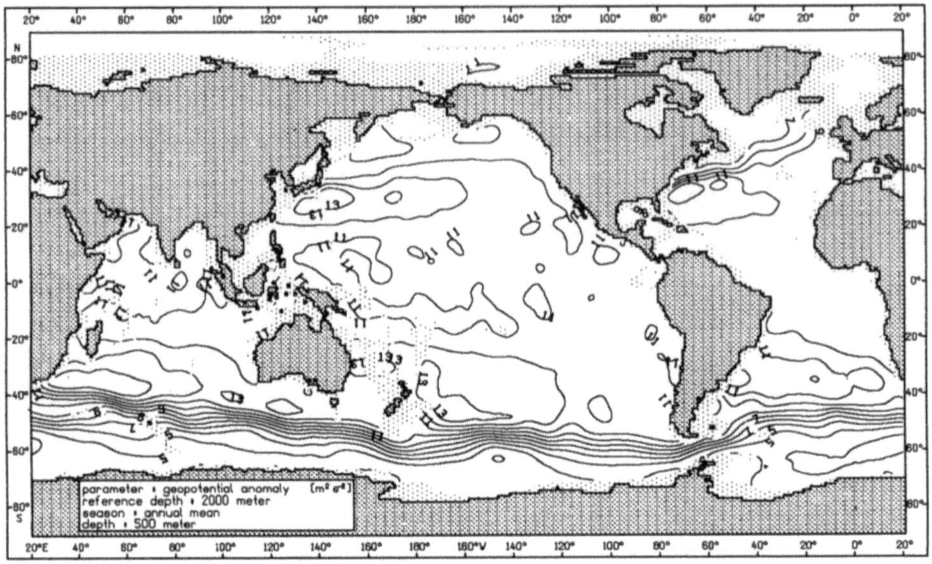

Fig. 3. Geopotential anomaly at 500 m depth. Annual mean for a reference depth of 2000 m

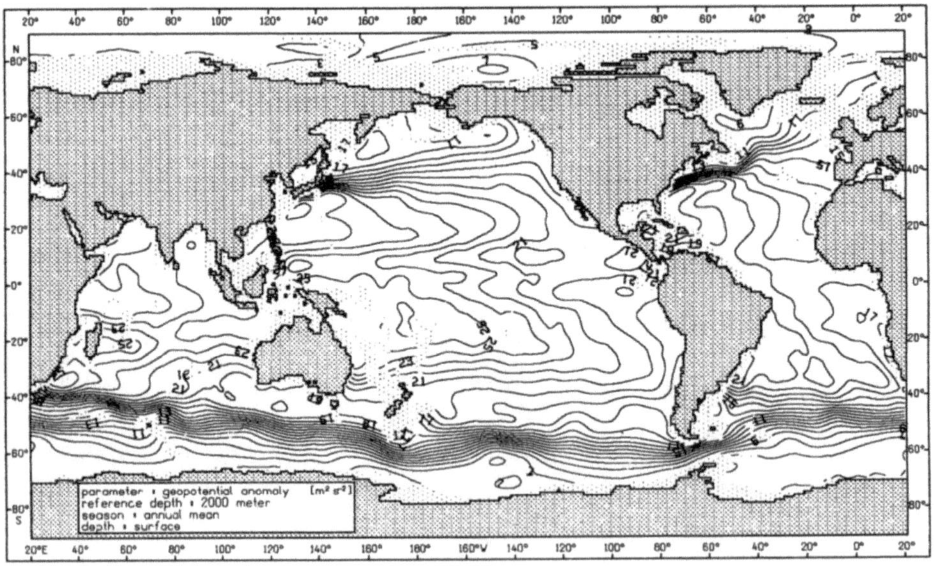

Fig. 4. Geopotential anomaly at the surface. Annual mean for a reference depth of 2000 m

3.2 Seasonal Variations

The seasonal variations of the oceanic circulation are caused by the variable atmospheric forcing; it is restricted therefore mainly to the upper 500 m of the water column. The major part again is formed by the wind-driven Ekman transport in the

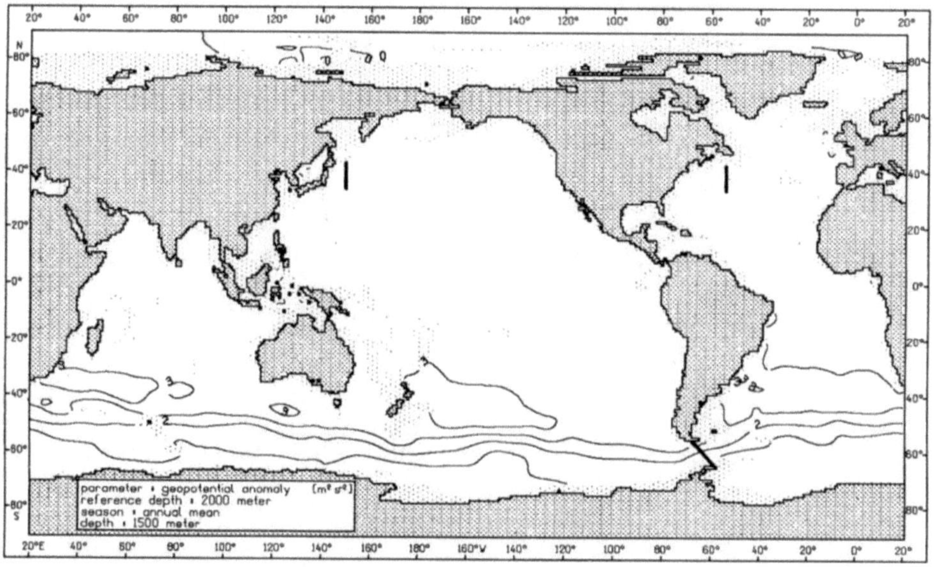

Fig. 5. Geopotential anomaly at 1500 m depth. Annual mean for a reference depth of 2000 m. Transport calculations have been done at the marked sections

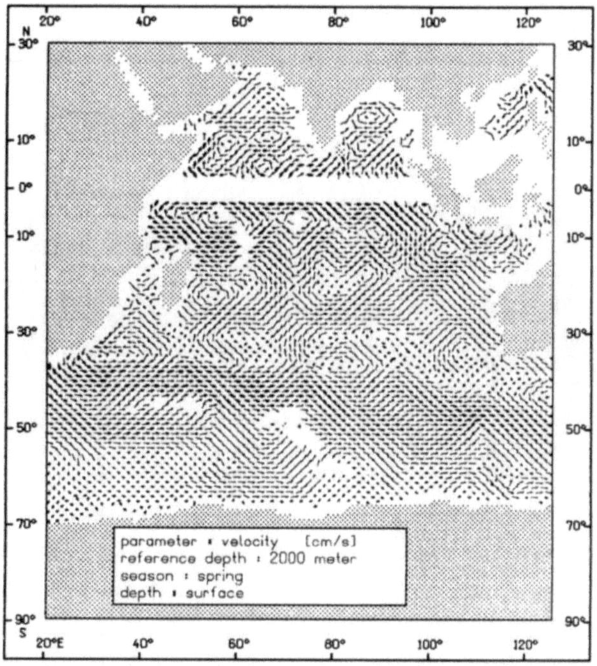

Fig. 6. Geostrophic surface currents of the Indian Ocean in spring. Reference level: 2000 m

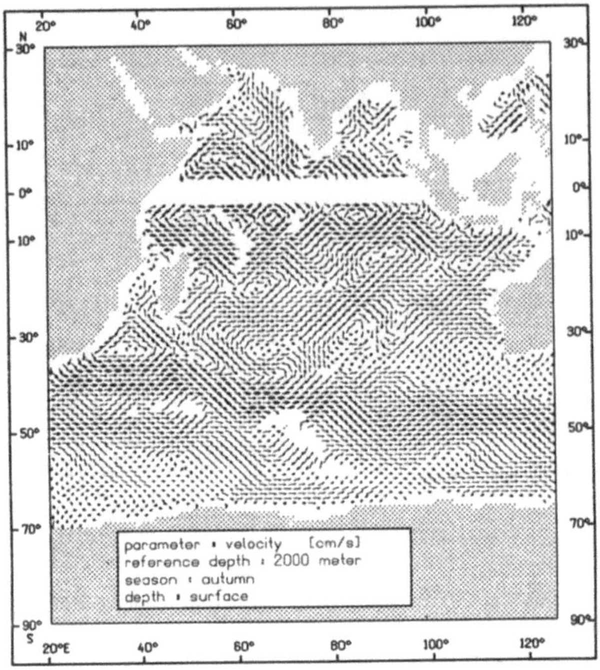

Fig. 7. Geostrophic surface currents of the Indian Ocean in autumn. Reference level: 2000 m

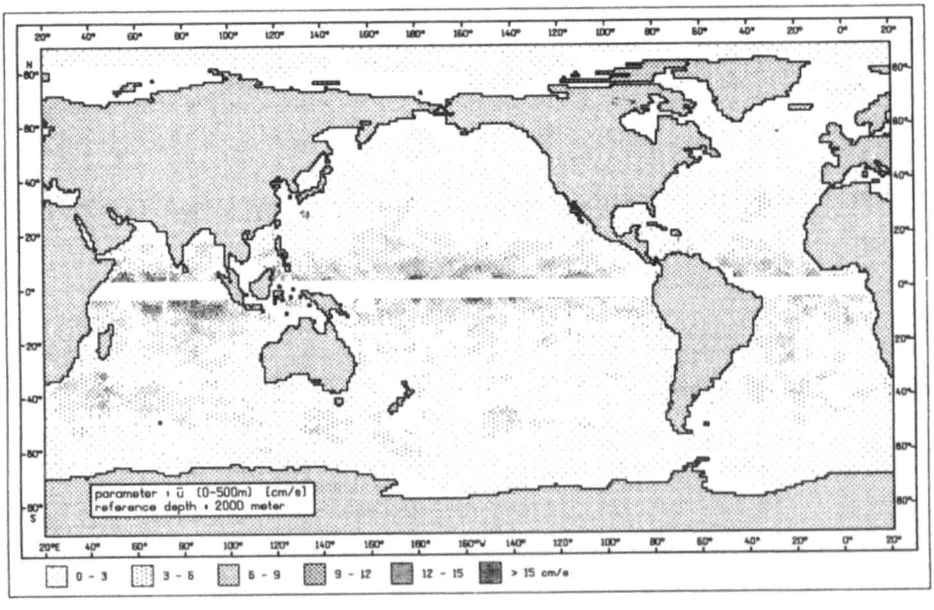

Fig. 8. Seasonal variability of the mean zonal geostrophic velocity for the upper 500 m. Reference depth: 2000 m

surface boundary layer which is not considered here. The variations in the thermohaline circulation produced by changes in the water mass distribution are relatively small.

The most significant variations are found in the Indian Ocean which is driven by the seasonal changing monsoon winds. Figures 6 and 7 show the thermohaline surface currents in the Indian Ocean in spring and in autumn. The most obvious feature is the reversal of the flow direction in the eastern South Equatorial Current and in the Somali Current.

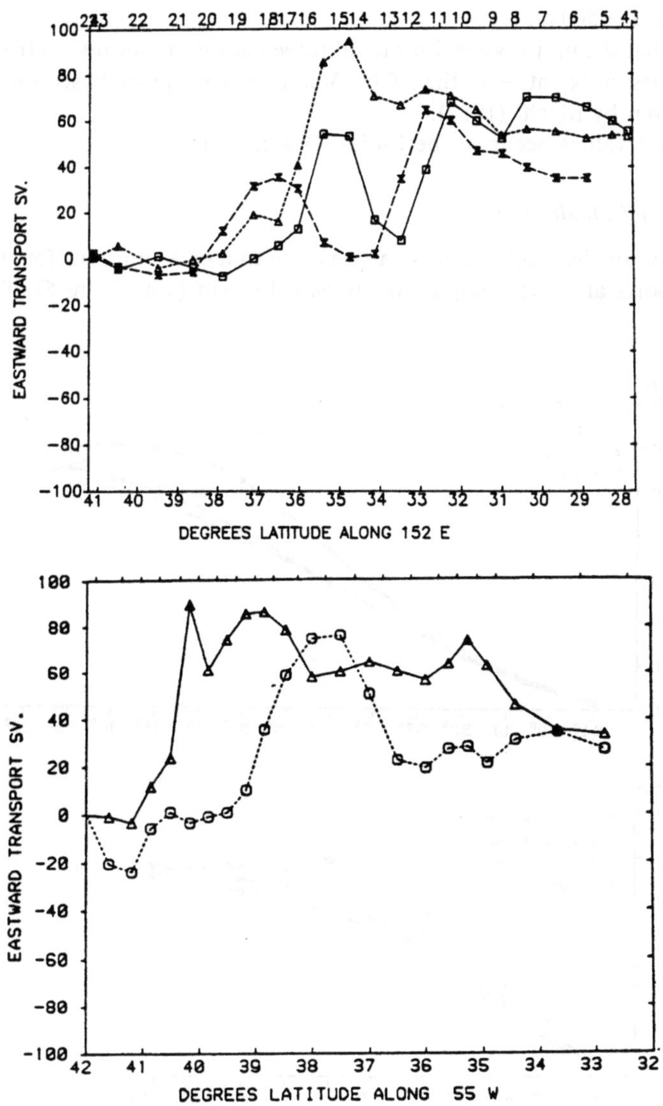

Fig. 9. Geostrophic transport in the Kuroshio (*upper panel*) and the Gulf Stream (*lower panel*) for different cruises according to Niiler et al. (1985)

Introducing a mean velocity $\bar{u}_i$ for a water layer of depth d and a season i

$$\bar{u}_i(\lambda, \varphi, d) = \frac{1}{d} \int_{-d}^{0} u_i(\lambda, \varphi, z)\, dz \;, \quad i = (1, 2, 3, 4)$$

and defining the expression

$$\Delta \bar{u}(\lambda, \varphi, d) = \max_i \bar{u}_i - \min_i \bar{u}_i$$

as a measure of seasonal variability, the geographical distribution of Figure 8 is obtained for d = 500 m.

It shows that the major variations in all three oceans are located in tropical regions and – to a lesser extent – in the ACC. A similar description based on observations has been given by Meehl (1982).

The highest values occur in the Indian Ocean (> 15 cm/s).

3.3 Transport Calculations

A verification of the model results can be done to a certain extent for the calculated mass transports at some sections of the world ocean (see Figure 5). The transport

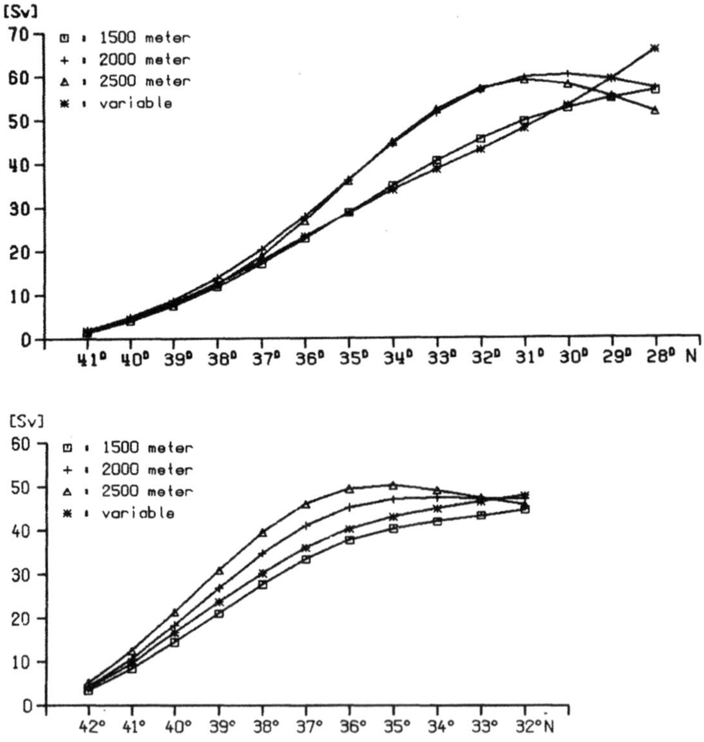

Fig. 10. Annual mean geostrophic transport in the Kuroshio (*upper panel*) and in the Gulf Stream (*lower panel*) for different reference levels

across a meridional section $(\lambda_1, \varphi_1, \varphi_2)$ is obtained by integration of the zonal component of the geostrophic velocity:

$$U(\lambda, \varphi_1, \varphi_2) = \int_{\varphi_1}^{\varphi_2} dU\, d\varphi = \int_{\varphi_1}^{\varphi_2} \int_{-h}^{\varphi_2} u(\lambda, \varphi, z)\, d\varphi\, dz$$

where dU denotes the transport per degree latitude.

For the sections across the Kuroshio and the Gulf Stream geostrophic calculations are given by Niiler et al. (1985), however only for an observation period of 14 days. The deduced transports are shown in Figure 9. They are compared with the authors' annual means (Figure 10).

The curves are qualitatively and quantitatively similar. The smaller distance of hydrographic stations in Niilers study allows far higher transport gradients which means a narrower flow. In our calculations all four reference layers lead to similar results. The values for the total transport vary from 56 to 66 Sv (1 Sv = 1 Sverdrup $= 10^6\,m^3\,s^{-1}$) for the Kuroshio and from 44 to 48 Sv for the Gulf Stream. They are in good agreement with Niilers estimates.

In Figure 11 the annual mean geostrophic transport in the Drake Passage for different reference levels is presented. The variable layer and the constant one with $z_0 = 2500$ m yield identical results, because they are in the same depth. The total transport (U = 69 Sv) is significantly higher as for $z_0 = 2000$ or 1500 m (45 Sv resp. 18 Sv). According to Whitworth (1983) and Nowlin and Klinck (1986) the baroclinic

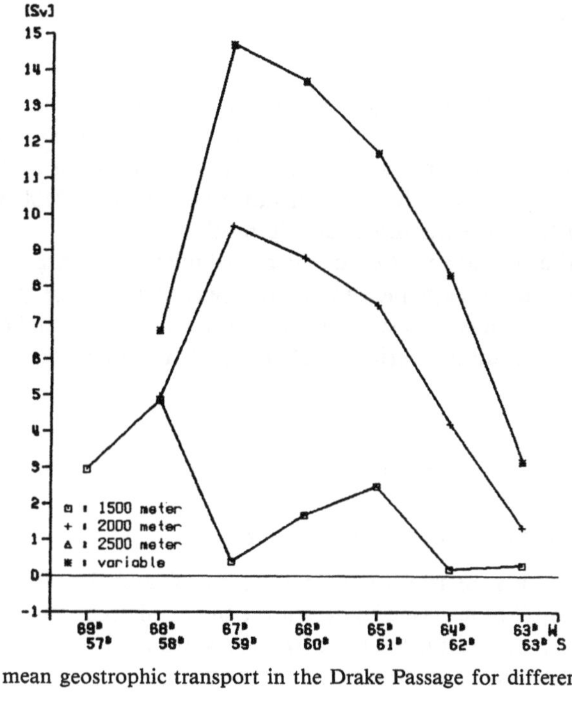

Fig. 11. Annual mean geostrophic transport in the Drake Passage for different reference levels

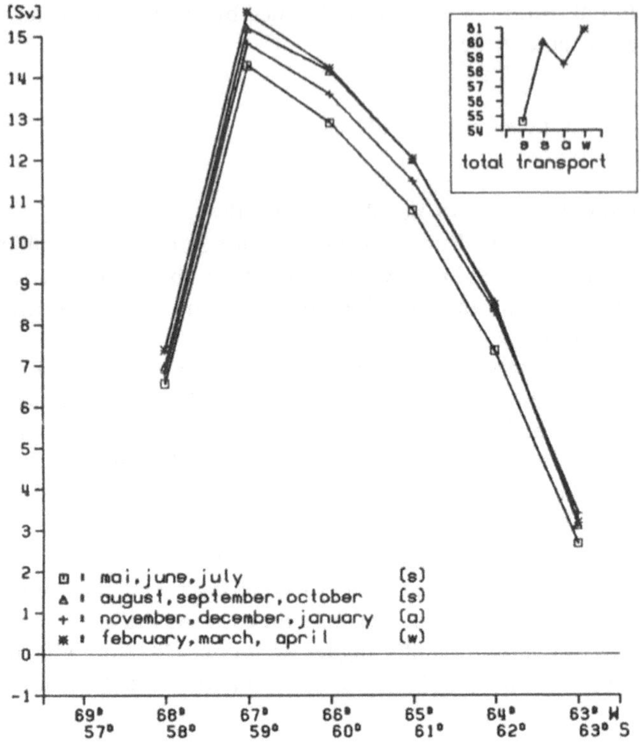

Fig. 12. Geostrophic transport in the Drake Passage for the four seasons and a variable reference level (*large picture*) and total transport during the year (*small inserted picture*)

(thermohaline) portion of the transport amounts to two-thirds of the total value. This would mean an annual mean of the total transport of 105 Sv which is in good agreement with observations. The conclusion from these comparisons is that the choice of a variable reference layer, respectively a constant one with $z_0 = 2500$ m, provides in this part of the world ocean the most reliable results.

Figure 12 shows the annual variation of the baroclinic mass transports in the Drake Passage for this case. In the large picture the transport is drawn for the four seasons, in the small one the total mass transport during the year. The seasonal curves are very close to each other. The annual variation of the total transport is ± 2 Sv which corresponds to $\pm 2\%$.

4 Angular Momentum Contained in Thermohaline Currents

4.1 Calculations Based on Geostrophic Currents

In the following the mass distribution calculated by (3) and the zonal geostrophical currents calculated by (6b) are used to determine the axial component of the relative angular momentum J according to (1). In the drawings again an area density δJ is shown, sometimes separately for the three oceans:

$$J(\lambda_1, \lambda_2, \varphi_1, \varphi_2) = \int_{\varphi_1}^{\varphi_2} \delta J(\lambda_1, \lambda_2, \varphi)\, d\varphi \ .$$

J contains already the zonal integration between the coasts (λ_1, λ_2).

Figures 13 and 14 show $\delta J(\varphi)$ for the world ocean $(\lambda_1 = -\pi, \lambda_2 = +\pi)$, for reference levels 2500 m and variable. Both pictures agree qualitatively. They demonstrate the large anticyclonic gyres in the world ocean (eastward flow in middle, westward flow in tropical latitudes) and the dominating ACC. A more detailed analysis shows however significant differences in terms of numbers: the variable reference layer produces higher positive (eastward) values of the total angular momentum. The numbers are given in Table 1: $10.9 \cdot 10^{24}\ \mathrm{kg\,m^2\,s^{-1}}$ compared with

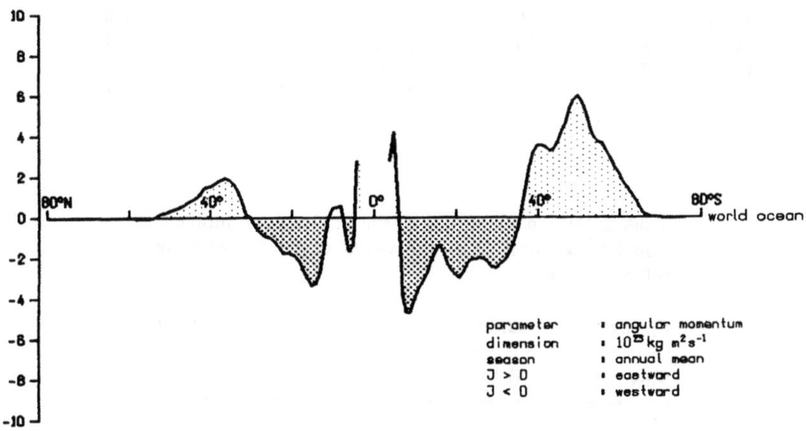

Fig. 13. Annual mean meridional density of the axial angular momentum of the world ocean. Reference level: 2500 m

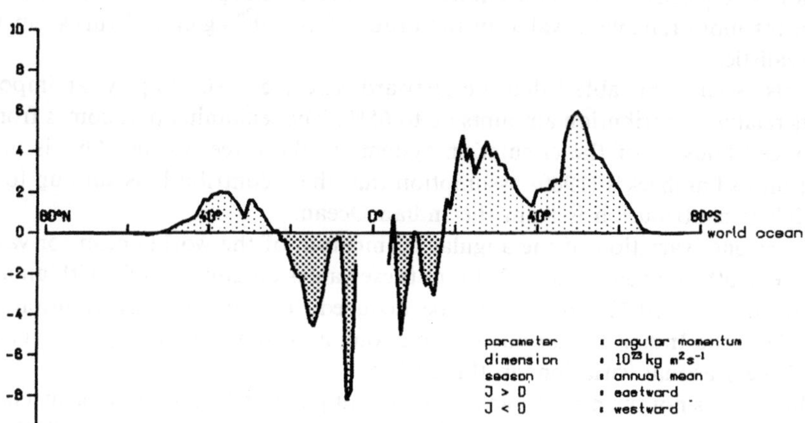

Fig. 14. Annual mean meridional density of the axial angular momentum of the world ocean. Reference level: variable

Table 1. Annual mean axial angular momentum (10^{24} kg m^2 s^{-1}) of the oceans between parallels of latitude for reference depth being (a) 2500 m and (b) variable

2500 m	Atlantic	Pacific	Indian	World
90 °N – 90 °S	1.76	–0.47	1.24	2.53
90 °N – 4 °N	0.64	–1.31	–0.05	–0.72
4 °S – 45 °S	–0.54	–1.84	–1.83	–4.21
45 °S – 90 °S	1.66	2.68	3.02	7.35
90 °N – 45 °S	0.10	–3.16	–1.77	–4.83
Variable	**Atlantic**	**Pacific**	**Indian**	**World**
90 °N – 90 °S	3.55	0.55	6.80	10.90
90 °N – 4 °N	0.66	–2.37	0.08	–1.63
4 °S – 45 °S	1.39	0.22	3.81	5.43
45 °S – 90 °S	1.50	2.71	2.89	7.11
90 °N – 45 °S	2.05	–2.15	3.90	3.81

Table 2. Seasonal variation of the axial angular momentum (10^{24} kg m^2 s^{-1}) of the world ocean for variable reference depth

Spring	Summer	Autumn	Winter
9.64	10.14	11.85	10.04

$2.5 \cdot 10^{24}$ kg m^2 s^{-1}. Thus, although the geostrophic currents for both reference depths obviously differ only slightly, the angular momentum varies by a factor of 4. It responds very sensitively to the choice of the reference depth. Since the variable option seems more reliable, a value in the order of $10 \cdot 10^{24}$ kg m^2 s^{-1} turns out to be most realistic.

It is also seen from Table 1 that the eastward ACC (45 °–90 ° S) plays an important role, its relative contribution amounts up to 65%. The remaining part comes from the unbalanced budgets of the circulation systems in the three oceans. This is in contradiction to Lambeck's (1980) assumption that these contributions sum up to zero. The highest contribution is from the Indian Ocean.

The seasonal variation of the angular momentum of the world ocean for variable reference depth is given in Table 2. From these values an annual cycle with an amplitude of $\Delta J = 1.1 \cdot 10^{24}$ kg m^2 s^{-1} can be deduced. The greatest contribution again comes from the Indian Ocean due to the strong monsoon signal. Figure 15 gives an idea of the seasonal variations in this region.

Although it is not the main topic of the present paper, a further oceanic mechanism generating an annual variation of the Earth's rotation should be mentioned: the variable moment of inertia due to the displacement of water masses in the sea. This contribution $J_\theta(t)$ is represented by the formula (Seiler 1988)

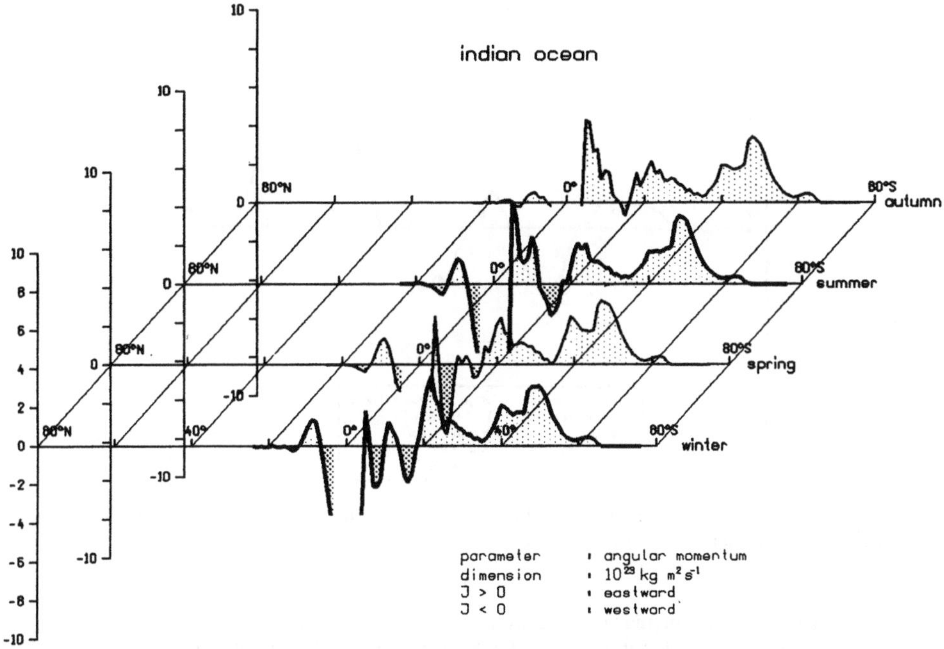

Fig. 15. Meridional density of the axial angular momentum of the Indian Ocean for the four seasons. Reference depth: variable

$$J_\theta(t) = \omega R^4 \int_{-\pi}^{+\pi} \int_{-\pi/2}^{+\pi/2} \int_{-h}^{\zeta} \varrho(\lambda, \varphi, z, t) \cos^3 \varphi \, d\lambda \, d\varphi \, dz$$

Its annual variation amounts up to $\Delta J_\theta = 6.0 \cdot 10^{24}$ kg m^2 s^{-1} which is about five times greater than the amplitude caused by the variable thermohaline ocean currents.

4.2 Calculations Based on a General Circulation Model

In order to evaluate the variability of the geostrophic calculations, Ernst Maier-Reimer of the Max-Planck-Institut für Meteorologie Hamburg has been asked to run his general circulation model (GCM) of the world ocean (Maier-Reimer 1985) with a subroutine calculating the axial angular momentum J (t). This model is based on the complete prognostic equations and is forced by the seasonal fluxes of momentum (wind), heat and salt at the surface. It has a grid resolution of 3.5 °×3.5 ° horizontally and 11 layers vertically.

The calculated meridional distribution of the area density of the annual mean axial angular momentum $\delta J(\varphi)$ is shown in Figure 16. This curve has to be compared with that of Figure 14. Qualitatively, the pattern of east- and westward contributions is very similar, also the relative importance of the different flow systems. Looking at the values between 4 °N and 4 °S omitted in the geostrophic calculations, their contributions cancel practically out. The corresponding error of the computations presented here is consequently small.

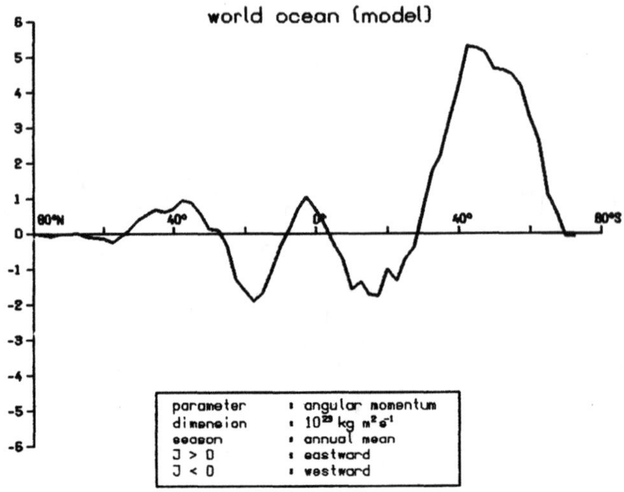

Fig. 16. Meridional density of the axial angular momentum of the world ocean from a GCM, annual mean (Maier-Reimer, personal communication)

Table 3. Axial angular momentum (10^{24} kg $m^2\,s^{-1}$) of the world ocean between parallels of latitude from a GCM, annual mean (Maier-Reimer, pers. comm.)

90°N–90°S	90°N–4°N	4°S–45°S	45°S–90°S	90°N–45°S
10.61	−0.74	0.35	10.32	−0.29

Table 4. Seasonal variation of the axial angular momentum (10^{24} kg $m^2\,s^{-1}$) of the world ocean from a GCM (Maier-Reimer, pers. comm.)

Spring	Summer	Autumn	Winter
10.84	10.85	10.41	10.33

Comparing the corresponding numbers (Table 3) with those of the geostrophic calculations (see Table 1) it is seen, that the values of the global mean and the northern hemisphere are in good agreement. In the southern ocean, however, major differences occur as a consequence of a stronger separation of the gyres in the GCM. The amplitude of the annual variation is $\Delta J = 0.26 \cdot 10^{24}$ kg $m^2\,s^{-1}$ with a maximum in summer (Table 4).

The annual variation of the variable moment of inertia is $\Delta J_\theta = 1.1 \cdot 10^{24}$ kg $m^2\,s^{-1}$. Similarly to the geostrophic calculations, the effect on the Earth's rotation due to the displacement of water masses is a multiple of the effect due to ocean currents. The absolute values of both calculations differ by a factor of 4. One possible reason for the lower GCM values seems to be that the windforcing and the density gradients are acting against each other.

4.3 Implications for Length of Day

A change in angular momentum ΔJ of the rotating Earth means a change in length of day

$$\Delta LOD = -\frac{\Delta J}{J_E} \times 1\,\mathrm{day}$$

with $J_E = 5.9 \cdot 10^{33}\,\mathrm{kg\,m^2\,s^{-1}}$ the mean angular momentum of the Earth. For a harmonic variation with an annual period by a running summation maximum time deviations are obtained of

$$\Delta UT = 0.94\,\mathrm{ms} \quad \text{for} \quad \Delta J = 1.1 \cdot 10^{24}\,\mathrm{kg\,m^2\,s^{-1}}$$

from geostrophic calculations and

$$\Delta UT = 0.22\,\mathrm{ms} \quad \text{for} \quad \Delta J = 0.26 \cdot 10^{24}\,\mathrm{kg\,m^2\,s^{-1}}$$

from the GCM.

As a result of the seasonal redistribution of water masses follows

$$UT = 0.94\,\mathrm{ms} \quad \text{for} \quad \Delta J_\theta = 1.1 \cdot 10^{24}\,\mathrm{kg\,m^2\,s^{-1}}\ ,$$

from the GCM

$$UT = 5.14\,\mathrm{ms} \quad \text{for} \quad \Delta J_\theta = 6.10 \cdot 10^{24}\,\mathrm{kg\,m^2\,s^{-1}}\ ,$$

from the geostrophic calculations.

5 Conclusions

The present paper provides an estimation of the axial angular momentum content of thermohaline currents in the world ocean and its variability based on the most complete hydrographical data set available today. Because of data gaps especially on the southern hemisphere and because of the uncertainty of the chosen reference level for the geostrophic calculations considerable errors of the absolute values of current velocities and corresponding axial angular momentum must be expected. By comparison with field data and results from a general circulation model it is concluded that this error is of the order of a factor 2.

The calculated seasonal variations of the axial angular momentum are expected to be in the same range of accuracy. With this restriction it can be stated that the maximum annual defect in Universal Time due to the thermohaline currents alone is in the order of 1 ms. The effect of the seasonal redistribution of ocean water masses is in the order of 5 ms. The corresponding relation from the GCM is 0.25 ms to 1 ms. This shows that a more detailed analysis of the variable moment of inertia, essentially its regional distribution, is necessary.

The analysis of the influence of the world ocean's variable thermohaline circulation on the Earth rotation presented here bridges a gap in the geophysical interpretation of astronomical data. The effects of ocean tides are sufficiently investigated as well (Seiler 1988 and this Vol.). A similar detailed investigation of wind-driven currents, its fluctuations and the corresponding variations of the Earth's angular momentum is still missing.

References

Brosche P, Sündermann J (eds) (1978) Tidal friction and the Earth's rotation. Springer, Berlin Heidelberg New York

Brosche P, Sündermann J (eds) (1982) Tidal friction and the Earth's rotation II. Springer, Berlin Heidelberg New York

Brosche P, Sündermann J (1985) The Antarctic Circumpolar Current and its influence on the Earth's rotation. Dtsch Hydrogr Z 38

Bryan K, Cox MD (1968) A non-linear model of an ocean driven by wind and differential heating. Part I and II. J Atm Sci 25

Defant A (1941) Die absolute Topographie des physikalischen Meeresniveaus und der Druckflächen sowie die Wasserbewegungen im Atlantischen Ozean. Meteor-Werk 6:2

Fomin LM (1964) The dynamic method in oceanography. Elsevier, Oceanogr Ser 2

Frische A (1989) Untersuchungen zum saisonalen Drehimpulsinhalt des Weltozeans. Diploma thesis, Univ Hamburg

Gill AE (1982) Atmosphere-ocean dynamics. Academic Press, London

Hide R (1984) Rotation of the atmospheres of the earth and planets. Phil Trans R Soc Lond A 313

Lambeck K (1980) The Earth's variable rotation. Univ Press, Cambridge

Levitus S (1982) Climatological Atlas of the World Ocean. NOAA Tech Pap

Maier-Reimer E (1985) A large-scale ocean circulation model. Int Rep, Max-Planck-Inst Meteorol, Hamburg

Meehl G (1982) Characteristics of surface current flow inferred from a global ocean current data set. J Phys Oceanogr 12

Millero FJ, Poisson A (1981) International one-atmosphere equation of state of seawater. Deep-Sea Res 28A

Munk WH, MacDonald GJF (1960) The rotation of the Earth. Univ Press, Cambridge

Niiler P, Schmitz WJ, Lee DK (1985) Geostrophic volume transport in high eddy-energy areas of the Kuroshio Extension and Gulf Stream. J Phys Oceanogr 15

Nowlin WD Jr, Klinck JM (1986) The physics of the Antarctic Circumpolar Current. Rev Geoph 24:3

Pedlosky J (1979) Geophysical fluid dynamics. Springer, Berlin Heidelberg New York

Seiler U (1988) Eine Untersuchung der Gezeiten des Weltozeans und ihres instantanen Drehimpulshaushalts. Doctoral thesis, Univ Hamburg

Whithworth T (1983) Monitoring the transport of the Antarctic Circumpolar Current at Drake Passage. J Phys Oceanogr 13

Atmospheric Effects on the Earth's Rotation

H. Volland

1 Introduction

The atmosphere as the outer volatile region of the Earth is directly exposed to solar radiation. It is also connected to the solar wind via interaction with the geomagnetic field and the magnetospheric plasma. The atmospheric wind blowing over the rough surface can exchange angular momentum with the solid Earth. On a short-term scale (days to years), fluctuations of the angular momentum vector give rise to variations in the length of the day (LOD) and changes in the polar motion. Secular variations of the LOD may be partly due to a solar gravitational torque acting on the thermally excited semidiurnal density bulge on the ground. The frictional force between the magnetopause and the solar wind may be capable of braking the Earth's rotation. In this paper, I estimate the various effects of atmospheric motions which may induce short-term and long-term changes in the Earth's rotation.

2 Outline of the Theory of Atmospheric Large-Scale Motions

The atmosphere behaves like a huge wave guide with respect ot large-scale waves. Only certain individual wavemodes can be excited by solar heating, orographic momentum forcing, or by turbulent interaction with smaller-scale waves. A wave component (e.g., the amplitude of the geopotential Φ) has the general form (e.g., Volland 1988)

$$\Phi = \theta_n^m(\phi) F_n^m(\zeta) \exp[i(m\lambda - \omega t)] \quad . \tag{1}$$

Here, $m = 0, 1, 2, \ldots$ is the zonal wave number, t the time, ω the wave angular frequency, ϕ the latitude, λ the longitude, $\zeta = z/2H$ a normalized height, z the geometric height, H the pressure scale height, θ_n^m the Hough function depending on latitude, zonal wave number m, and meridional wave number n, and F_n^m a height structure function, also depending on m and n, on the height ζ and on the heat input configuration. A free characteristic upgoing wave within an isothermal background atmosphere outside the source region has the height structure function

$$F_n^m(\zeta) = \exp[(1 - i k_n^m)\zeta] \tag{2}$$

Radioastronomisches Institut, Universität Bonn, Auf dem Hügel 71, D-5300 Bonn 1, FRG.

with $k_n^m = [\varepsilon_n^m/\varepsilon_c - 1]^{1/2}$ a normalized vertical wavenumber, ε_n^m an eigenvalue, and $\varepsilon_c \simeq 10$ a constant. ε_n^m is a separation constant, called the Lamb parameter, which connects the Hough function with the height structure function. If $\varepsilon_n^m > \varepsilon_c$, the corresponding wave is an internal wave with finite vertical wavelength. Waves with $\varepsilon_n^m < \varepsilon_c$ are external waves (or evanescent waves) with no wave characteristics in the vertical. Since $k_n^m = -i|k_n^m|$ in this case, this parameter can partially or totally compensate for the exponential growth of the wave mode propagating into a region of exponentially decreasing density (see Eq. (2)]. Such wave growth follows from the strict linearization of the hydrodynamic and thermodynamic equations and indicates that the wave energy of a non-dissipative wave remains constant. Waves with $\omega > 0 (\omega < 0)$ propagate to the east (west). Large-scale waves are waves with small m ($m < 3$, in general). Only wave modes with low wave numbers can be filtered out from the meteorological noise. Higher order modes interfere destructively and contribute to the meteorological noise.

3 Internal Friction

One of the most difficult problems in atmospheric dynamics is the interaction between waves of all scales. Being a notoriously non-linear problem, several attempts have been made to parameterize this interaction in a crude way. The turbulent interaction between large-scale and smaller-scale waves is often parameterized by an eddy viscosity force:

$$F_{edd} \simeq K_h(\partial^2 \mathbf{V}/\partial x^2 + \partial^2 \mathbf{V}/\partial y^2) + K_v \partial^2 \mathbf{V}/\partial z^2 \tag{3}$$

with $K_h \simeq 10\ m^2/s$ and $K_v \simeq 10^5\ m^2/s$ the coefficients of horizontal and vertical eddy viscosity (e.g., Pedlosky 1979), $\mathbf{V} = (u; v)$ the horizontal wind velocity vector with its components u, directed to the east (x), and v, directed to the north (y). Since vertical transport of eddy momentum renders the hydrodynamic equations non-separable and prevents the concept of wave modes outlined in Sect. 2, one can use an alternative concept of parameterization, introducing a Rayleigh friction coefficient ν_R and an eddy momentum force $\mathbf{F}_e$:

$$F_{edd} \simeq -\Omega \nu_R \mathbf{V} + \mathbf{F}_e \tag{4}$$

with $\Omega \nu_R \simeq -K_h(\partial^2 \mathbf{V}/\partial x^2 + \partial^2 \mathbf{V}/\partial y^2).\ \mathbf{V}^{-1}; \mathbf{F}_e \simeq K_v \partial^2 \mathbf{V}/\partial z^2$, and $\Omega = 7.29 \times 10^{-5}\ s^{-1}$ the angular frequency of the Earth's sidereal rotation. ν_R and K_h are generally positive. Therefore, the corresponding terms act as sinks of kinetic energy of the large-scale waves. K_v and the component of $\mathbf{F}_e$ may be positive or negative, indicating that energy can be exchanged in both directions. If $K_v > 0$, energy is transferred from large-scale to smaller-scale waves.

Including ν_R in the equations of horizontal momentum and assuming a harmonic time variation as in (1), one obtains a complex frequency

$$\partial \mathbf{V}/\partial t + \Omega \nu_R \mathbf{V} = -i\omega_R \mathbf{V} \tag{5}$$

with $\omega_R = \omega + i\Omega\nu_R$. Introducing this complex frequency into the eigenvalue equation, both the eigenvalue ε and the vertical wave number k become complex ($k = k_r - ik_i$) so that the height structure function of the free wave in (2) can be expressed as

$$F = \exp[(1-k_i)\zeta - ik_r\zeta] \ . \tag{6}$$

Dissipation now reduces or even reverses the wave growth.

In a similar manner, one can introduce a Newtonian cooling coefficient ν_N and an eddy heat source Q_e to replace the concept of eddy heat conduction. This leads to a second complex frequency, $\omega_N = \omega + i\Omega\nu_N$, to be included in the equation of energy. The ratio $P = \nu_R/\nu_N$, called the Prandtl number, is of the order of unity. Since $1/\nu_R \simeq 20$ in the lower atmosphere, internal friction is important only for waves with periods larger than about 10 days.

4 Surface Friction

An exchange of momentum between atmosphere and Earth takes place in the boundary layer of the atmosphere. This layer is of variable size depending on orography and latitude. Typically, its thickness is about 1 km at mid-latitudes. In the free atmosphere above the boundary layer, the horizontal large-scale and long-periodic wind is supposed to be almost in geostrophical equilibrium with the pressure gradient. The prototype of a boundary layer is the Ekman layer (e.g., Pedlosky 1979). The change of the wind direction with height follows the well-known Ekman spiral. The wind is zero on the ground, increases with height and turns to the right (in the Northern Hemisphere) until it reaches the wind of the free atmosphere. The exchange of momentum is via eddy momentum transport described by the second term in (3). At the bottom of the layer is the surface layer with a thickness of a few decameters depending on the roughness of the surface (e.g., von Mieghem 1973).

If the driving force to maintain the geostrophic wind would suddenly stop, the wind would decay due to surface friction after a characteristic time (the spin-down time) of

$$t_s = H/(2fK_v)^{1/2} \simeq 2 \text{ days} \tag{7}$$

at mid-latitudes with $H \simeq 7$ km the pressure scale height, $f = 2\Omega \sin\phi$ the Coriolis parameter, and K_v from (3). The vertical momentum flux is given by

$$\mathbf{f} = -K_v \partial\mathbf{V}/\partial z \ . \tag{8}$$

In order to prevent any exchange of momentum between Earth and atmosphere, the condition

$$u = \partial u/\partial z = 0 \quad \text{at} \quad z = 0 \tag{9}$$

must be fulfilled at the surface. The Earth's reaction to a change of atmospheric surface stresses is expected to occur over time scales larger than the spin-down time in (7).

In the surface layer, the frictional stress resulting from (8) is nearly constant. The zonal component of this stress is usually parameterized by (e.g., Lorenz 1967; Lambeck 1980)

$$f_\lambda = c_D V u \qquad (10)$$

with V the magnitude of the horizontal wind vector, u the zonal wind component at anemometer heights (a few meters above the ground), and c_D an empirical drag coefficient that depends on orography, varying between about 0.001 over the oceans to about 0.004 over the continents. c_D also depends on V, and probably on the wave period, too.

The wind blowing against a mountain range can effectively push the mountains and the rest of the Earth with it. This mountain torque can be as effective as the surface stress due to the winds (e.g., Wahr and Oort 1984). For a rough estimate, this effect can be incorporated into an effective zonally averaged drag coefficient c_D.

5 Atmospheric Angular Momentum

The axial angular momentum of the atmosphere (AAM) consists of two components. The first component is due to the zonal wind u:

$$M_w = a \int \varrho_0 u \cos\phi \, dV \ . \qquad (11)$$

The second component is due to pressure loading on the surface:

$$M_p = \frac{\beta a^2 \Omega}{g} \int p(0) \cos^2\phi \, dA \qquad (12)$$

with $dA = a^2 \cos\phi \, d\phi \, d\lambda$ the area element, $dV = dA\,dz$ the volume element, p(0) the pressure amplitude at the ground, a = 6371 km the Earth's radius, $g = 9.81\ m/s^2$ the gravitational acceleration, ϱ_0 the mean density, and $\beta \simeq 0.7$ a Love number accounting for the compensating deformation of the Earth in response to atmospheric loading (e.g., Lambeck 1980; Barnes et al. 1983).

The meridional structure of the zonally averaged zonal wind and ground pressure can be developed into series of spherical functions $P_n^m(\phi)$ (in Neumann's normalization):

$$u = \sum_r u_r P_r^1(\phi) \ ; \quad P_s = \sum_r p_r P_r(\phi) \ . \qquad (13)$$

Evidently, only the first symmetric terms in (13) with the coefficients u_1 and p_2, respectively, can contribute to the globally averaged AAM. All other terms of the wind and pressure fields, including zonal wave numbers $m > 0$, can only redistribute angular momentum in latitude and longitude. The wind component with the coefficient u_1 is responsible for a superrotation of the atmosphere like a rigid body moving faster (provided $u_1 > 0$) than the underlying solid Earth.

No net exchange of AAM with the solid Earth is expected to take place for time scales larger than a few years (e.g., Palmen and Newton 1969). Therefore, condition (9) must be applied to the first component or the zonal wind in (13). One expects then that the second term with P_3^1 dominates near the ground. $P_3^1(\phi)$ is negative in the tropics giving rise to easterlies. It is positive at middle and higher latitudes with westerlies in these regions. This is in fair agreement with the observations (e.g., Lorenz 1967; Newell et al. 1974).

The mean zonally averaged circulation within the lower atmosphere can be simulated to a first approximation by two symmetric wave modes of zero zonal wave number m = 0: a Rossby-Haurwitz wave having a zonal wind and a geopotential of the form

$$u_{RH} = -U_1 P_1^1(\phi)\Phi_1(\zeta) \; ; \quad \Phi_{RH} = 2a\Omega U_1 P_2(\phi)\Phi_1(\zeta)/3 \tag{14}$$

and an external class II-wave (0; −2) having the wave components

$$\begin{aligned} u_{II} &= U_{-2}[B_1 P_1^1(\phi) + B_3 P_3^1(\phi) + \ldots]\,\Phi_{-2}(\zeta) \\ \Phi_{II} &= -2a\Omega U_{-2}[C_2 P_2(\phi) + C_4 P_4(\phi) + \ldots]\,\Phi_{-2}(\zeta) \end{aligned} \tag{15}$$

with $B_1 = 1.305$; $B_3 = 1.060$; $C_2 = 1.041$; $C_4 = 0.544$; and the higher order terms smaller by at least one order of magnitude. The velocities U_1 and U_{-2} are determined from the observations. The wind of the RH-wave ($U_1 \simeq 5$ m/s) is mainly responsibly for the tropical easterlies. Applying the boundary condition (9), one obtains a constraint on U_1 and U_{-2}:

$$\begin{aligned} &-U_1\Phi_1(0) + U_{-2}B_1\Phi_{-2}(0) = 0 \\ &-(U_1\,d\Phi_1/d\zeta - U_{-2}B_1\,d\Phi_{-2}/d\zeta)\,|_{\zeta=0} = 0 \;. \end{aligned} \tag{16}$$

Together with a realistic choice of the vertical heat input configuration for both waves, one obtains a zonal mean flow as function of height and latitude which fairly well simulates the basic features of the observations (Volland 1989a).

The mean values for M_w and M_p in (11) and (12) are calculated in this case to be

$$\begin{aligned} M_w &= (16/3)\pi a^3 \varrho_s H \int_0^\infty [-U_1\Phi_1(\zeta) + B_1 U_{-2}\Phi_{-2}(\zeta)]\exp(-2\zeta)\,d\zeta \\ &\simeq 9.9\times 10^{25}\ \mathrm{J\,s} \\ M_p &= 16\beta\pi a^5 \varrho_s \Omega^2 U_1 \Phi_1(0)\,[3C_2/B_1 - 1]/45\,g \simeq 4.0\times 10^{25}\ \mathrm{J\,s} \end{aligned} \tag{17}$$

with $\varrho_s = 1.29$ kg/m^3 the mean density on the ground, and $p(0) = \varrho_s\Phi(0)$.

An isothermal background atmosphere with scale height $H \simeq 7$ km has been applied for the calculation of M_w.

There remains one inconsistency in the model just described. If (10) is valid, then the second component of u_{II} in (15) with the coefficient B_3, introduced into (10) and integrated over the sphere, would give rise to a torque on the solid Earth because of the quadratic dependence on u in (10). This inconsistency can be resolved either by relaxing condition (16) so that the integrated product Vu of all components in (14) and (15) becomes zero, or by using two different values for c_D in (10) at low and middle latitudes such that

$$c_{D2} \int_{\phi_0}^{\pi/2} |P_3^1| P_3^1 dA \simeq c_{D1} \int_0^{\phi_0} |P_3^1| P_3^1 dA \tag{18}$$

with $\phi_0 = 26.6°$ the latitude where $P_3^1(\phi_0) = 0$, and $c_{D1} \simeq 2.3\, c_{D2}$. The meridional velocity is smaller by a factor of about 10 than the zonal velocity and can therefore be neglected.

The fluctuations of the AAM and the length of the day (LOD) contain an annual component of about 2.4×10^{25} J s (corresponding to a change of LOD of 0.4 ms) and a somewhat smaller semiannual component (Eubanks et al. 1985). The solar heat input which drives the class II-wave varies semiannually with an amplitude of about 1/5 of the climatic mean. The tropical easterlies display an annual variation with a similar amplitude (e.g., Newell et al. 1974). Both these variations are consistent with the corresponding changes in M_w and M_p in (17).

A tropical semiannual wave within the middle atmosphere (e.g., Belmont et al. 1974) is of a type corresponding to the RH-waves with $m = 0$ (Volland 1989a). Its largest component is of the form (14) with a maximum zonal wind velocity of the order 15 m/s near 50 km altitude. One can estimate, however, that its contribution to M_w is at most $\exp(-3) = 0.05$ where $\zeta \simeq 3$ corresponds to the region above the 50 hPa level ($z = 3H \simeq 21$ km), a height where the waves considered in (14) have already decayed to small amplitudes.

A small quasi-biennial oscillation (QBO) in LOD as well as 40–60 day quasi-periods, both attributed to changes in the AAM, have also been observed (Eubanks et al. 1985). The QBO with a typical period of about 28 months has maximum wind velocities of the order of 10–30 m/s in the tropical stratosphere near 25–30 km altitude (e.g., Naukojat 1986). This wave is also of the type of series of RH-waves with $m = 0$ (Volland 1989a). The first term of this series having the form of (14) contributes to the AAM. Its zonal wind can be expressed as

$$u_{QBO} \simeq U_1 P_1^1(\phi)(\zeta - \zeta_0)^2 \exp[(1 - \wedge)(\zeta - \zeta_0)] \cos(\omega_{QBO} t + k\zeta) \tag{19}$$

valid in the height region $\zeta > \zeta_0$, with $\zeta_0 \simeq 1$ ($z_0 \simeq 14$) the lower boundary of the QBO, $\zeta_m \simeq 2$ ($z_m \simeq 28$ km) the height of maximum QBO, $\omega_{QBO} \simeq \Omega/840$ its angular frequency, $\wedge = 1 + 2/(\zeta_m - \zeta_0) \simeq 3$, $k \simeq 3.14$, and $U_1 \simeq 10$ m/s. We can now estimate the contribution of the QBO to the AAM as

$$(M_w)_{QBO} \simeq \frac{32\pi H a^3 U_1 \varrho_s \exp(-2\zeta_0)}{3[(\wedge + 1)^2 + k^2]^{3/2}} \simeq 1.0 \times 10^{24} \text{ J s} \tag{20}$$

corresponding to an LOD oscillation of 0.02 ms. The analysis of Eubanks et al. (1985) suggests a QBO amplitude of 0.05 ms.

The atmosphere cannot create new angular momentum in the closed system Earth – atmosphere. If the Sun could be turned on at time $t = 0$, a Hadley cell would develop initially in each hemisphere with westerlies in the upper branch of the cell and easterlies within its lower branch so that the total AAM would remain zero. Surface friction allows the atmosphere to pick up angular momentum from the solid Earth in the course of time. The final stage is reached when no net exchange of angular momentum takes place, as required by condition (16), and the AAM exactly equals the angular momentum loss of the solid Earth (Mayr and Harris 1983).

6 Polar Motion

The polar motion of the Earth's rotation axis contains an annual component with a maximum displacement from the poles of almost 100 milliarcseconds (mas). This component very likely is mostly of atmospheric origin (Hide 1984). The vector of the excitation function for this forced oscillation consists of two terms (e.g., Lambeck 1980; Moritz and Mueller 1987): one due to atmospheric motion

$$\psi_w = -\frac{a\beta_w}{\Omega(C-A)} \int (u \sin\phi + i v) e^{i\lambda} \varrho_0 dV \tag{21}$$

and one due to pressure loading

$$\psi_p = -\frac{a^2 \varrho_s \beta_p}{3g(C-A)} \int \Phi(0) P_2^1(\phi) e^{i\lambda} dA . \tag{22}$$

The meaning of the symbols is the same as in (11) and (12). In addition, v is the meridional velocity, $\beta_w \simeq 1.43$ and $\beta_p \simeq 1.0$ are two Love numbers, $C = 8.04 \times 10^{37}$ kg m^2 is the polar moment of inertia, $A = C/(1+\omega_r/\Omega)$ its mean equatorial moment, and $\omega_r = \Omega/304$ is the Euler frequency of free nutation.

Only those atmospheric wavemodes contribute to the annual component of the polar motion which are antisymmetric with respect to the equator, have a zonal wave number $m = 1$, and oscillate at frequencies $l\omega_A$ (with $l > 0$, and $\omega_A = \Omega/366.24$ the angular frequency of 1 year). The only waves of this type are the quasi-stationary internal Rossby waves (1; −3) that propagate within the westerly background wind at mid- and high latitudes. Their geopotential and wind components can be approximated by (Volland 1988)

$$\left.\begin{aligned} \Phi &= \theta^1_{-3}(\phi) \\ u &= u^1_{-3}(\phi) \\ v &= v^1_{-3}(\phi) \end{aligned}\right\} c g z_0 \cos(\lambda-\lambda_0) \cos(l\omega_A t + k\zeta) \exp(\zeta) \tag{23}$$

with $k \simeq 1.73$ a vertical wave number, z_0 the maximum geopotential height on the ground, λ_0 the longitude of this maximum, and c a normalization factor. For a rough estimate, one can approximate the meridional structure functions by those of the corresponding Rossby-Haurwitz wave (1; −3):

$$\begin{aligned} &\theta^1_{-3} = -[16P^1_2/3 + 9P^1_4/4]/7; \quad u^1_{-3} = \frac{3}{4a\Omega} \sin\phi(15\cos^2\phi - 4) \\ &v^1_{-3} = \frac{3i}{4a\Omega}(5\cos^2\phi - 4); \quad c = 0.564 . \end{aligned} \tag{24}$$

The wind of this wave is not in geostrophic equilibrium.

One finds from (21) to (24) (Volland 1989b)

$$\begin{aligned} \boldsymbol{\Psi}_w &= z_0 \mathbf{B}_w \cos(l\omega_A t + \gamma_w) \\ \boldsymbol{\Psi}_p &= z_0 \mathbf{B}_p \cos(l\omega_A t) \end{aligned} \tag{25}$$

with $\mathbf{B}_w = -B_w \exp(i\lambda_0)$; $\mathbf{B}_p = B_p \exp(i\lambda_0)$

$$B_w = \frac{4\pi H a^2 \varrho_s g \beta_w c}{\Omega^2 (C-A)(1+k^2)^{1/2}} \simeq 1.30 \times 10^{-8}\,\mathrm{m}^{-1}$$

$$B_p = \frac{64\pi a^4 \varrho_s \beta_p c}{105(A-C)} \simeq 0.87 \times 10^{-8}\,\mathrm{m}^{-1}$$

and the two terms of polar motion become (j = "w" or "p") (e.g., Wahr 1988)

$$\mathbf{m}_j = \frac{\mathbf{B}_j z_0}{(\alpha^2 l^2 - 1)} [\cos(l\omega_A t + \gamma_j) + \alpha l i \sin(l\omega_A t + \gamma_j)] \tag{26}$$

with $\gamma_w = \arctan(k) \simeq 60°$, $\gamma_p = 0$, $\alpha = \omega_A/\omega_c \simeq 1.19$, and $\omega_c \simeq \Omega/436$ the Chandler eigenfrequency. With $B_w/B_p \simeq 1.5$ and $\lambda_0 = 0°$, one obtains finally for the amplitudes of the total **m**

$$m_1 = D\cos(l\omega_A t - \delta);\; m_2 = \alpha l D \cos(l\omega_A t - \delta) \tag{27}$$

with $D \simeq \sqrt{28}/6\, B_w z_0/(\alpha^2 l^2 - 1)$ and $\delta \simeq \arctan(\sqrt{27}) = 79.1°$.

This is an equation of the ellipse with a ratio between major axis and minor axis of

$$\frac{m_2}{m_1} = \alpha l\;. \tag{28}$$

For the annual wave (l = 1), one obtains as ratio of the axes $\alpha = 1.19$ which agrees reasonably with the observations (e.g., Lambeck 1980). Using $z_0 = 14$ m (corresponding to a maximum ground pressure amplitude of $|p(0)| = c z_0 g \varrho_s \simeq 1.0$ hPa and a maximum horizontal wind of $|v(0)| = 3cgz_0/a\Omega \simeq 50$ cm/s), the amplitude of the major axis arising from both wind and pressure loading becomes 4.6×10^{-7} radians corresponding to 95 mas. Such number of z_0 is not unrealistic (McAveney et al. 1978). The atmosphere is therefore responsible for a major part of the annual polar motion (Hide 1984). The semiannual component (l = 2) is smaller by at least a factor of 5.

Eubanks et al. (1988) have recently attributed rapid polar motions to the atmospheric antisymmetric normal mode (1; −3). This mode has essentially the same pressure and wind field as in (24) with a period of about 10 days, moving to the west. Its effect on the polar motion can be estimated from (26) if one replaces the frequency by $-\Omega/10$ and accounts for the retrograde propagation of the wave.

7 Solar Gravitational Tidal Torque

Migrating diurnal tides are excited by the diurnally varying solar heat input into the atmosphere. They depend on local time τ and travel westward with the Sun. The tidal component of relevance for this section is the symmetric semidiurnal tidal wave (2; 2). Its geopotential can be calculated by assuming a simple heat input configuration (constant heat per mass Q_s at all heights) (Volland 1988)

$$\Phi = \theta_2^2(\phi) \frac{2i\kappa Q_s}{\Omega(1+k^2)} \left[1 + \frac{4\kappa - 1 - k^2}{2(1-2\kappa+ik)} \right] \exp\left[(1+ik)\zeta\right] \exp(2i\tau) \tag{29}$$

with $\kappa = c_p/c_v = 0.29$ the ratio of the specific heats, $k \simeq 0.3$ the vertical wave number of this wave, $\tau = \Omega t + \lambda$ the local time, and

$$\theta_2^2 = \sum_r C_{2r} P_{2r}^2(\phi) \tag{30}$$

the Hough function with its two first coefficients $C_2 = 0.838$ and $C_4 = -0.216$. The pressure amplitude on the gound is given by

$$D \exp(-2i\tau_m) = \frac{i\kappa \varrho_s (1+ik) Q_s}{\Omega(1-ik)(1-2\kappa+ik)} \tag{31}$$

with $D \simeq 116$ Pa and $\tau_m \simeq 9.05$ h if $Q_s = -12$ mW/kg peaking at local noon.

The observations give the same amplitude and a slightly different time of maximum ($\tau_m = 9.73$ h) (Chapman and Lindzen 1970). A pressure bulge thus leads the Sun by about 2.3 h.

In these calculations, the Earth's surface was considered to be frictionless so that no exchange of angular momentum takes place. With strong friction (V = 0 at the surface), the pressure variation would disappear on the ground. Evidently, the time rate of change of the surface wind is too large to reach the steady state assumed in Sect. 4. Nevertheless, a small amount of friction certainly must occur. This may be the reason for the difference between the theoretically predicted and the observed phase on the ground.

The gravitational tidal force of the Sun generates a much smaller pressure bulge on the ground given by

$$|p(0)| \left| \frac{c_2 S_{20}(1+ik)\varrho_s}{(1-2\kappa+ik)} \right| \simeq 1 \text{ Pa} . \tag{32}$$

Here, $S_{20} = 0.37$ is the amplitude of the principal solar tidal potential

$$S_2 = -S_{20}(r/a)^2 P_2^2(\phi) \cos(2\tau) \tag{33}$$

and $c_2 \simeq 1$ a coefficient from the expansion of S_2 into a series of Hough functions (2; 2n). The values given by (32) may be underestimated by a factor of about 2 due to the influence of the tidally deformed ocean. The observed lunar semidiurnal tidal wave has a maximum surface pressure amplitude of 5.5 Pa (Chapman and Lindzen 1970).

The solar gravitational tidal force cannot significantly modulate the thermally excited pressure bulge in (31). It can, however, exert a torque L on this bulge (e.g., Gold and Soter 1969; Kundt 1977), given by

$$\begin{aligned} L &= -\frac{1}{g}\int \frac{\partial S_2}{\partial \lambda}\, \theta_2^2 D \exp\left[2i(\tau-\tau_m)\right] dA \\ &= \frac{96 a^2 S_{20} D}{5g} \sin(2\tau_m) \simeq (3.2 \pm 0.3) \times 10^{15}\,\mathrm{Nm} \ . \end{aligned} \tag{34}$$

This torque tries to move the bulge toward local noon and thus pulls on the whole atmosphere via internal friction. Surface friction transfers this torque to the solid Earth so that the whole system solid Earth – atmosphere is accelerated. This gives rise to an increase of Ω at a rate

$$d\Omega/dt = L/C \simeq 3.9\; 10^{-23}\,\mathrm{s}^{-2} \tag{35}$$

corresponding to a present-day decrease of the LOD by about 0.15 ms/100 yr. Tidal braking of the solid Earth is responsible for an increase of LOD by about 2 ms/100 yr (e.g., Brosche and Sündermann 1978). The gravitational tidal torque therefore reduces this effect by almost 10%.

8 Magnetospheric Friction

A steady electrically neutral flux of low energy particles, mainly protons and electrons, flows outward from the Sun. The velocity of this "solar wind" varies between about 300 and 900 km/s depending on the degree of solar activity. The number density of the solar wind particles in the vicinity of the Earth is of the order of 10^6 to $10^7\,\mathrm{m}^{-3}$. Magnetic fields are drawn out of the solar surface and carried by the solar wind into space. The field strength of this interplanetary magnetic field (IMF) at one AU is a few nanoteslas.

Most of the solar wind particles cannot penetrate into the Earth's atmosphere. They are deflected by the geomagnetic field, thereby creating a cavity called the magnetosphere. Only within the polar regions of the magnetosphere does the solar wind have direct access. It is also here that the field lines of the IMF merge with those of the Earth, sweeping the rotating geomagnetic field lines to the night side where they are again separated from the IMF and reconnected with each other. This process is accompanied by large-scale plasma convection, downstream at high latitudes and upstream at lower latitudes.

A non-rotating observer fixed in the frame of reference of the magnetosphere measures an electric field **E** in interplanetary space that is orthogonal to the solar wind velocity **v** and the IMF **B**:

$$\mathbf{E} = -\mathbf{v}\times\mathbf{B} \ . \tag{36}$$

This field is accompanied by electric polarization charges along the IMF. Since the electric conductivity orthogonal to the IMF lines is almost zero, no discharging current can flow in the undisturbed solar wind. When the solar wind reaches the magnetosphere, field line merging maps the electric field (36) into the polar regions of the well-conducting ionospheric dynamo layer near 100 km height. Different IMF lines are now short-circuited via the ionosphere, and a discharging current flows from the solar wind into the ionosphere at dawn and out of the ionosphere at dusk. This current is of the order of a few megaamperes and is largest within the aurora zones near $\pm 70°$ latitude (e.g., Volland 1984, and references within).

The electric field **E** mapped into the magnetosphere has a configuration such that magnetospheric plasma velocity and geomagnetic field are related to **E** according to (36). It is directed mainly from dawn to dusk. The vertical component of the IMF (in the direction parallel to the geomagnetic dipole axis) enhances the plasma convection if it is directed to the south, or is suppresses convection if it is directed to the north. The magnitude of the magnetospheric electric convection field is therefore a measure of the strength of the coupling of solar wind energy into the magnetosphere. The total dissipation of solar wind energy within the magnetosphere varies between about 0.1 TW during quiet conditions (low solar activity; IMF directed to the north) and 10 TW during disturbed conditions (high solar activity; IMF directed to the south) (Akasofu 1981).

Viscous-like interaction between solar wind and magnetosphere, which supports the magnetospheric plasma convection and maintains the electric field, occurs at the boundary of these two regions, the so-called magnetopause. Even if the vertical component of the IMF is directed to the north, the plasma convection, and therefore the dawn-to-dusk electric convection field, is maintained by this viscous-like interaction.

The inner magnetosphere corotates with the Earth's atmosphere with almost the same angular velocity Ω, up to distances of about 5 to 10 Earth radii. The role of the atmosphere is to provide a viscous transfer of momentum from the rotating surface of the Earth into the ionosphere. Here, the plasma is set into motion by collisional friction between the ions and the neutral particles (Hines 1960, 1974). The rotating ionospheric plasma produces a corotating electric field (as seen by an observer, fixed in the frame of reference of the magnetosphere) given by

$$\mathbf{E} = -(\boldsymbol{\Omega}\times\mathbf{r})\times\mathbf{B} \tag{37}$$

where $\boldsymbol{\Omega}$ is the angular velocity vector of the Earth, $\mathbf{r}$ is the radius vector from the spin axis, and $\mathbf{B}$ is the geomagnetic field.

Assuming again that the magnetic field lines are perfect conductors, the electric field (37) is then transmitted outward to enforce the corotation of the magnetospheric plasma. An observer moving with the Earth would not measure this corotation field.

The superposition of the electric corotation field and the convection field results in two fundamentally different regions within the magnetosphere. The first is an area of closed electric equipotential shells within the inner magnetosphere, called the plasmasphere, where the corotation field dominates and thermal ions, mainly protons, remain trapped. The ionosphere provides this thermal plasma so that a relatively high proton density is maintained ($>10^8\,m^{-3}$). The convection field dominates outside the plasmapause, which is the last closed equipotential near 5 to 10 Earth radii distance from the Earth, and the protons can move freely along the open equipotential shells to the magnetopause where they are permanently lost in the solar wind. The proton density in this region therefore remains low ($<10^7\,m^{-3}$) (Banks 1979). In the polar regions situated outside the plasmapause, the ionospheric outflow, also called the polar wind, transports part of the Earth's angular momentum to the solar wind. It is this mass transport which limits the extent of corotation, even in the absence of a solar wind (Hill 1979).

The solar wind thus exerts a torque on the rotating Earth with the magnetospheric plasma acting as a transmission medium. We estimate an upper limit to this torque, assuming that the total energy dissipation of the solar wind is used for braking the Earth's rotation. This torque is

$$L = P\Delta t \simeq 10^{15}\,\mathrm{Nm} \tag{38}$$

with $P \simeq 10^{12}$ W the average dissipated power from the solar wind, $\Delta t \simeq b/v \simeq 10^3$ s the residence time of a solar wind particle within the magnetosphere of length b, and v its velocity. The time rate of change of the Earth's angular velocity is then given by

$$\frac{d\Omega}{dt} = \frac{L}{C} \simeq 1.2\times 10^{-23}\,s^{-2} \tag{39}$$

with C the polar moment of inertia of the Earth.

The secular increase of the length of the day τ (LOD) due this effect is then

$$\frac{d\tau}{dt} = -\frac{\tau^2}{2\pi}\frac{d\Omega}{dt} \simeq 1.5\times 10^{-14} \tag{40}$$

corresponding to 46 μs in 100 years.

A similar estimate has been derived by Kundt (1977) in a somewhat different way. If a magnetic shear pressure of ($B^2/2\mu$) acts tangentially on a spherical rotating magnetopause of radius R, a torque of

$$L \simeq 4\pi R^3 B^2/2\mu \simeq 5\times 10^{14}\,\mathrm{N\,m} \tag{41}$$

is expected, where $R \simeq 15\ a = 10^8$ m, $B \simeq 10$ nT, and $a = 6371$ km the Earth radius (see also Hirshberg 1972).

The real braking effect is probably smaller by at least one order of magnitude because of the small internal viscous dissipation within the magnetospheric plasma. The change of the LOD due to magnetospheric braking is therefore almost negligible.

Even if the dissipative solar wind power were to increase to a maximum value of $\delta P \simeq 10^{13}$ W during one day, the change of the LOD would then be only at most of the order of

$$\delta\tau \simeq -\frac{\tau^2}{2\pi}\frac{\delta L}{C} \simeq 13\,\mathrm{ns}\ . \tag{42}$$

We furthermore estimate the influence of the polar wind on the change of the LOD. Taking the magnetic coupling within the magnetosphere into account, the change of the angular momentum due to such outflow is (Mestel 1968)

$$\frac{d}{dt}(C\Omega) \simeq \bar{r}^2 \Omega \frac{dm}{dt} \tag{43}$$

with $\bar{r} < 10$ a the range of magnetospheric corotation, a the Earth's radius, $C = 0.33\, m_E\, a^2$ the principal moment of inertia of the Earth, $m_E = 6 \times 10^{24}$ kg its mass, and $dm/dt < 0.1$ kg/s the polar wind mass transport (Hill 1974). From (43) we derive

$$\frac{d\Omega}{dt} \simeq 3\left(\frac{\bar{r}}{a}\right)^2 \frac{\Omega}{m_E} \frac{dm}{dt} < 4 \times 10^{-28}\, s^{-2} \tag{44}$$

which is again insignificant.

Finally, we estimate the influence of the IMF on the precession of the Earth's axis. The solar magnetic field is a dipole field, to a first approximation, which reverses its dipole axis every 11 years in the course of the solar cycle. The solar magnetic north pole is in the southern hemisphere since 1979 and is expected to turn to the north in 1990 (Hoeksema 1983 et al.). The solar magnetic field lines are carried by the solar wind into space and become the open interplanetary field lines, essentially parallel to the heliospheric equator, a quasi-neutral plane, also referred to as the interplanetary current sheet. Higher order magnetic poles, varying in time erratically, give rise to departures from this idealized picture (e.g., Smith 1979).

The solar equator is inclined by 7° with respect to the ecliptic plane. Therefore, the IMF exerts a torque on the geomagnetic dipole that reverses over the course of a solar cycle. The torque maximizes during equinox and becomes zero during solstices. The geomagnetic dipole axis, and thus the Earth's axis, precesses in one direction during one solar cycle and in the opposite direction during the following cycle. The amount of precession during one cycle can be approximated by

$$\Delta\phi \simeq \frac{m_p B \cos\gamma \Delta t}{2M} < 10^{-9} \text{ degrees} \tag{45}$$

with $m_p = 8 \times 10^{22}$ A m^2 the magnetic moment of the geomagnetic field (approximated by a coaxial dipole field), $B < 10$ nT an upper limit of the IMF at the Earth's orbit, $M = 5.9 \times 10^{33}$ kg m^2/s the Earth's angular momentum, $\gamma = 7.25°$ the angle between solar equator and ecliptic plane, and $\Delta t = 11$ years the average duration of the solar cycle.

We conclude from these estimates that the interaction of the ionospheric-magnetospheric system with the solar wind and IMF has no measurable influence on the total angular momentum budget of the Earth.

Acknowledgment. I thank for the valuable discussions with W. Kundt.

References

Akasofu SI (1981) Energy coupling between the solar wind and the magnetosphere. Space Sci Rev 28:121

Banks PM (1979) Magnetospheric, ionospheric, and atmospheric interaction. In: Kennel CF et al. (eds) Solar systems plasma physics II. North Holland, Amsterdam, p 57

Barnes RTH, Hide R, White AA, Wilson CA (1983) Atmospheric angular momentum fluctuations, length-of-the-day changes and polar motions. Proc R Soc Lond A387:31

Belmont AD, Dart DG, Nastrom GD (1974) Periodic variations in stratospheric zonal wind from 20 to 65 km, at 80°N to 70°S. Quart J R Soc 100:203

Brosche P, Sündermann J (eds) (1978) Tidal friction and the Earth's rotation. Springer, Berlin Heidelberg New York

Chapman S, Lindzen RS (1970) Atmospheric tides. Reidel, Dordrecht

Eubanks TM, Steppe JA, Dickey JD, Callahan PS (1985) A spectral analysis of the earth's angular momentum budget. J Geophys Res 90:5385

Eubanks TM, Steppe JA, Dickey JD (1988) The atmospheric excitation of rapid polar motions. In: Babcock AK, Wilkins GA (eds) The Earth's rotation and reference frames for geodesy and geodynamics. Kluwer, Dordrecht

Gold T, Soter S (1969) Atmospheric tides and the resonant rotation of Venus. Icarus 11:356

Hide R (1984) Rotation of the atmospheres of the Earth and planets. Phil Trans R Soc Lond A313:107

Hill TW (1974) Origin of the plasma sheet. Rev Geophys Space Phys 12:379

Hill TW (1979) Inertial limit of corotation. J Geophys Res 84:6554

Hines CO (1960) On the rotation of the polar ionospheric regions. J Geophys Res 65:141

Hines CO (1974) Solar wind torque as an inhibitor of terrestrial rotation. J Geophys Res 79:1543

Hirshberg J (1972) Upper limit of the torque of the solar wind on the Earth. J Geophys Res 77:4855

Hoeksema JT, Wilcox JM, Scherrer PH (1983) The structure of the heliospheric current sheet: 1978–1982. J Geophys Res 88:9910

Kundt W (1977) Spin and atmospheric tides of Venus. Astron Astrophys 60:85

Lambeck K (1980) The Earth's variable rotation. Univ Press, Cambridge

Lorenz EN (1967) The nature and theory of the general circulation of the atmosphere. WMO, Geneva

Mayr HG, Harris I (1983) Quasi-axisymmetric circulation and superrotation in planetary atmospheres. Astron Astrophys 121:124

McAvaney BJ, Bourke W, Puri K (1978) A global spectral model for simulation of the general circulation. J Atm Sci 35:1557

Mestel L (1968) Magnetic breaking by a stellar wind. Astron Soc Lond 138:359

Moritz H, Müller II (1987) Earth rotation. Ungar, New York

Naukojat B (1986) An update of the observed QBO of the stratospheric winds over the tropics. J Atm Sci 43:1873

Newell RE, Kidson JW, Vincent DG, Boer GJ (1974) The "General Circulation" of the tropical atmosphere, Vol 1. MIT, Cambridge, MA, p 371

Palmen E, Newton CW (1969) Atmospheric circulation systems. Academic Press, New York

Pedlosky J (1979) Geophysical fluid dynamics. Springer, Berlin Heidelberg New York

Smith EJ (1979) Interplanetary magnetic fields. Rev Geophys Space Phys 17:610

van Mieghem J (1973) Atmospheric energetics. Clarendon, Oxford

Volland H (1984) Atmospheric electrodynamics. Springer, Berlin Heidelberg New York Tokyo

Volland H (1988) Atmospheric tidal and planetary waves. Kluver, Dordrecht

Volland H (1989a) Rossby-Haurwitz waves of zero zonal wavenumber. Contr Atm Phys 62:77

Volland H (1989b) On the seasonal components of polar motion. Geophys Res Lett 16:303

Wahr JM (1988) The earth's rotation. Ann Rev Earth Planet Sci 16:231

Wahr JM, Oort AH (1984) Friction and mountain torque estimates from global atmospheric data. J Atm Sci 190

The Solar Torque – A Leak for the Angular Momentum of the Earth-Moon System

P. Brosche and J. Wünsch

1 Introduction

The tides in the terrestrial oceans raised by the Moon lead to a transfer of angular momentum from the rotation of the Earth into the orbit of the Moon. Likewise, the tides raised by the Sun transport angular momentum from the rotation of the Earth into the "orbit of the Sun" which is the orbit of the Earth around the Sun. Due to the smaller tidal constant [mass/(distance)^3] of the Sun, the solar effect is smaller than the lunar one. More important is the difference in the measurable manifestations. Today our best values stem from observations of the lunar orbit (lunar laser ranging and stellar occultations by the Moon); therefore they purely represent the *lunar* part of the tidal interchange. In contrast, the deceleration of the Earth being caused by both effects, does contain the solar signal as well. Unfortunately, the Earth underlies many other influences and hence the terrestrial value is not as accurate as the lunar one. Consequently, the standard procedure consists in a *theoretical* estimate of the ratio κ = lunar/solar torque. This is applied to the *observed* lunar torque in order to obtain the *total* torque acting on the Earth (except for non-tidal torques; see Volland, this Vol.). Finally, this "hybrid" torque is compared with the *observed* tidal or secular change in the Earth's rotation – as accurately as it can be isolated from the manifold of changes.

2 The Ratio

2.1 The Schematic View

The following considerations represent a stepwise deviation from an equilibrium picture. A global ocean of uniform depth covering the whole surface of the Earth was assumed by Newton to react in an equilibrium manner such that the sea level is identical with an equipotential surface. In this case a disturbing outer point mass would cause a two-bulge deformation of the surface, the amplitude of the deformation being proportional to the amplitude of the tidal force, that is, to the disturbing mass m divided by the third power of its distance r. There would be no phase lag between force and deformation and hence no net torque.

As a first deviation, we can assume the existence of a constant phase lag and a constant Love number such that the deformations are a constant fraction of the variation

Observatorium Hoher List der Universitäts-Sternwarte, Universität Bonn, D-5568 Daun, FRG.

of the equipotential surface. Then a non-vanishing horizontal force exists at the bulges. There the torque is the axial distance times the incremental mass times the horizontal acceleration. The two latter are *both* proportional to the tidal force, hence the torque is proportional to the *square* of the tidal force (of course, an exact integration is necessary; it leads to the same result as our heuristic explanation).

Third, we could admit that the complexity of the real oceans leads to far-from-equilibrium tides but we could expect that the response is linear for tides with nearly identical periods. That would mean: (a) their phase lags are nearly the same, (b) their amplitudes are proportional to the amplitudes of the tidal force. In other words, we use the foregoing arguments in a *relative* manner and for "*neighbour*" tides only. For solar and lunar tides with adjacent periods, one would predict a ratio of the average tidal torques L (Munk and MacDonald 1960):

$$\kappa = L_{☾}/L_{\odot} = \left(\frac{m_{☾}}{r_{☾}^{3}} : \frac{m_{\odot}}{r_{\odot}^{3}}\right)^{2} = 4.6 \ . \tag{1}$$

The instantaneous value varies because of the eccentricities of the orbits; the value above corresponds to mean distances. Munk and MacDonald (1960) use a slightly deviating 5.1. Burša (1987) has 4.7. Lambeck (1980) starts from ocean models and satellite data but arrives at a similar value of 4.9.

2.2 *Periodic Variations*

As we know today, the main tidal transfer of angular momentum is mediated by the tides of the oceans. They are far from equilibrium. The secular transfer values are averages over tidal periods, the variations within one period being much larger than the averages. So one could surmise: if our schematic arguments above are not valid for the *periodic* variations, they are even less applicable to the time *averages*.

The treatment of the periodic variations in the angular momentum of the oceans by a recent hydrodynamical model (Seiler, this Vol.) leads to phase differences of $\sim 30°$ in the oceanic angular momentum variation due to tides of the same tidal band, namely (M2, S2) and (K1, O1). Moreover, the amplitude ratios of that variations are certainly not represented by Eq. (1) (Table 1).

These results look quite ominous for the justification of the schematic view. On the other hand, the periodic variations are mainly due to transfers between the oceans and the solid Earth, therefore the relations for the exchange between oceans and Moon do not apply. It could be that the *average* torques do follow Eq. (1) nevertheless.

2.3 *Average Torques*

The average tidal torques of the most important oceanic partial tides were kindly provided by Dr. U. Seiler and are exhibited in Table 2. We have sorted the tides according to their origin in lunar and solar ones. The K1 tides were treated as one tide and we have split up the result using Eq. (1) into a solar and lunar part. Even in the case of a completely agreeing period, the response may not be exactly linear but this does not influence our conclusions seriously.

Table 1. Periodic angular momentum variations

Tides of the same band	κ_E	Angular momentum content	
		Amplitude ratio (first divided by second)	Phase difference (first minus second)
M2, S2	4.6	2.0	−33°
K1, O1	2.0	0.5 (!)	+30°

Table 2. Average negative torques (unit 10^{22} dyn cm = 10^{15} Nm)

Tidal band	Partial tide caused by			
	Moon		Sun	
Semidiurnal	M2	37.66	S2	4.64
	N2	1.85		
Diurnal	K1m	2.22	K1s	0.48
	O1	2.44	P1	0.29
Fortnightly	Mf	0.00		
	Mf′	0.00		
Monthly	Mm	0.00		
Sum		44.17		5.41
Ratio			$\kappa = 8.16$	

The outcome is clear and simple: already the two major constituents on each side – M2 and S2 – and also the sums of all treated partial tides lead to a ratio $\kappa \simeq 8$ – about a factor of two away from the quasi-equilibrium value κ_E. In our view the reason for this outcome lies in the great sensitivity of the oceans to the forcing periods. Their eigenperiod spectrum is rich and the "width" of the spectral lines is narrow (Sündermann 1982; Brosche and Hövel 1982). Hence the small difference in angular velocity of ~3% between M2 and S2 is enough to alter the response as shown in the table.

Besides the problem discussed here, our result is also relevant to other intelligent approaches to the tidal friction problem, such as the one of Mignard (1982), who assumes a constant (effective) Love number and a constant time-lag.
If this assumption is not valid for tides of the same tidal band and for the present oceans, it is even more doubtful for tides of various bands and of ancient oceans. Therefore it seems not possible to estimate the relative influence of the various partial tides by their equilibrium parameters.

3 The Balance

As pointed out in the introduction, the precisely observed changes in the lunar orbit can be converted into a lunar part $\dot{\omega}_{TL}$ of the secular deceleration of the rotation of the Earth. This part can be augmented by the solar part $\dot{\omega}_{TS}$ with the aid of an appropriate κ. If there are non-tidal changes $\dot{\omega}_{NT}$, e.g., by non-tidal torques or as reaction on a changing moment of inertia θ of the Earth, they have to be added. The sum can then be confronted with a directly observed value $\dot{\omega}$ for the Earth:

$$\dot{\omega}_{TL} + \dot{\omega}_{TS} + \dot{\omega}_{NT} \overset{?}{=} \dot{\omega} \ . \tag{2}$$

(The symbol "$\overset{?}{=}$" means that the equality has to be checked, of course, within error limits.)

According to Burša (1987):

$$\begin{aligned} \dot{\omega}_T &= \dot{\omega}_{TL} + \dot{\omega}_{TS} = -(6.3 \pm 0.5) \cdot 10^{-22}\,\mathrm{rad\,s^{-2}} \\ \dot{\omega}_{NT} &= \qquad\qquad\quad +(1.2 \pm 0.3) \cdot 10^{-22}\,\mathrm{rad\,s^{-2}} \\ \dot{\omega} &= \qquad\qquad\quad -(5.4 \pm 0.5) \cdot 10^{-22}\,\mathrm{rad\,s^{-2}} \ . \end{aligned}$$

The second term stems from the decrease in the zonal geopotential harmonic J_2 observed by the *Lageos* satellite and transformed into a change of the Earth's largest moment of inertia. Burša (1987) argues that the use of the $\dot{\omega}_{NT}$ belonging to this effect produces a perfect balance and this in turn supports the meaning of the observed change in J_2.

With Burša's $\kappa_E = 4.7$ we can derive the constituents of his $\dot{\omega}_T$ to be

$$\left.\begin{aligned} \dot{\omega}_{TL} &= -5.2 \\ \dot{\omega}_{TS} &= -1.1 \end{aligned}\right\} 10^{-22}\,\mathrm{rad\,s^{-2}} \ .$$

If we keep the $\dot{\omega}_{TL}$ but introduce our $\kappa = 8.16$ and the corresponding

$$\dot{\omega}_{TS} = -0.6 \cdot 10^{-22}\,\mathrm{rad\,s^{-2}}$$

the new $\dot{\omega}_T$ becomes

$$\dot{\omega}_T = -5.8 \cdot 10^{-22}\,\mathrm{rad\,s^{-2}} \ .$$

Augmented by the same $\dot{\omega}_{NT}$ as above, the left side of the questionable Eq. (2) becomes $-4.6 \cdot 10^{-22}\,\mathrm{rad\,s^{-2}}$. Thus the introduction of a realistic non-schematic κ degrades the fulfillment of Eq. (2). This does not call in question the *Lageos* results but it shows that the balance has to be revised and that at present it is not possible to use it as a support for the reality of the observed dJ_2/dt.

Acknowledgment. We acknowledge gratefully support by the Deutsche Forschungsgemeinschaft.

References

Brosche P, Hövel W (1982) Tidal friction for times around the presence. In: Brosche P, Sündermann J (eds) Tidal friction and the Earth's rotation II. Springer, Berlin Heidelberg New York, p 175

Burša M (1987) Secular tidal and nontidal variations in the Earth's rotation. Studia Geoph Geod 31:219

Lambeck K (1980) The Earth's variable rotation. Univ Press, Cambridge, Chap 10, Table 10.3

Mignard F (1982) Long time integration of the Moon's orbit. In: Brosche P, Sündermann J (eds) Tidal friction and the Earth's rotation II. Springer, Berlin Heidelberg New York, p 67

Munk WH, MacDonald GJF (1960) The rotation of the Earth. Univ Press, Cambridge, p 200ff

Sündermann J (1982) The resonance behaviour of the world ocean. In: Brosche P, Sündermann J (eds) Tidal friction and the Earth's rotation II. Springer, Berlin Heidelberg New York Tokyo, p 165

Tectonic Consequences of the Earth's Variable Rotation on Geological Time Scales

C. Denis[1] and P. Varga[2]

1 Introduction

The Earth-Moon system evolved quite considerably since its formation, both as regards its orbital parameters and the kinetic and dynamic properties of its individual components. The evolution in time of the interior and surface features of the Earth depends on a number of endogenous and exogenous factors. Exogenous agents of geological evolution are, for instance, meteorite impacts and luni-solar interactions causing tidal deformations. Primary endogenic evolutionary factors are the decay of abundant radioactive elements, as well as physical segregation and chemical fractionation processes, producing gravitational energy release and compositional change. Climatic mechanisms, such as glaciation-deglaciation cycles, which have both exogenous and endogenous causes, quite evidently also produce geological change which, however, seems to be of a more or less cyclic nature and probably does not affect the Earth's deep interior. The purpose of this paper is to consider some aspects of the evolution over geological time spans which are, or may be, a consequence of the Earth's tidal despinning.

The evolution of the Earth on a global scale caused by tidal friction is a very slow process, and for computational purposes can be considered as a time sequence of near-equilibrium states. Moreover, because the characteristic time scales range from a 100 million years to over a 1000 million years, it seems reasonable to use an Earth model with quite simple rheological properties. In this paper, we consider a model consisting of an interior incompressible bulk with no shear strength, and of a very thin elastic outer shell, the effective lithosphere, which cannot exert any appreciable bending stress and thus cannot sustain a significant non-hydrostatic bulge. Hence, if this model represents the actual situation with a sufficient degree of approximation, we may use the Clairaut-Laplace-Lyapunov theory of equilibrium figures to calculate the flattening of the Earth's internal strata and related kinetical parameters for different palaeorotation rates and, from the results of these calculations, estimate the incremental strains and stresses produced in the effective lithosphere by the varying centrifugal force. Using then Anderson's theory of faulting, it is possible to give a tentative tectonic interpretation of the incremental stress patterns as a function of latitude, and speculate about the early evolution of the Earth's lithosphere.

[1] Institut d'Astrophysique, Université de Liège, 5, Av. de Cointe, B-4200 Ougrée, Belgium.
[2] Loránd Eötvös Geophysical Institute, Columbus u., 17–23, H-1145 Budapest, Hungary.

2 Outline of the Theory of Equilibrium Figures

Before we proceed to investigate the Earth's palaeokinetic and palaeodynamic parameters, it may be worthwhile sketching briefly the Clairaut-Laplace-Lyapunov, or CLL, theory of figures. A more comprehensive account of the latter may be found in Zharkov and Trubitsyn (1978) or in Denis (1985, 1989). The CLL theory assumes that the distance, r, joining any point $P(r, \theta, \lambda)$ of a level surface

$$U(r, \theta, \lambda) = \text{constant}$$

to the geocentre, may be written explicitly in terms of Lyapunov variables, s and θ, as follows

$$r(s,\theta) = \left[1 + \sum_{n=0}^{\infty} s_{2n}(s) P_{2n}(\cos\theta)\right] s \ .$$

The scalar function U denotes the effective gravity potential, which is the sum of the gravitational potential, $Y(r, \theta)$, and of the centrifugal potential,

$$Z(r,\theta) = -\tfrac{1}{3}\omega^2 r^2 [1 - P_2(\cos\theta)] \ .$$

The variables r, θ, λ are the spherical coordinates (radius, colatitude, longitude) of the point P in a geocentric reference system. We consider here an Earth model in relative hydrostatic equilibrium. It can be shown that such a model is necessarily axisymmetric and possesses an equatorial plane of symmetry (see Denis 1985, 1989). The actual deviations of the real Earth from this state of symmetry are of the order of 10^{-6} or less, and will be neglected throughout this paper. The symbol P_k stands, as usually, for a Legendre polynomial of degree k. All the geometric and kinetic quantities, such as the oblateness of internal strata, the moments of inertia, or other characteristic parameters associated with the equilibrium shape of a rotating body, can be expressed in terms of the functions s_k ($k = 0, 2, 4, \ldots$). The latter are called figure functions. The Lyapunov variable s is a parameter which unambiguously labels the level surface associated with a given internal physical stratum. It may represent the polar radius $c = r(s, \theta = 0°)$ or the equatorial radius $a = r(s, \theta = 90°)$ of a physical stratum, but here we shall use s to represent the equivolumetric radius, i.e., the radius of the sphere which includes the same volume as the level surface associated with the stratum. Thus, we have the defining relation

$$s = \sqrt[3]{\frac{1}{2}\int_{-1}^{+1} r^3(s,z)\,dz} \ .$$

The CLL theory considers the rotating body to be incompressible. This means that the total average radius of the model Earth is assumed to be constant for different rotation velocities, and the labelling parameter of the outer level surface will always assume the value of the present epoch, R = 6371 km. The effects of the Earth's finite compressibility will be discussed briefly in a later section, and are also considered by Hinderer and Legros (this Vol.).

The symbols used here have their usual meaning (cf. Moritz and Mueller 1988), i.e., $G = (6673 \pm 1) \times 10^{-14}\,m^3\,s^{-2}\,kg^{-1}$ is the gravitational constant[3], ϱ denotes mass density inside the Earth, ω is the Earth's angular rotation speed at any epoch and has for the present epoch the mean value $\Omega = 7292115 \times 10^{-11}\,rad\,s^{-1}$, correct to the number of digits provided.

The CLL theory is expressed in terms of a small parameter m, which represents the ratio of the centrifugal acceleration to the gravitational acceleration at the equator of the mean surface, i.e., $m = \omega^2 R^3/GM$, where M is the Earth's total mass. It leads to a very efficient numerical code for computing successive figure functions to any desired accuracy because the order of magnitude of the maximum absolute value of s_{2n} is m^n, for $n>0$, and m^2 for $n = 0$. The procedure used in this paper considers quantities up to the third order in m, and solves iteratively a system of three complicated integro-differential figure equations, which may be written in the following general form

$$s_2^{(k+1)}(s) = F_2[m, s_2^{(k)}(s), s_4^{(k)}(s), s_6^{(k)}(s)] \ ,$$

$$s_4^{(k+1)}(s) = F_4[m, s_2^{(k+1)}(s), s_4^{(k)}(s), s_6^{(k)}(s)] \ ,$$

$$s_6^{(k+1)}(s) = F_6[m, s_2^{(k+1)}(s), s_4^{(k+1)}(s), s_6^{(k)}(s)] \ .$$

The explicit forms of the expressions F_2, F_4, F_6 are provided in the references already mentioned (Zharkov and Trubitsyn 1978; Denis 1989). We start the iterations with

$$s_2^{(1)}(s) = F_2[m, 0, 0, 0] \ ,$$

$$s_4^{(1)}(s) = F_4[m, s_2^{(1)}(s), 0, 0] \ ,$$

$$s_6^{(1)}(s) = F_6[m, s_2^{(1)}(s), s_4^{(1)}(s), 0] \ .$$

and stop when convergence over the whole interval $0 \le s \le R$ has been achieved within given tolerances.

To third order approximation, the expressions for the equatorial radius, a(s), the polar radius, c(s), and the geometrical flattening,

$$f(s) = \frac{a(s) - c(s)}{a(s)} \ ,$$

of any internal level surface of average radius s, are as follows:

$$\frac{a(s)}{s} = 1 + s_0 - \frac{1}{2} s_2 + \frac{3}{8} s_4 - \frac{5}{16} s_6 \ ,$$

$$\frac{c(s)}{s} = 1 + s_0 + s_2 + s_4 + s_6 \ ,$$

$$f(s) = -\frac{3}{2} s_2 - \frac{5}{8} s_4 - \frac{3}{4} s_2^2 - \frac{21}{16} s_6 - \frac{27}{40} s_2^3 + \frac{1}{4} s_2 s_4 \ .$$

[3] In our computations we adopted the value $G = 6.672 \times 10^{-11}\,m^3\,s^{-2}\,kg^{-1}$.

We readily find, to the same order of approximation, the following expressions for the principal moments of inertia A and C about an equatorial axis and a polar axis, respectively:

$$A = \frac{2}{5}\int_0^R \varrho(s)\frac{d}{ds}\left[s^5\left(1+\frac{1}{2}s_2+\frac{9}{7}s_2^2+\frac{53}{105}s_2^3+\frac{4}{77}s_2s_4\right)\right]ds \ ,$$

$$C = \frac{2}{5}\int_0^R \varrho(s)\frac{d}{ds}\left[s^5\left(1-s_2+\frac{3}{7}s_2^2+\frac{44}{105}s_2^3+\frac{8}{7}s_2s_4\right)\right]ds \ .$$

The trace of the inertia tensor, 2A+C, is provided by the following expression, which involves only quantities of the second order in m and higher:

$$\mathrm{tr}\, I = 2A+C = \frac{2}{5}\int_0^R \varrho(s)\frac{d}{ds}\left[s^5\left(3+3s_2^2+\frac{10}{7}s_2^3-\frac{80}{77}s_2s_4\right)\right]ds \ .$$

Thus, once the figure functions s_0, s_2, s_4, s_6 have been computed for a given m, i.e., for a given rotation rate ω, it is possible to derive all relevant kinetic quantities, such as the dynamic shape factor

$$J_2 = \frac{C-A}{Ma^2} \ ,$$

or the precessional constant

$$H = \frac{C-A}{C} \ .$$

Notice that the latter parameter plays an essential role in many geodynamic discussions (Khan 1983), as well as in the astronomical theory of palaeoclimates (Milankovitch 1920; Berger 1976).

3 Tidal Despinning of the Earth

The observed present-day tidal despinning of the Earth is, according to the final report of the IAG Special Study Group 5-99 on *Tidal friction and the Earth's rotation* (Burša 1987)

$$\left(\frac{d\omega}{dt}\right)_{\mathrm{tidal}} = -(6.4\pm0.3)\times10^{-22}\,\mathrm{rad\,s^{-2}} \ ,$$

about 83% of this value stemming from tidal interaction with the Moon, and 17% from tidal interaction with the Sun[4]. The absolute amount of this value, inferred

[4] In a paper presented at this meeting, Brosche and Wünsch (this Vol.) argue quite convincingly that the ratio κ of the average lunar tidal torque to the average solar tidal torque is almost twice the quasi-equilibrium value, $\kappa = 4.9$, given here. Using Brosche and Wünsch's value, $\kappa = 8.2$, some 11% of the rate of decrease of the Earth's rotation speed caused by tides are due to the Sun, and some 89% are caused by the Moon.

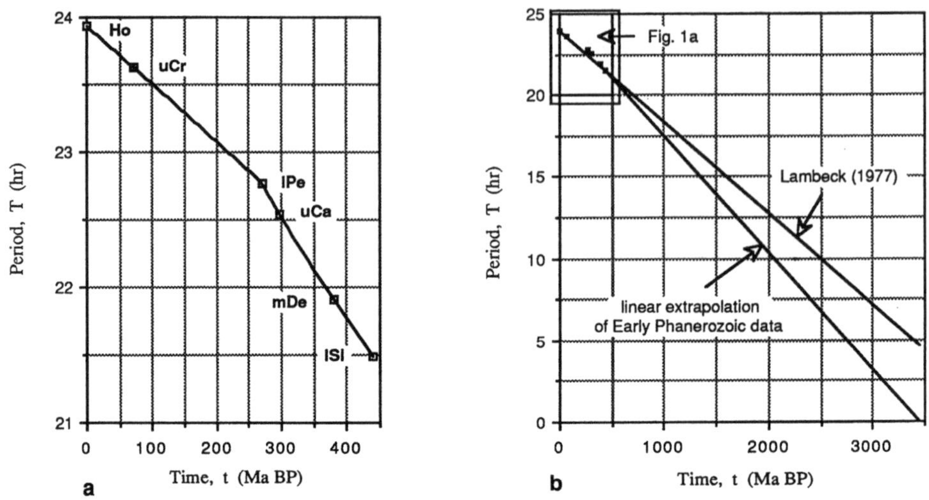

Fig. 1a. Estimated values of Earth's rotation period (in hours) in the geological past, according to Stoyko (1970). The notations used are as follows: *Ho* Holocene; *uCr* Upper Creaceous; *lPe* Lower Permian; *uCa* Upper Carboniferous; *mDe* Middle Devonian; *lSi* Lower Silurian. The figure suggests that the rate of despinning might have been more important before the Permian (270 to 225 Ma BP) than afterwards, which would imply stronger tidal dissipation in the Early Phanerozoic than in the Late Phanerozoic. However, the large error bounds, which are not indicated here, do not permit to draw definite conclusions about time changes in the tidal despinning rate. **b** Extrapolation of Lambeck's (1977) linear regression line and of the Early Phanerozoic data into the Proterozoic and Archaean. If Lambeck's law is used, the Earth would have made one turn around its polar axis in 5 h about 3.4 Ga ago. If the Early Phanerozoic data are used, the moment when the rotation period was 5 h would be reduced to about 2.7 Ga BP. Both these values would favour the theory that the Moon was captured by the Earth some 2.7 to 3.4 Ma ago, a scenario which is neither supported by the dated morphological features on the Moon's surface, nor by geological observations on the Earth

from astronomy and celestial mechanics, is about 15% smaller than the value obtained by Lambeck (1977) from a regression analysis performed on paleontological data which cover most of the last 5×10^8 years, i.e., the Phanerozoic Aeon:

$$\frac{d\omega}{dt} = -(5.4 \pm 0.5) \times 10^{-22}\, \text{rad s}^{-2} \, .$$

It is presently not clear whether the difference between these values is meaningful, despite the fact that the indicated error intervals do not overlap. It is of course quite possible that the present area and geographical distribution of shallow seas, brought about by recent plate tectonics, favours tidal friction and thus leads to a despinning rate which is larger than the mean despinning rate inferred from growth elements in ancient corals and bivalves.

Figure 1 a shows estimations, provided by Stoyko (1970), of the length of day (l.o.d.) in the geological past. These data are in essential agreement with more recent determinations reviewed by Scrutton (1978) and Lambeck (1980). They suggest that the rate of tidal despinning would have been larger than the present-day value between the lower Silurian (440 Ma BP) and the lower Permian (270 Ma BP), amounting to

-8.5×10^{-22} rad s^{-2} on the average, and lower than the present-day value between the Permian and recent times, amounting to -4.4×10^{-22} rad s^{-2} on the average. Clearly, such an inference does not look unreasonable if we associate the smaller rate with the existence of a supercontinent, Pangaea, and a reduced number of shallow seas. Nevertheless, considering the rather large uncertainties underlying the fossil data, it seems too early to draw definite conclusions (see also Krohn and Sündermann 1982).

All in all, fossil coral data and ancient bivalve data allow us to trace the Earth's rotational history backwards for about 500 Ma. Trying to get deeper into the past necessarily involves, at present, some speculation. However, geological observations

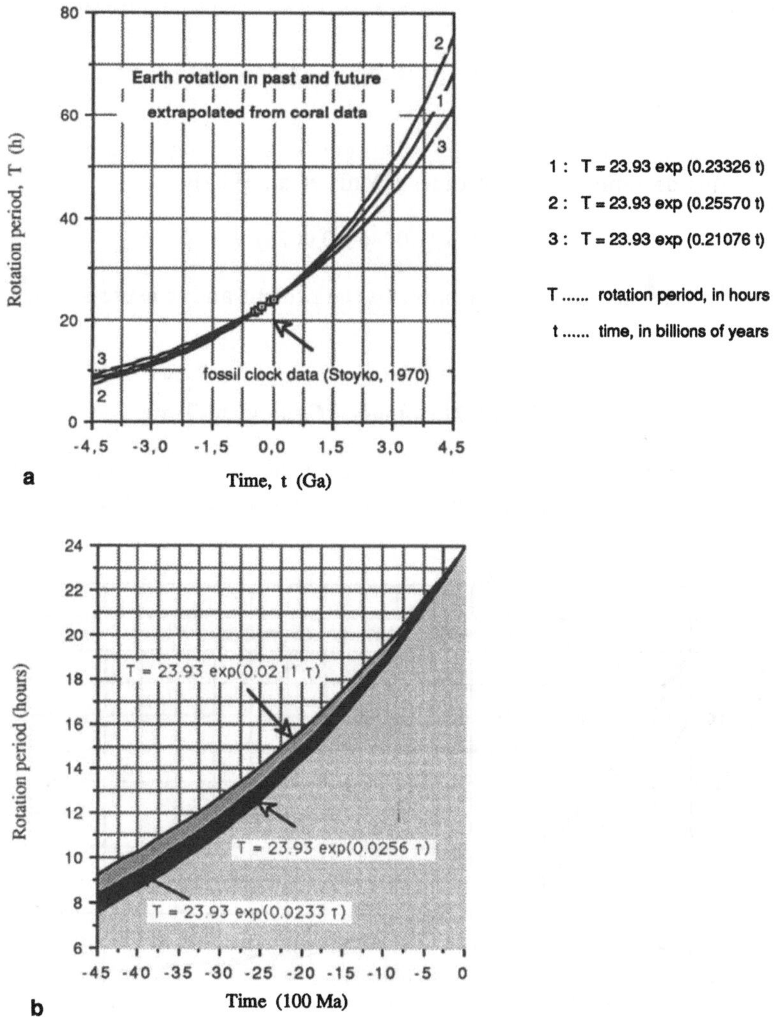

Fig. 2 a. Earth's rotation period (in hours) between 4.5 Ga BP to 4.5 Ga AP inferred from the rotation law (see text): $\omega^{-1}d\omega/dt = -(0.233\pm0.022)\,Ga^{-1}$. **b** Closeup of past values of the rotation period according to the same law, in terms of the time τ expressed in units of 100 Ma

do not support the idea of dramatic changes in the Earth–Moon system, such as the occurrence of a Gerstenkorn event, for the last 2.5 billion years. Moreover, the relatively smooth evolution of the lunar surface during the last three billion years is evidence that close approach must have occurred more than three billion years ago, if it occurred at all (Lambeck 1980). This argument indicates that the more or less linear decrease of the Earth's rotation speed suggested by Phanerozoic fossils cannot be extrapolated very far into the Proterozoic (Fig. 1 b; see also Lambeck 1988). A tentative assumption is that the actual rate of despinning is not constant over geological time spans, but that it is proportional to the spin rate itself. In all the subsequent computations, we adopt therefore the following despinning law, which agrees well with Phanerozoic estimations of l.o.d., and leads to plausible values of l.o.d. in the very distant geological past (Fig. 2):

$$\frac{1}{\omega}\frac{d\omega}{dt} = -(7.4 \pm 0.7) \times 10^{-18}\,s^{-1} = -0.23 \pm 0.02\,Ma^{-1} \ .$$

In terms of l.o.d., which we do not distinguish here from the rotation period $T = 2\pi/\omega$, this despinning law becomes (Denis et al. 1987)

$$T = 23.93 \times (1.0236 \pm 0.0023)^{\tau} \ ,$$

where the time τ is measured with respect to the present and is expressed in units of 100 million years.

4 Kinetic Parameters of the Earth Throughout Geological Time

In a previous paper, one of us (Denis 1986a) used the palaeorotation rates provided by Stoyko (1970) to derive the time changes of major kinetic parameters throughout

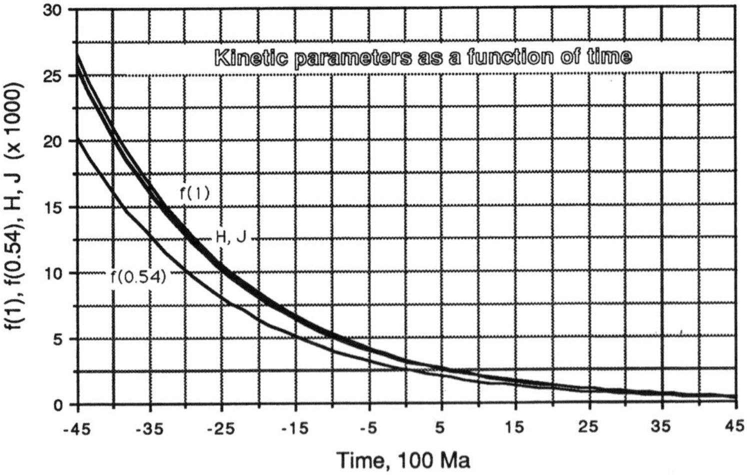

Fig. 3. Kinetic parameters $f_s = f(1)$, $f_c = f(0.54)$, H and $J = 3J_2$, expressed in units of 1/1000, as a function of past and future epochs

the Phanerozoic Aeon, assuming that the internal structure of the Earth had remained essentially unchanged during the last 500 Ma. Notice that unessential changes of the Earth's constitution would only have minor effects on integrated kinetical parameters, such as J_2, H, etc. The present paper extends these results to most of the Earth's history, and extrapolates tentatively the computations into the distant future, assuming that the rotation law given above and illustrated in Figure 2, holds from 4.5 Ga BP ($\tau = -45$) to 4.5 Ga AP ($\tau = +45$).

Clearly, spin rates beyond the Phanerozoic (i.e., for $\tau < -5$) are uncertain, and thus the association of a given set of kinetic parameters with a specific age is somewhat speculative. It is possible that stromatolites may soon be used successfully as palaeontological clocks (Walter 1976). In this way, Proterozoic l.o.d. values may become available which would possibly make necessary some rescaling of the ages indicated here. Data provided for the distant future are even less certain, and should not be considered too seriously. Nevertheless, we believe that the values of the ancient kinetic parameters given here are sufficiently constrained by our model calculations to yield meaningful conclusions about some global evolutionary aspects.[5]

From these model calculations we derived close formulae, which fit our numerical results almost perfectly, for the following parameters (Fig. 3):

geometrical flattening of the outer surface:

$$f_s(\tau) = 0.00334 \times 10^{-0.02\tau} \; ;$$

geometrical flattening of the core-mantle boundary:

$$f_c(\tau) = 0.00255 \times 10^{-0.02\tau} \; ;$$

dynamical flattening:

$$H(\tau) = 0.00324 \times 10^{-0.02\tau} \; ;$$

dynamic shape factor:

$$J_2(\tau) = 0.00107 \times 10^{-0.02\tau} \; .$$

We notice that the evolutionary time scale of all these parameters is typically 2.2 Ga, i.e., about four to five times the duration of the whole Phanerozoic Aeon. An interesting parameter to consider, which will be used in the next section, is the decrement of flattening in time measured from a given reference state. We shall assume that a solid crust was formed soon after the Earth's formation, and use as a reference state

[5] The authors are grateful to Prof. Brosche for having drawn their attention, after submittal of the manuscript, to a result published very recently by Williams (1989) stemming from late Precambrian glaceogenic formations in South Australia. If the sedimentary rythmites of siltstone and fine sandstone observed there are interpreted as ancient ebb-tidal deposits, and taking as a premise that the observed first-order peak is a yearly, non-tidal signal, it may be inferred that c. 650 Ma ago the day lasted 21.9 ± 1.2 h. For $\tau = -6.5$, the despinning law assumed in this paper yields $T = 20.6 \pm 0.3$ h. Considering all the uncertainties involved in the datum provided by Williams (1989), we believe that there is still no need to revise the despinning law used here.

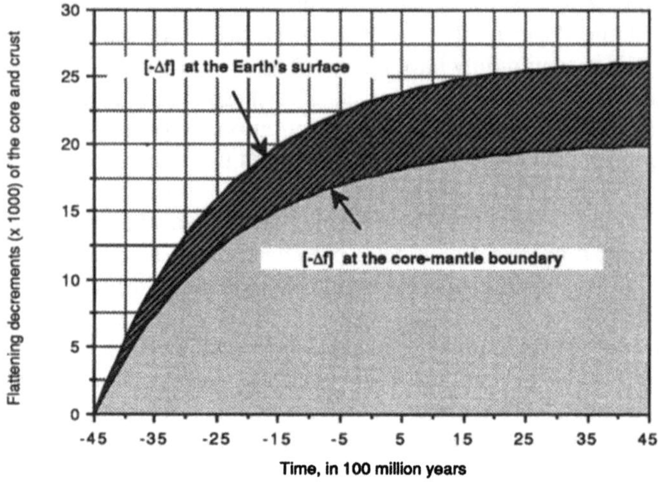

Fig. 4. Decremental flattening of the Earth's outer surface and core-mantle boundary throughout time

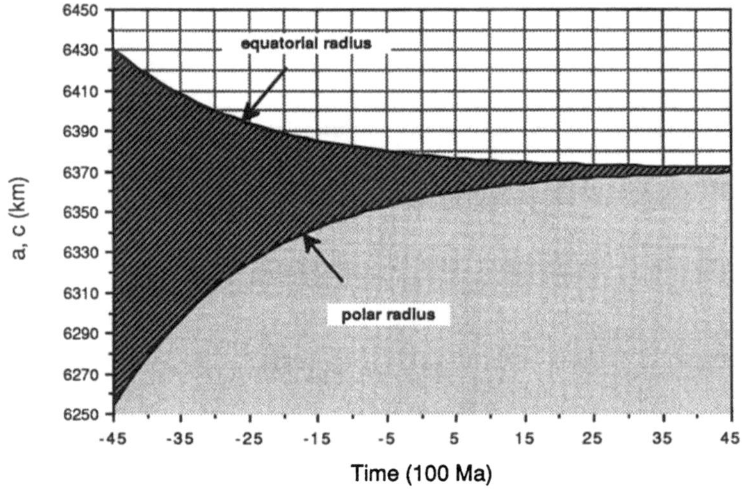

Fig. 5. Inferred variation throughout time of the Earth's equatorial and polar radii

the flattening corresponding to $\tau = -45$, i.e., $f_s(-45) = 0.02653$. With respect to this initial state, the flattening has decreased throughout time by an amount Δf according to the law (Fig. 4):

$$\Delta f(\tau) = 0.00334 \times 10^{-0.02\tau} - 0.02653 \ .$$

Figure 5 shows the variation with time of the equatorial and polar radii. Since the early epochs of Earth's history, tidal friction has reduced the equatorial radius by about 0.8%, and increased concomitantly the polar radius by about 1.6%, excepting an

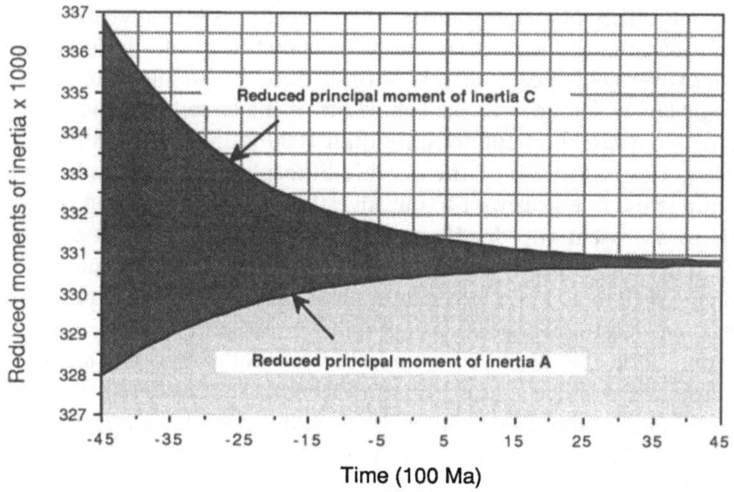

Fig. 6. Inferred variation throughout time of the Earth's principal moments of inertia

overall decrease of the Earth's mean radius due to the radial part of the centrifugal potential.

Figure 6 shows the variation with time of the principal moments of inertia. Since the early Archaean, the moment of inertia around the polar axis has decreased by about 1.6%, and the moment of inertia around an equatorial axis has increased concomitantly by about 0.8%, again supposing that there is no general decrease of the Earth's volume and that core formation was essentially completed very soon after the Earth's formation. The variation of inertia moment is being investigated more thoroughly in a paper by Varga and Denis (this Vol.).

5 Lithospheric Stresses Caused by Tidal Despinning

The computation of the Earth's shape by means of the CLL theory assumes hydrostatic equilibrium, which implies that all shear stresses are relaxed. Strictly speaking, this assumption is in conflict with the assumption that tidal spindown could build up incremental lithospheric shear stress which might give rise to tectonic faulting. However, as stated in the introductory section, we consider that for the time spans involved here, the Earth can adequately be modelled as a two-layered body consisting of a thin outer elastic shell, the effective lithosphere, and an inner bulk in which rocks behave like a fluid and thus do not sustain shear strength.

It is fashionable to believe at present that over geological times all shear stresses in solid rock are getting relaxed by creep phenomena, thus excluding the existence of elastic-brittle behaviour on time scales ranging from millions to billions of years. It should be realized, however, that this belief is not based on experiment or theory, but rather on general ideas about creep processes in the mantle extrapolated to the upper lithosphere. The mere observation that very ancient surface features have been preserved for hundreds of millions of years on Earth, and more distinctly on the

Moon, Mars and other planetary bodies where erosion is less intensive, point to the fact that the cool outer lithospheres of earthlike planets are indeed capable of sustaining shear stress over geological intervals of time. Actually, we ought to admit that the exact rheologic law which governs the behaviour of the terrestrial lithosphere on time scales ranging from less than 100 to more than 1000 million years is not yet known.

Nevertheless, flexure studies of the oceanic lithosphere seem to indicate that over time spans and stress levels relevant to our investigation, the transition from an essentially elastic to an essentially ductile regime occurs in the Earth at an isotherm of about 400 °C (see, e.g., Turcotte and Schubert 1982), yielding a thickness of the effective lithosphere of 10 to 20 km, quite smaller (by a factor of 5 to 10) than the thickness of the lithosphere determined from seismological investigations. Such a thin shell cannot exert appreciable bending stress and thus cannot support a significant nonhydrostatic rotational bulge at the equator. Hence, we may use hydrostatic figure theory as a good approximation to infer the overall shape of the effective lithosphere at any moment, and use thin elastic shell theory (Kraus 1967) to compute incremental strains and stresses produced by a change of curvature.

Turcotte (1979) states that there is extensive observational evidence that the surface plates behave elastically on geological times, and provides evidence that stresses relax in rocks for geological time scales of 10^6 to 10^9 years only at temperatures greater than 600 °C.

Laboratory studies on common rocks by Goetze and Evans (1979) show that these materials cannot sustain shear stress exceeding 10^8 to 10^9 Pa (1 to 10 kb). The exact value of this yield stress depends strongly on temperature. Beyond the yield stress, the mechanical behaviour of rocks ceases to be elastic. Thus, Goetze and Evans suggest a rheological model of the lithosphere which agrees with the model used in this paper: an upper, cooler part which is elastic-brittle, and a lower, hotter part which is ductile. Bodine et al. (1981), McNutt and Menard (1982), Watts and Ribe (1984), Diament (1987), among many others, have used in situ observations such as geoid heights and bathymetric profiles around seamounts and volcanic island chains to study flexure of the oceanic lithosphere under load. Their results give support to the idea that the long-term rheologic behaviour of the Earth's lithosphere may indeed be represented as a thin, continuous and homogeneous, elastic-brittle plate on top, a visco-elastic material at the bottom. The effective elastic thickness decreases with the age of the load, but tends asymptotically to a finite limit somewhere between 10 and 20 km (Fig. 7).

Amalvict and Legros (1986, in press) have derived sets of formulae for incremental lithospheric stresses in simple gravito-elastic models. Their results confirm and extend earlier results obtained by Vening Meinesz (1947) and Melosh (1977). Let β and λ denote latitude and longitude, respectively, μ the effective shear modulus of the elastic outer shell, $\Delta \mathrm{f}$ the change of flattening with respect to a reference value, f_0, at a reference epoch, τ_0. Then, the incremental meridional stress is given by

$$\sigma_{\beta\beta} = -\frac{\mu \Delta \mathrm{f}}{11}[5-3\cos 2\beta-4\varepsilon(3+7\cos 2\beta)] \quad ,$$

where ε denotes the normalized lithospheric thickness, i.e., the thickness of the effective lithosphere divided by the Earth's total radius. The incremental azimuthal stress is

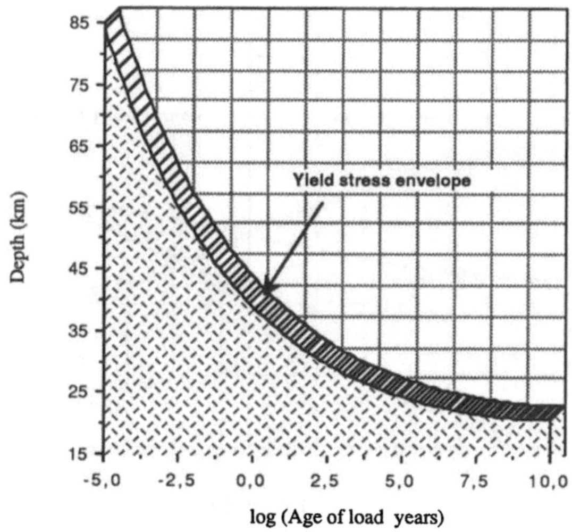

Fig. 7. Plot of the thickness of the effective lithosphere against the decimal logarithm of the age of an applied load, expressed in millions of years, according to the yield stress model supported by a number of observations (based on data by Bodine et al. 1981)

$$\sigma_{\lambda\lambda} = +\frac{\mu \Delta \mathrm{f}}{11}[1+9\cos 2\beta - 4\varepsilon(5+\cos 2\beta)] ,$$

and the resulting incremental stress difference becomes

$$\Delta\sigma = \sigma_{\beta\beta} - \sigma_{\lambda\lambda} = -\frac{\mu \Delta \mathrm{f}}{11}(6-32\varepsilon)(1+\cos 2\beta) .$$

Notice that this stress pattern is symmetric with respect to the equator, as it should be according to Curie's Law of Symmetry (cf. Denis 1989). The value of ε does not exceed 0.003, and thus the finite thickness of the effective lithosphere may safely by neglected in the formulae above.

Figure 8 shows plots of the incremental stresses $\sigma_{\beta\beta}$ and $\sigma_{\lambda\lambda}$, and of the incremental stress difference, $\Delta\sigma$, in the effective lithosphere as a function of latitude. These plots are drawn for a flattening decrement $\Delta \mathrm{f} = -0.001$, and for an effective shear modulus $\mu = 1$ Mb, which is a typical value encountered in surface rocks. We conclude from these plots that $\Delta\sigma$ is always positive. It reaches a maximum at the equator and becomes zero at the poles ($\beta = \pm 90°$). The meridional stress, $\sigma_{\beta\beta}$, is positive, i.e., tensional, at all latitudes, whereas the azimuthal stress, $\sigma_{\lambda\lambda}$, is negative, i.e., compressional, up to mid-latitudes ($\beta_{\mathrm{crit}} = \pm 48.2°$ of latitude in our idealized planetary model), poleward of which it becomes positive, that is tensional.

If we assume that σ_{rr}, $\sigma_{\beta\beta}$ and $\sigma_{\lambda\lambda}$ are principal stresses, and moreover consider that σ_{rr} is almost zero at the Earth's surface, we may interpret the former stress pattern in terms of the theory of near-surface faulting (Anderson 1951; Burns 1976; Melosh 1977; Denis 1986b). Thus, the stress pattern points to the existence of two

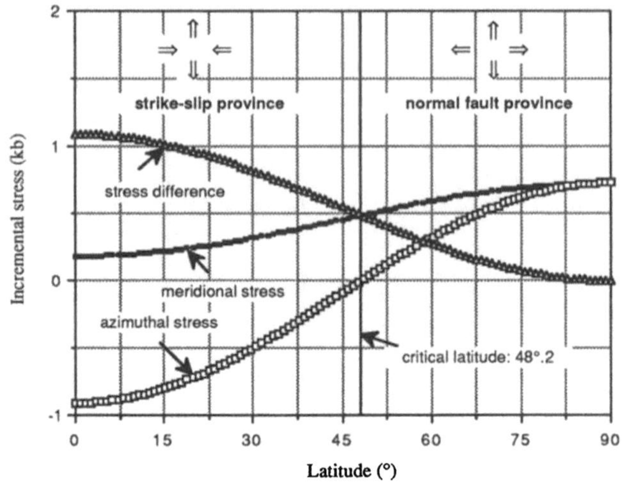

Fig. 8. Incremental stress for $\Delta f = -0.001$ and $\mu = 1$ Mb

major tectonic provinces: a strike-slip fault province ($\sigma_{\beta\beta} > \sigma_{rr} > \sigma_{\lambda\lambda}$ (which extends from the equator to mid-latitudes, and a normal fault province ($\sigma_{\beta\beta} > \sigma_{\lambda\lambda} > \sigma_{rr}$) which extends from mid-latitudes to the poles.

6 Tidal Shrinkage

Until now, we have assumed that the Earth behaves essentially like an incompressible body, thus taking into account only the quadrupole term of the rotational perturbation. The radial part has actually led to a continuous contraction and renumbering of the level surfaces,

$$s_0 = -\frac{1}{5}s_2^2 - \frac{2}{105}s_2^3 < 0 \ ,$$

which we did not associate, however, with a contraction of the material body.

Stoneley (1924) was probably the first to consider shrinkage of the Earth's volume caused by tidal despinning in an attempt to explain mountain building. He found, with the then admitted parameters describing the Earth's compressibility, that a slowing down of the rotation speed from a 12-h day to a 24-h day would reduce the Earth's radius by about 1.6 km. For the same change in the rotation period, our results lead to a decrease of the equatorial radius of about 16 km, and a concomitant increase of the polar radius by about 32 km.

Hinderer and Legros (this Vol.) find a contraction of about 470 m during the last 400 Ma. The latter is quite compatible with Stoneley's result obtained for a much longer time interval. Our computations yield, for the last 400 Ma, $\Delta a = -1.5$ km and $\Delta c = +3$ km.

Using experimental values for the change of density of common rocks with pressure, and an increase of Earth's gravity (i.e., gravitational plus centrifugal acceleration) by about 3% over its history, we may infer a value of $|\Delta R|/R$ comprised between 0.1% and 1%. We derive this estimate from a paper by Birch (1968), which shows that a doubling of the present value of gravity would lead to a decrease of the Earth's radius by at most 400 km. Hence, the overall increase of gravity by about $0.3\ \mathrm{m\,s^{-2}}$ resulting from an increase of l.o.d. from 8.4 to 24 h would lead to $\Delta R \cong -12$ km, as compared to $\Delta a \cong -50$ km and $\Delta c \cong 100$ km.

We conclude, therefore, that the results obtained in the previous section should not be changed dramatically when accounting for the Earth's finite compressibility. The effects of tidal spindown caused by the radial term of the centrifugal potential are quite smaller than the effects due to the quadrupole term, but can possibly lead to a thrust province in a restricted neighbourhood of the equator. It should be realized, however, that the Earth's volume may not have changed at all in any significant way, or many have increased during geological terms instead of decreasing. Indeed, dilatation due to radioactive heating, dilatation due to tidal heating, dilatation due to a possible decrease of the gravitational constant, are some effects that oppose contraction due to tidal despinning. Thus, if any global dilatation has occurred, the strike-slip province at low latitudes may extend to slightly higher latitudes.

7 Tectonic Evolution of the Earth Due to Tidal Despinning

We are now in a position to discuss, in general terms, the dynamics in the remote past of the uppermost layers of our planet. The geological record of the Earth's early history is sparse and uncertain, and therefore gives rise to much controversy. However, there seems to be a more or less general agreement that the Earth's crust was mainly granitic more than 3.5 billion years ago. Larger quantities of sedimentary and volcanic rocks were added only later, between ca. 3.5 and 2.7 Ga BP (Lambeck 1980). Stable continental blocks, i.e., the present continental cratons, formed probably from 2.6 to 2.0 Ga ago. During this time interval of 700 million years, a gradual transition from great tectonic mobility and crustal thickening to the present style of global plate tectonics seems to have occurred. The continental drift we watch today seems to be essentially governed by convective motions deep inside the Earth's mantle. However, because the asthenosphere uncouples to a large extent motions in the lower part of the mantle from motions of the lithosphere, it does not seem evident that mantle convection can account for the original breakup of the lithosphere into a number of plates (Ichiye 1971). If lithospheric breakup is due to tidal despinning, it is easy to understand why no evidence of plate tectonics has been found on Mars and Venus, two planets with no massive companion. Liu (1974) has suggested that lithospheric breakup may be due to polar wandering, by which the lithosphere experiences a change of curvature, similar to the one caused by tidal despinning.

Shear strength, S, of granite measured in the laboratory under confining pressures and temperatures relevant to the effective lithosphere is typically of the order of 0.1 to 0.3 kb. For basalt, S generally ranges from 0.3 to 0.5 kb. The angle of internal friction, Φ, is comprised between 56° and 58° for granite, and between 48° and 50° for basalt (Jaeger 1978). According to the Coulomb-Navier-Mohr theory, shear failure in brittle materials such as granite and basalt occurs if

$$\Delta\sigma \geq 2\mathrm{S}\tan\gamma + (\tan^2\gamma - 1)\sigma_{\lambda\lambda}\ , \quad \gamma = 45^\circ + \frac{\Phi}{2}\ .$$

Failure takes place in a vertical plane at an angle γ to the longitudinal stress. For granite, tensile strength is about 0.04 kb, whereas compressive strength is much larger, of the order of 1.4 kb. Thus, shear fracture occurs in granite if

$$\Delta\sigma - 10.4\sigma_{\lambda\lambda} \geq 1.35\ \text{(kb)}\ ,$$

assuming that S = 0.2 kb and $\Phi = 57°$. It occurs in basalt if

$$\Delta\sigma - 6.15\sigma_{\lambda\lambda} \geq 2.14\ \text{(kb)}\ ,$$

assuming that S = 0.4 and $\Phi = 54°$. Typical shear moduli for granite and basalt are 243 kb and 307 kb, respectively (Clark 1966).

The latter criteria, and the expressions derived in the previous sections for Δf, $\sigma_{\lambda\lambda}$, and $\Delta\sigma$, indicate that an original crust mainly made up of granite must have undergone severe strike-slip faulting at low latitudes about 4.05 billion years ago. If the original crust had been mainly composed of basalt, then the same events would have occurred somewhat later, about 3.91 billion years ago. Dip-slip faulting at high latitudes must have happened quite earlier, maybe as soon as 4.48 Ga BP, only 20 million years after the consolidation of the original crust. The fact that normal faulting occurred so early in Earth's history, shortly after its formation, may have caused a lasting asymmetry between the Northern and Southern hemispheres. Indeed, at that epoch core formation probably was not completed yet, thus allowing whole-mantle cellular convection of dipolar type (Runcorn 1965; Fairbridge 1972). Hot ascending magma may thus have extruded at high latitudes in one hemisphere, but neither at low latitudes on both sides of the equator, nor at high latitudes in the other hemisphere, where cool material was descending toward the center.

On the other hand, reverse faulting, and in particular thrust faulting, is not typical of a tidal spindown regime, and is presumably associated with collisions of moving plates. This suggests that folded orogenic belts can only have occurred much later, probably only 2.5 billion years ago or less. Volcanic mountain chains, however, must have occurred on Earth about 20 million years after the original granitic crust had been formed.

It is not yet possible to ascertain whether faulting due to stress buildup in the effective lithosphere by tidal despinning was episodic or more or less continuous in time, after the original crust had been fractured into several pieces. Because tidal despinning is a very slow but continuous process, and because rupture occurs suddenly for a certain stress level, there is a definite possibility that major strike-slip fault systems have occurred episodically before an actual plate tectonic regime was established. On the other hand, the creation rate of major rift systems has a more or less stationary character.

Drifting of plates, a phenomenon quite unrelated directly with tidal deceleration, may give rise to stress buildup much in the same way tidal despinning does. Indeed, a plate possessing a northward or southward component of drift, such as the African plate presently has, will experience a change in curvature which may possibly lead to some kind of intraplate tectonics. Membrane stress tectonics of the African plate has been investigated by Freeth (1979), but much of the theoretical formulation of the theory is due to Turcotte (1974). Because of larger values of the Earth's flattening, intraplate tectonics was presumably more active in the Cryptozoic than in more recent times.

Acknowledgments. The authors wish to express their sincerest thanks to the organizers of this meeting. C.D. is grateful to the Fonds National de la Recherche Scientifique (Brussels) for financial support. P.V. gratefully acknowledges support from the Alexander von Humboldt Stiftung.

References

Amalvict M, Legros H (1986) Lithospheric stresses: gravito-elastic model and geometric deformation. Manuscripta Geodaetica 11:197–206

Amalvict M, Legros H (in press) Stresses in the lithosphere generated by a changing shape. In: Teisseyre R (ed) Physics and Evolution of the Earth's Interior, Vol 5. PWN, Warszawa and Elsevier, Amsterdam

Anderson EM (1951) The Dynamics of Faulting. Oliver and Boyd, Edinburgh

Berger AL (1976) Obliquity and precession for the last 5000000 years. Astron Astrophys 51:127–135

Birch F (1968) On the possibility of large changes in the Earth's volume. Phys Earth Planet Int 1:141–147

Bodine JH, Steckler MS, Watts AB (1981) Observations of flexure and the rheology of the oceanic lithosphere. J Geophys Res 86(B5):9905–9918

Burns JA (1976) Consequences of the tidal slowing of Mercury. Icarus 28:453–458

Burša M (1987) Final Report of IAG Special Study Group 5-99 "Tidal Friction and the Earth's Rotation". XIX Gen Ass IUGG, Vancouver, August 1987

Clark SP Jr (ed) (1966) Handbook of Physical Constants (revised edition). Geol Soc Am Mem 97

Denis C (1985) The Hydrostatic Figure of the Earth. Geophys Rep Publ IAL, Réf 85/2, Univ Liège

Denis C (1986a) On the change of kinetical parameters of the Earth during geological times. Geophys J R Astr Soc 87:559–568

Denis C (1986b) Lithospheric deformation caused by the secular change of the Earth's rotation speed. Terra Cognita 6:315 (abstr)

Denis C (1989) The hydrostatic figure of the Earth. In: Teisseyre R (ed) Physics and Evolution of the Earth's Interior, Vol 4, Gravity and Low-Frequency Geodynamics, Chap 3. PWN, Warszawa and Elsevier, Amsterdam, pp 111–186

Denis C, Karafistan A, Amalvict M, Legros H (1987) Changing curvature and tectonic activity of planetary bodies. In: Proc CNES Scient Workshop Comparative Planetology and Earth Sciences, La Londe-les-Maures. Geophys Rep Publ IAL, 7:3–15

Diament M (1987) Isostasie, réponses mécanique et thermique de la lithosphère. Applications à la Géodynamique. Thèse de doctorat d'Etat, Univ. Paris-Sud, Centre d'Orsay

Fairbridge RW (1972) Planetary spin-rate and evolving cores. Ann NY Acad Sci 187:88–107

Freeth SJ (1979) Deformation of the African plate as a consequence of membrane stress domains generated by post-Jurassic drift. Earth Planet Sci Lett 45:93–104

Goetze C, Evans B (1979) Stress and temperature in the bending lithosphere as constrained by experimental rock mechanics. Geophys J R Astr Soc 59:463–478

Ichiye T (1971) Continental breakup by nonstationary mantle convection generated with differential heating of the crust. J Geophys Res 76:1139–1153

Jaeger JC (1978) Elasticity, Fracture and Flow, with Engineering and Geological Applications. Chapman and Hall, London

Khan MA (1983) Revised value of the dynamical ellipticity of the Earth. Geophys J R Astr Soc 72:327–331

Kraus H (1967) Thin Elastic Shells. Wiley, New York

Krohn J, Sündermann J (1982) Paleotides before the Permian. In: Brosche P, Sündermann J (eds) Tidal Friction and the Earth's Rotation II. Springer, Berlin Heidelberg New York Tokyo, pp 190–209
Lambeck K (1977) Tidal dissipation in the oceans: astronomical, geophysical and oceanographic consequences. Phil Trans R Soc London A287:545–594
Lambeck K (1980) The Earth's Variable Rotation: Geophysical Causes and Consequences. Univ Press, Cambridge
Lambeck K (1988) Geophysical Geodesy: The Slow Deformations of the Earth. Univ Press, Oxford
Liu HS (1974) On the breakup of tectonic plates by polar wandering. J Geophys Res 79:2568–2572
McNutt M, Menard HW (1982) Constraints on yield strength in the oceanic lithosphere derived from observations of flexure. Geophys J R Astr Soc 71:363–394
Melosh HJ (1977) Global tectonics of a despun planet. Icarus 31:221–243
Milankovitch M (1920) Théorie mathématique des phénomènes thermiques produits par la radiation solaire. Gauthier-Villars, Paris
Moritz H, Mueller II (1987) Earth Rotation – Theory and Observation. Ungar, New York
Runcorn SK (1965) Changes in the convection pattern in the Earth's mantle and continental drift: evidence for a cold origin of the Earth. Phil Trans R Soc Lond A258:228–251
Scrutton CT (1978) Periodic growth features in fossil organisms and the length of the day and month. In: Brosche P, Sündermann J (eds) Tidal friction and the Earth's Rotation. Springer, Berlin Heidelberg New York, pp 154–196
Stoneley R (1924) The shrinkage of the Earth's crust through diminishing rotation. Mon Not R Astr Soc Geophys Suppl 1:149–155
Stoyko A (1970) La variation séculaire de la rotation de la Terre et les problèmes connexes. Ann Guébhard (Neuchâtel) 46:293–316
Turcotte DL (1974) Membrane tectonics. Geophys J R Astron Soc 36:33–42
Turcotte DL (1979) Flexure. Adv Geophys 21:51–86
Turcotte DL, Schubert G (1982) Geodynamics. Wiley, New York
Vening Meinesz FA (1947) Shear patterns of the Earth's crust. Trans Am Geophys Un 28:1–61
Walter MR (ed) (1976) Stromatolites. Elsevier, Amsterdam
Watts AB, Ribe NM (1984) On geoid heights and flexure of the lithosphere at seamounts. J Geophys Res 89(B13):11152–11170
Williams GE (1989) Late Precambrian tidal rhythmites in South Australia and the history of the Earth's rotation. J Geol Soc Lond 146:97–111
Zharkov VN, Trubitsyn VP (1978) Physics of Planetary Interiors. Pachart, Tucson

Secular Variations of the Earth's Moment of Inertia and Related Quantities

P. Varga[1] and C. Denis[2]

1 Introduction

In investigations of long-term variations of the length of the day (l.o.d.) and of the evolution of the Earth-Moon-Sun system, use is often made of the law of conservation of angular momentum written as

$$C\frac{d\omega}{dt} + \omega \frac{dC}{dt} = L + N , \tag{1}$$

where C is the moment of inertia about Earth's polar axis, ω is Earth's angular rotation speed, t is time, L is the tidal torque exerted on the Earth by Moon and Sun ($L<0$), and N is a correction term which accounts for a possible imbalance between L and $d(C\omega)/dt$. Imbalance can arise from the fact that other external torques besides tidal torques may act on the Earth, leading to non-tidal accelerations or decelerations of the Earth's spin. Imbalance can also arise if the Earth as a whole does not rotate like a solid body. Then, exchange of angular momentum between mantle, core, oceans and atmosphere rotating at different speeds may occur, and such processes have not been considered in the left-hand side (l.h.s.) of Eq. (1). Finally, if we suppose that only tidal friction leads to important effects over long time scales, there is still the possibility of a numerical imbalance between $d(C\omega)/dt$ and L due to our imperfect knowledge of the exact values of the involved quantities. In the latter case, N represents just a numerical parameter with no physical meaning. Of course, conservation of angular momentum does not just involve the polar component C and, in precise studies, Eq. (1) must be replaced by more sophisticated expressions discussed by Burša and by Kostelecký in this volume, which involve a projection factor taking account of the fact that the obliquity is not zero.

One of us (P.V.) has estimated, from Schwiderski's (1980) cotidal charts, an oceanic tidal torque $L_{ot} = -5.0\times10^{16}$ N.m. The earth tidal torque, L_{et}, as estimated from what seem to be the most reliable determinations of the phase lag of the gravimetric body tide M_2, is -1.3×10^{16} N.m The contribution of the atmospheric tides, L_{at}, amounts approximately to $+0.4\times10^{16}$ N.m and has an overall accelerating effect. Our estimation thus leads to a value of about -5.9×10^{16} N.m. for the tidal torque L. However, we wish to stress that the exact present value of L ist still open to debate,

[1] Lorand Eötvös Geophysical Institute, Columbus u., 17–23, H-1145 Budapest, Hungary.
[2] Institute d'Astrophysique, Université de Liège, 5, Av. de Cointe, B-4200 Ougrée, Belgium.

although there is little doubt that $-L$ is comprised between 10^{16} and 10^{17} N.m (cf. Denis 1986).

It is important to realize that these values can only be considered as indicative, and that the error bounds may be quite large. A number of papers presented in this volume, as well as discussions during the conference itself, have more or less convinced us that the state-of-the-art presently reached in the modelling of oceanic and atmospheric large-scale motions induced by the lunar and solar tidal force fields is advanced enough to provide accurate values of L_{ot} and L_{at}. The contribution of the earth tides, which depends strongly on the debatable Q-value for tidal deformations of the interior parts of the Earth, is only known within a factor, say, of 5, and constitutes the largest source of uncertainty as regards the value of $L = L_{ot} + L_{et} + L_{at}$. Zschau (1977, 1978) has pointed out that the phase lag of bodily load tides is ruled by the rheology of the crust and upper mantle, whereas the phase lag of the direct bodily tides results mainly from anelasticity of the lower mantle. This fact largely complicates the measurement of a meaningful effective phase lag of tidal waves, and thus the derivation of the actual value of L_{et}, from gravity recordings at unevenly distributed earth tidal stations. Zschau (1983) has attempted to derive an effective tidal Q-value without actually using tidal observations, by assuming that the seismic absorption band model (cf. Anderson and Minster 1979) remains applicable to straining at tidal and Chandler wobble frequencies. The latter assumption had previously been used by Jeffreys (1972), who on this basis suggested that Q varies approximately as the quintic root of frequency, i.e., $Q \simeq \nu^{0.2}$. If this law is assumed to hold, we may argue that for the semi-diurnal tides the relevant Q is 350 ± 40, for the diurnal tides it is 300 ± 30, and for the Chandler wobble it is 90 ± 10.

However, the uncertainties in the Q models themselves as well as in the numerical parameters which characterize these models are presently too great to allow a reliable estimate of dissipation caused by bodily tides. Zschau (1986) has tried to avoid these difficulties by arguing that the rate of dissipation due to earth tides can be related linearly to the period of the Chandler wobble [435.5 ± 0.7 days, according to Höpfner (1983)], the relation being nearly independent of the Q model and its defining parameters.

We believe that the actual tidal torque is not known with sufficient accuracy to use Eq. (1), or a somewhat more complicated expression involving the intricacies of celestial mechanics, to derive from the observation of $d\omega/dt$ and from the knowledge of ω and C any meaningful value of dC/dt.

This, however, should be an important goal to reach, because dC/dt contains information about geodynamic processes which may or may not occur inside the Earth (convective motions in the mantle and core, chemical fractionation and gravitational settling, volume changes due to heating or cooling, or to a variable gravitational constant, mass redistribution caused by seismic and volcanic activity, or by climatic factors). Such processes do not give rise to a net change of angular momentum, but may change the polar moment of inertia and concomitantly speed up or slow down the Earth rotation.

In problems concerning the evolution of the Earth-Moon-Sun system, and in particular in studies concerning the secular accelerations of Moon and Sun (Jeffreys 1970, p. 301), it is common practice to disregard the rate of change of the polar moment of inertia, dC/dt, by omitting in Eq. (1) the second term of the l.h.s., i.e.,

$\omega\, dC/dt$, without justification. Denis (1986) has shown that the latter term can indeed be neglected with respect the term $C\, d\omega/dt$, as long as it may be interpreted as a feedback effect to changes in l.o.d. caused by tidal friction (see Sect. 3). Of course, for changes in l.o.d. which are a direct consequence of mass redistribution without implying any external torque, we have $\omega\, dC/dt = -C\, d\omega/dt$. Thus, in the realistic case where both tidal friction and mass redistribution processes are acting simultaneously, it may become necessary to take account of the term $\omega\, dC/dt$ in the angular momentum balance. However, as mentioned above, Eq. (1) by itself does not provide reliable information at present. Thus, one may speculate about possible non-tidal changes of C (Sect. 4), using theoretical arguments or arguments based on the value of dJ_2/dt recently derived from orbit-tracking data of the geodetic satellite LAGEOS (Sect. 5). Most theoretical discussions assume, either implicitly or explicitly, that variations of C can be derived by assuming that the trace of the inertia tensor remains constant. Section 2 will address this topic, and provide some new insight.

2 Rotational Deformation and Trace Changes of the Inertia Tensor

Let $\mathrm{tr}\,\mathfrak{Z} = \mathfrak{Z}_{11} + \mathfrak{Z}_{22} + \mathfrak{Z}_{33}$ denote the trace of the inertia tensor, $\mathfrak{Z}$. It is generally assumed that this quantity is invariant, not only under a rigid body rotation of the reference frame, a fact which follows at once from the tensorial character of $\mathfrak{Z}$, but also under a wide class of deformations of the Earth itself. Rochester and Smylie (1974) have discussed this assumption in some detail, and have shown in particular that *to the first order* in the arbitrary displacement field expanded into spherical vector harmonics,

$$\mathbf{u}(\mathbf{r},t) = \sum_{n=0}^{\infty} \sum_{m=-n}^{n} \left[u_n^m(r,t)\,\frac{\mathbf{r}}{r} + r\, v_n^m(r,t)\,\nabla - w_n^m(r,t)\,\mathbf{r}\times\nabla \right] Y_n^m(\theta,\phi) \ ,$$

the Lagrangian perturbation of the trace of the inertia tensor of a body which is spherically symmetric in the undeformed state and has a density distribution $\varrho(r)$, is given by

$$\delta\,\mathrm{tr}\,\mathfrak{Z} = 16\pi \int_0^R \varrho(r)\, r^3 u_0^0(r)\, dr \ . \tag{2}$$

The notations used here are standard: r, θ, ϕ are spherical coordinates, $\mathbf{r}$ is the radius-vector joining any point $P(r,\theta,\phi)$ to the geocentre, n and m are the degree and order, respectively, of the spherical harmonic function $Y_n^m(\theta,\phi) = P_n^m(\cos\theta)\, e^{im\phi}$. Integration is performed from the Earth's centre $r = 0$ to the Earth's surface $r = R$.

Munk and MacDonald (1960) assume that $\delta\,\mathrm{tr}\,\mathfrak{Z} = 0$ for the deformation caused by the body force potential arising from the Earth's rotation. Rochester and Smylie (1974) argue, by means of Eq. (2), that this assumption is wrong for a compressible Earth model, and note that the variations of the components of the inertia tensor produced by rotational deformation can be expressed as

$$\delta\mathfrak{Z}_{ij} = \frac{kR^5}{3G}(\omega_i\omega_j - \omega^2\delta_{ij}) + \frac{1}{3}\,\delta\,\mathrm{tr}\,\mathfrak{Z}\,\delta_{ij} \ ,$$

where k is the Love number of degree 2 describing the elastic yielding of the Earth due to the quadrupole term in the centrifugal potential. More precisely, k is defined in such a way that a disturbing potential U of degree 2 would lift a fluid equipotential surface covering the globe, subject to a gravitational acceleration g, by (1+k) U/g relative to the Earth's centre, and leave it an equipotential surface. The additional term, $1/3\,\delta \mathrm{tr}\,\mathfrak{Z}$, is caused by the radial term in the centrifugal potential which, for a compressible Earth model, must lead to an overall contraction if the Earth is spinning down, and to a global dilatation if the Earth is spinning up. Hinderer and Legros (this Vol.) find a total decrease of about 500 m of the Earth's radius for an increase of l.o.d. by some 10^4 s, a result which is in agreement with the estimations of Denis and Varga (this Vol.) comprised between 125 and 1250 m. Hence, using Eq. (2), we can estimate the order of magnitude of the decrease of $\mathrm{tr}\,\mathfrak{Z}$ during the Phanerozoic [for which $\delta(\mathrm{l.o.d.}) \cong 10^4$ s] by assuming the Earth to be homogeneous ($\varrho = 5500\ \mathrm{kg\,m^{-3}}$) and the displacement to be proportional to the radius, i.e. $u_0^0(r) = \alpha r$, with $\alpha \cong \delta R/R \cong 8\times 10^{-5}$. This estimate yields $\delta \mathrm{tr}\,\mathfrak{Z} = 4.6\times 10^{34}\ \mathrm{kg\,m^2}$.

On the other hand, for a radial deformation δR of a spherical Earth model of radius R, the changes in inertia are isotropic, i.e., $\delta\mathfrak{Z}_{11} = \delta\mathfrak{Z}_{22} = \delta\mathfrak{Z}_{33} = 1/3\,\delta \mathrm{tr}\,\mathfrak{Z} = \delta I$, and thus amount to

$$\delta I = \frac{8\pi}{3}\,\delta \int_0^R \varrho(r)\, r^4\, dr \;.$$

For a homogeneous Earth model of mean density $3M/4\pi R^3$ in the undeformed state, of mean density $3M/4\pi(R+\delta R)^3$ in the deformed state, the change in the inertia moment is (assuming $\delta R = 500$ m)

$$\delta I \cong \frac{4}{5}\,\frac{\delta R}{R}\, MR^2 \cong 1.5\times 10^{34}\ \mathrm{kg\,m^2} \;.$$

Within truncation errors, the latter value is in agreement with the value derived from Eq. (2), which yields $\delta I = 1/3\,\delta \mathrm{tr}\,\mathfrak{Z} \cong 1.5\times 10^{34}\ \mathrm{kg\,m^2}$. It is possible to take account of the Earth's actual heterogeneity by using for the inertia coefficient the observed value 0.331 instead of the value 0.4 corresponding to a homogeneous model. Therefore, the value, given above, of the change in the inertia moment during the Phanerozoic should be multiplied by the factor $0.331/0.4 \cong 0.83$, yielding $\delta I \cong 1.2\times 10^{34}\ \mathrm{kg\,m^2}$. Let the symbols M and R stand for the Earth mass and present average radius, respectively: $M = 5.973\times 10^{24}$ kg, $R = 6.371\times 10^6$ m, hence $MR^2 = 2.424\times 10^{38}\ \mathrm{kg\cdot m^2}$. Thus, for a variation of δP seconds of the present value of the Earth's rotation period, P, the change of the moment of inertia caused by the radial term in the centrifugal potential, is approximately given by

$$[\delta I]_{\mathrm{radial}} = 1/3\,[\delta \mathrm{tr}\,\mathfrak{Z}]_{\mathrm{radial}} = [\delta A]_{\mathrm{radial}} = [\delta B]_{\mathrm{radial}} = [\delta C]_{\mathrm{radial}}$$
$$= -1.2\times 10^{30}\,\delta P = -5.0\times 10^{-9}\, MR^2\,\delta P \;. \qquad (3)$$

with the usual notations $A = \mathfrak{Z}_{11}$, $B = \mathfrak{Z}_{22}$, $C = \mathfrak{Z}_{33}$.

As mentioned above, changes in the trace of the inertia tensor vanish for the linear effects associated with the perturbing action of the second degree harmonic term in the centifugal potential. However, the changes of $\operatorname{tr}\mathfrak{Z}$ caused by non-linear effects need not necessarily vanish, and indeed seem not to vanish. Although a study of $\operatorname{tr}\mathfrak{Z}$ considering such non-linear terms in the perturbation approach appears to be lacking, it is interesting to note that the theory of relaxed equilibrium figures (Denis 1989), which implicitly assumes incompressibility, leads to a small, but non-zero expression for $\operatorname{tr}\mathfrak{Z}$, which stems from terms of order higher than the first in the surface flattening $f_s \cong 3\times 10^{-3}$ (cf. Denis and Varga, this Vol.). Specifically, for a variation of δP seconds of the present value of the Earth's rotation period, T, we find by means of the figure theory that

$$[\delta \operatorname{tr}\mathfrak{Z}]_{\text{quadrupolar}} \cong -2\times 10^{-10}\, MR^2 \delta P \ , \tag{4}$$

a value which happens to be about one hundredth of the corresponding change in the polar moment of inertia, $[\delta C]_{\text{quadrupolar}}$, and agrees with the fact that $\delta \operatorname{tr}\mathfrak{Z}/\delta C = 3\times O(f_s) \cong 0.01$. We conclude from this result that, at an accuracy level of 1%, we may in general safely neglect changes of $\operatorname{tr}\mathfrak{Z}$ associated with the quadrupolar part of the rotational deformation, but not the changes associated with the radial part of the rotational deformation.

3 Variations of the Polar Moment of Inertia and Related Quantities Caused by Changes of l.o.d.

For the present epoch, a fluctuation of δP seconds in the rotation period, P, leads to a change of the polar moment of inertia, C, given by

$$\begin{aligned}\delta C = [\delta C]_{\text{quadrupolar}} + [\delta C]_{\text{radial}} &\cong -(1.7\zeta + 0.5)\times 10^{-8} MR^2 \delta P \\ &\cong -(4.1\zeta + 1.2)\times 10^{30} \delta P \ ,\end{aligned} \tag{5}$$

where ζ is a time-dependent rheologic reduction factor. The variations, for the present epoch, of the related parameters, geometrical surface flattening (f_s), dynamic shape factor (J_2), dynamic oblateness (H), equatorial radius (a), are given by similar expressions:

$$\delta f_s \cong [\delta f_s]_{\text{quadrupolar}} \cong -7.8\times 10^{-8} \zeta \delta P \ , \tag{6}$$

$$\delta J_2 \cong [\delta J_2]_{\text{quadrupolar}} \cong -2.5\times 10^{-8} \zeta \delta P \ , \tag{7}$$

$$\delta H \cong [\delta H]_{\text{quadrupolar}} \cong -7.5\times 10^{-8} \zeta \delta P \ , \tag{8}$$

$$\delta a \cong [\delta a]_{\text{quadrupolar}} + [\delta R]_{\text{radial}} \cong -(0.17\zeta + 0.05)\delta P \,(\text{metres}) \ . \tag{9}$$

By definition, the reduction factor ζ is unity for an incompressible fluid Earth model. On geological time scales, the Earth behaves indeed nearly like a fluid and ζ should be close to 1, if we disregard the Earth's small but finite compressibility. On short and intermediate[3] time scales, the rotational distortion of the Earth's shape will be reduced to some extent by shear stress, and the factor ζ will be smaller than unity. An estimate of ζ for short time scales is provided by the ratio k/k_f of the actual Love number, k, to the fluid, or secular, Love number, k_f. For rotational deformations in the range of hours to days, this ratio is about 0.3 (Munk and MacDonald 1960). For increasing durations of the applied stimulus, ζ should come closer and closer to 1. We suggest to use the following simple relaxation law for mantle material, except in the upper cool part of the lithosphere which may retain shear strength over hundreds of millions of years (cf. Denis and Varga, this Vol.):

$$\zeta(t) = 1 - \beta \exp\left(-\frac{t}{\tau}\right) . \tag{10}$$

Here, the variable t denotes the time elapsed since the onset of the stimulus, $\beta = 1 - k/k_f \cong 0.7$, and the parameter τ is a typical relaxation time, measured by the ratio η/μ of a typical dynamic viscosity, η, to a typical shear modulus, μ. We take $\eta \cong 2\times10^{21}$ Pa s and $\mu \cong 2\times10^{11}$ Pa as representative values of mantle rocks, and thus end up with a relaxation time τ of 10^{10} s, or roughly 300 years.

Considering that presently P increases on the average by 2 ms every century, i.e.,

$$[dP/dt]_{secular} \cong 6.3\times10^{-13} , \tag{11}$$

we find the secular rate of change of C due to variations of l.o.d. by combining Eq. (5) with Eq. (11) and setting $\zeta = 1$. The result is

$$[dC/dt]_{secular\,\delta l.o.d.} \cong -3.4\times10^{18}\,kg\,m^2\,s^{-1} \cong -1.1\times10^{26}\,kg\,m^2\,yr^{-1} , \tag{12}$$

as well as a concomitant change of J_2,

$$[dJ_2/dt]_{secular\,\delta l.o.d.} \cong -1.6\times10^{-20}\,s^{-1} \cong -5.0\times10^{-13}\,yr^{-1} . \tag{13}$$

The latter value corresponds to the instantaneous present rate of secular change. The mean rate of change of J_2 over the last 500 million years, based on l.o.d. data derived from paleontological clocks (cf. Denis and Varga, this Vol., and Table 1) and attributed to tidal friction, is

$$\langle dJ_2/dt\rangle_{tidal\ friction} \cong -2.1\times10^{-20}\,s^{-1} \cong -6.6\times10^{-13}\,yr^{-1} . \tag{14}$$

[3] According to Scheidegger (1963, p. 160), it appears that the behaviour of the material of the Earth's crust and mantle is substantially different in the three following time ranges: It is essentially that of an elastic solid with some strength limit in the *short range*, i.e., up to 4 h, with a typical time scale of 3 s. From 4 h to 15000 years (typical time scale: 1 year), the *intermediate range*, the behaviour may be suitably characterized as that of a body with elastic afterworking and some undetermined strength limit, i.e., by a firmoviscous solid. In the *long range*, i.e., for durations longer than 15000 years, the behaviour is that of a viscoelastic liquid with some low elastic yield stress in the cooler upper parts of the Earth's lithosphere. For geological time scales, which are typically of the order of 100 million years or longer, most of the tectonic stress is relaxed and the material may adequately be modelled as a viscous liquid.

Table 1. Values of some kinetic quantities of the Earth as a function of the period of rotation

Period(s)	Time BP	$10^6 f_s$	$10^6 J_2$	a(km)	c(km)	A/MR^2	C/MR^2
86164	0	3337	1072	6378.11	6356.82	0.330448	0.331523
84142	100	3543	1138	6378.55	6355.94	0.330426	0.331568
82119	200	3749	1204	6378.99	6355.07	0.330404	0.331612
80097	300	3956	1270	6379.43	6354.19	0.330382	0.331657
78074	400	4162	1336	6379.87	6353.31	0.330361	0.331701
76052	500	4368	1403	6380.31	6352.44	0.330339	0.331746

This rate of decrease averaged over 5×10^8 years is about 10% larger than the present rate of decrease due to tidal friction suggested by astronomical observations, which amounts to

$$[dJ_2/dt]_{\text{tidal friction}} \cong -(6.0\pm0.3)\times10^{-13}\,\text{yr}^{-1} \quad . \tag{15}$$

Note that the actual value to be used for Earth's rotational deceleration, $d\omega/dt$, over long time intervals ($\Delta t > 3\times10^3$ years), is still open to debate, although only little doubt exists that on long-term averages it has remained nearly constant for the last 500 million years. Applying linear regression on palaeontological data, Lambeck (1977) found the value $d\omega/dt = -(5.4\pm0.5)\times10^{-22}\,\text{rad}\cdot\text{s}^{-2}$. The latter, which the authors favour as the actual rate of tidal spindown on geological time scales, does not differ significantly from Brosche's (1987) value $-5.6\times10^{-22}\,\text{rad}\cdot\text{s}^{-2}$. However, it seems to be distinct from the present-day tidal deceleration value $-(6.4\pm0.3)\times10^{-22}\,\text{rad}\cdot\text{s}^{-2}$ given by Burša (1987). The latter is in essential agreement with the value $-6.07\times10^{-22}\,\text{rad}\cdot\text{s}^{-2}$ obtained from recent ranging to artificial satellites (Christodoulis et al. 1988), or with the Earth's spindown value $-5.98\times10^{-22}\,\text{rad}\cdot\text{s}^{-2}$ derived from recent lunar ranging (Newhall et al. 1986). Thus, there exists presently a difference of about 10% between short-term and long-term observations of $[d\omega/dt]_{\text{secular}}$, but it is not quite clear yet whether this difference is actually significant or not.

4 Some Causes of Secular Change of the Polar Moment of Inertia

The Earth's inertia tensor, and in particular the moments of inertia, undergo changes on time scales which range from a fraction of a second to several billion years. These changes are associated with deformations of the Earth caused by both external forces and internal processes which result in a redistribution of mass. Here, we shall be interested mainly in changes with time of the moment of inertia, C, about the polar axis, that is in the quantity dC/dt. Among the possible causes that give rise to secular changes of C, we may consider the following phenomena which are roughly ordered according to decreasing time scales: (1) tidal friction; (2) volume changes in response to a number of possible causes, including thermal effects and change in the gravitational constant; (3) core growth and chemical fractionation processes; (4) polar wandering and plate motions; (5) glacial loading and unloading; (6) large-scale motions in the Earth's fluid parts; (7) straining of the Earth by long-period tides with zonal symmetry; (8) magnetohydrodynamic and magnetoelatic interactions.

The Earth as a whole is orbiting and spinning in the gravity fields of the Moon and the Sun, and the ensuing tidal motions are damped by frictional and non-linear processes occurring inside the Earth's bulk, within the oceans and atmosphere, and on the boundaries. Although these frictional effects are small on short time scales, energy supplied by rotation and gravitation is being dissipated continuously. Thus, the cumulative effect of tidal friction becomes a very important factor of evolution if very long time intervals are considered (cf. Denis and Varga, this Vol.).

Tidal friction builds up torques which redistribute angular momentum between the Earth's crust and mantle and other parts of the Earth-Moon-Sun system. This results in the long run in a net transfer of angular momentum from the spinning Earth to orbital motion. As a consequence, Earth rotation is slowed down and the centrifugal force field is decreased, causing the Earth's shape to become less flattened and the polar moment of inertia to decrease with time. A typical time scale associated with tidal friction is about 2 billion years. Denis (1986) has computed plausible values of the principal moments of inertia and related kinetic and geodynamic parameters in the geological past. Table 1 shows some of his results interpolated for a set of values of the sidereal rotation period, P, throughout the Phanerozoic Aeon. These data are based on the Clairaut-Laplace-Lyapunov theory of equilibrium figures.

It seems to make little sense to calculate the changes in the inertia tensor that may be associated with mass transfer due to plate tectonic motion and mantle convection, while neither the mechanism driving the plates nor the mantle flow at depth is known (Lambeck 1980, p 93). Absolute motion of plates has been investigated more than a decade ago (Minster and Jordan 1978; see also Moritz and Mueller 1987, p. 554), but does not allow to reliably estimate the rate of change of the inertia tensor caused by tectonic processes. A detailed picture of the dynamic state of the mantle which is emerging slowly by means of seismic tomography may change this situation in the coming years.

Denis (1985, 1986) has discussed the possibility that core growth would be going on at the present time at a sufficient rate to contribute significantly to time changes in the moments of inertia. Again, the mechanisms of core formation and core growth are too speculative to make calculations of dC/dt very meaningful. Most authors believe that core formation was nearly completed during the early stages of the formation of the Earth as a whole (Oversby and Ringwood 1971), and that the Earth's rotational history was ruled, throughout geological times, by tidal friction, not by segregation processes. Nevertheless, the possibility of such segregational processes should be borne in mind when discussing long-term changes of the Earth's moments of inertia. Birch (1965) has estimated that the average moment of a hot, undifferentiated Earth would be around 9.09×10^{37} kg$\cdot$m^2, that for a cold, undifferentiated Earth would be about 8.79×10^{37} kg$\cdot$m^2, values which are respectively some 13% and 10% larger than the present value, 8.02×10^{37} kg$\cdot$m^2. It is generally admitted, in theories starting with an undifferentiated proto-Earth, that core formation was nearly completed in 100 to 500 million years after the formation of the Earth as a planet. If this is so, the early average rate of change of C must have been comprised roughly between $+1 \times 10^{29}$ and $+1 \times 10^{28}$ kg$\cdot$m$^2\cdot$yr^{-1}, leading to an acceleration of Earth rotation which must have outweighed tidal deceleration by a factor of more than 100. Again, this idea of an early spin-up due to core formation may be considered somewhat speculative, because we cannot rule out the possibility that the

Earth had been differentiated right from the start (see, e.g., Anderson and Hanks 1972).

Many authors have investigated the possiblity of large volume changes and concomitant changes in the moment of inertia caused by a secular decrease of the gravitational constant, G, postulated by Dirac (1937). Reviews of this hypothesis, with references to the original papers, may be found in the works of Dicke (1964), Jordan (1966), Runcorn (1969), and Carey (1976). Birch (1968) has investigated the consequences of a possible, though speculative, decrease of G over geological times, and has concluded that a *plausible upper limit* to the variation of the Earth's radius is about 100 km, or 0.016 R. Such a variation would give rise to a secular increase of C by the rate $[dC/dt]_{cosmological} \cong +4\times10^{26}\ kg\cdot m^2\cdot yr^{-1}$. Yabushita (1986) has discussed the hypothesis of a variable G, and found that a value (dG/dt)/G comprised between -1.4×10^{-11} and -3.3×10^{-11} per year is compatible with the observed lunar accelerations. Runcorn (1964) suggested that $[dC/dt]_{cosmological}$ was comprised between $+3.6\times10^{27}$ and $+3.4\times10^{28}\ kg\cdot m^2\cdot yr^{-1}$, but did not rule out the possibility that G did in fact not vary at all, thus making $[dC/dt]_{cosmological} = 0$. A recent discussion about this topic has been made by Bretterbauer (1987). We believe that the whole matter concerning cosmological variations of the gravitational constant is still too speculative to consider it as a plausible factor causing variations of the inertia moment but, of course, such a possibility should be kept in mind.

Presently, besides tidal friction, only post-glacial rebound is considered to bring about significant changes of the moments of inertia on time scales longer than a few decades. The phenomenon of post-glacial rebound, and the related geophysical problem of the variation with depth of the mantle viscosity, has been investigated recently by a number of authors (Nakiboglu and Lambeck 1980; Peltier 1983, 1985, 1986; Sabadini and Peltier 1981, a.o.), as a consequence of constraints becoming available from data contributed by the International Latitude Service, by the Bureau International de l'Heure, by lunar laser ranging, and lastly by the geodetic satellite LAGEOS.

Changes of the moments of inertia occur also on time scales ranging from several hours to several decades (Rochester 1984). They are caused by deformations due to earthquakes (Anderson 1974), tides (Yoder et al. 1981), atmospheric and oceanic loading (Cazenave 1986). Significant changes of the angular momentum of the Earth's crust and mantle, and concomitant feedback changes in the polar moment of inertia (see below, Sect. 5), may also arise, e.g., from circulations in the atmosphere (Hide 1984; Brosche and Sündermann 1985), from solar wind effects (Djurovic and Paquet 1988; Mazzarella and Palumbo 1988), from core-mantle interactions (Rochester 1970).

5 Long-Term Variations of the Geopotential Coefficient J_2

Fits of the LAGEOS tracking data covering a period of several years to sophisticated dynamical models leave significant residual signatures in orbital parameters which reflect long-term variations in the zonal gravitational coefficient J_2 (Yoder et al. 1983). According to Burša (1987), the most recent determination of the rate of these variations yields

$$[dJ_2/dt]_{LAGEOS} = -(2.8\pm0.3)\times10^{-11}\ yr^{-1}\ . \qquad (16)$$

This rate is about 50 times in excess of the rate of decrease of J_2 which can reasonably be expected from tidal friction, cf. Eq. (15) (Denis 1986).

Yoder et al. (1983) derive from the LAGEOS value of dJ_2/dt a non-tidal rate of decrease of the largest principal moment of inertia.

$$[dC/dt]_{LAGEOS} = -(4.5 \pm 0.5) \times 10^{27}\ \mathrm{kg\,m^2\,yr^{-1}}\ , \tag{17}$$

which again is about 50 times larger than the rate of decrease of C associated with tidal friction.

The general definition of J_2:

$$J_2 = \frac{C-(1/2)(A+B)}{Ma^2}\ , \tag{18}$$

where $A, B, C\,(A \leq B < C)$ are the principal moments of inertia, M is the mass and a is the equatorial radius of the Earth, allows us to write

$$J_2 = \frac{3C - \mathrm{tr}\mathfrak{Z}}{2Ma^2}\ . \tag{19}$$

Yoder et al. (1983), followed in that respect by a number of other authors (f.i. Burša 1987), state that a variation, δJ_2, of J_2 is directly proportional to the concomitant variation, δC, in the polar moment of inertia C. They obtain this result by interpreting the quantity a in Eq. (19) as a constant scale factor, and by noting that for deformations that preserve volume such as tidal strain or mass redistribution over the Earth's surface, the trace of the inertia tensor, $\mathrm{tr}\mathfrak{Z} = A+B+C$, remains constant as well. Yoder et al. (1983) put the quantity "a" equal to the Earth's mean radius, R, whereas Burša (1987) takes for "a" the *present* value of the equatorial radius, $a_0 = 6378140$ m. However, to achieve consistency at any epoch in the definitions of the inertia factor, $J_2/H = C/Ma^2$, or of the very definition of J_2 by means of the Earth's internal mass distribution (see, e.g., Teisseyre et al. 1989), it is necessary to consider "a" as the Earth's equatorial radius which varies along with J_2 and the inertia moments. Thus, neglecting $\delta\mathrm{tr}\mathfrak{Z}$ (cf. Sect. 2), we obtain from Eq. (19):

$$\delta J_2 = \frac{3}{2}\frac{\delta C}{Ma^2} - 2J_2\frac{\delta a}{a}\ . \tag{20}$$

The second term on the right hand side amounts to about 0.2% of the first term, and thus Yoder's result,

$$\delta J_2 \cong \frac{3}{2}\frac{\delta C}{Ma^2}\ ,$$

holds as a quite fair approximation.

Assuming then conservation of angular momentum in the Earth, Yoder et al. (1983) derive a non-tidal acceleration of the Earth's rotation which amounts to

$$[d\omega/dt]_{nontidal} = -(\omega/C)\,dC/dt = (1.3 \pm 0.14) \times 10^{-22}\ \mathrm{rad\,s^{-2}}\ . \tag{21}$$

Thus, the total observed change of rotation is at present (Burša 1987)

$$d\omega/dt = -(5.1 \pm 0.33) \times 10^{-22}\, \mathrm{rad\, s^{-2}} \ . \tag{22}$$

According to Yoder et al. (1983), the rate of decrease of J_2 obtained from the LAGEOS data is consistent with both historical observations of the non-tidal acceleration of Earth rotation and models of viscous rebound of the solid Earth following the removal of large polar ice caps during the last deglaciation. Peltier (1983, 1986) argues that the LAGEOS data require a lower-mantle viscosity comprised between 2.7×10^{21} Pa s and 4.4×10^{21} Pa s, and thus reinforce the view that the Earth's mantle has no large-scale contrasts in its radial variation of viscosity. Peltier (1986) notes that this inference is of crucial importance in understanding the thermal convection pattern which is responsible for plate tectonics.

We believe that the LAGEOS data may indeed be used to set constraints on the variation of viscosity with depth, via the theory of isostatic uplift occurring as a delayed response to the melting of huge polar ice caps, some ten to six thousand years ago. It is interesting to note, moreover, that the melting of polar ice leads, without delay, to an *increase* of the polar moment of inertia as a consequence of the redistribution of large water masses towards lower latitudes, but that the uplift of the depressed crustal layers leads, with delay, to a *decrease* of the polar moment of inertia. Nevertheless, before using the LAGEOS data for quantitative estimates of lower mantle viscosity, we must make sure that the dynamic modelling includes all important effects. For instance, Yoder and Ivins (1985) suggested that the retreat of mid-latitude glaciers, as we observe it today, should cause an increase in J_2, comparable to the one caused during the major deglaciation period. The predicted value for dJ_2/dt from the melting of present-day glaciers is $+3.5 \times 10^{-11}$ per year, whereas the expected value for dJ_2/dt caused by post-glacial crustal uplift phenomena is near $-3 \times 10^{-11}\,\mathrm{yr}^{-1}$ for a lower mantle viscosity of 2×10^{21} Pa·s or 5×10^{23} Pa·s, and reaches a maximum value of $-6 \times 10^{-11}\,\mathrm{yr}^{-1}$ for a lower mantle viscosity of some 1.5×10^{22} Pa·s. The combination of these two effects would yield the observed rate of change of J_2 only if the contribution from mantle rebound is near its maximum possible value (Mignard 1986).

In their original investigation, Yoder et al. (1983) fit the LAGEOS tracking data to sophisticated dynamical models which include the tidal deformations and the associated variations of Earth rotation and UT1 (Yoder et al. 1981). However, as far as we understand, they do not consider rotational deformation produced as a feedback effect to a change of l.o.d., which we may evaluate by means of Eq. (7).

The tidally induced elastic change in the polar moment of inertia is given by (Yoder et al. 1981)

$$\delta C = k \frac{M'}{M} \left(\frac{a}{r'} \right)^3 M a^2 \left(\tfrac{1}{3} - \sin^2 \delta' \right) \ , \tag{23}$$

where M', r', δ' are respectively the mass, distance, and declination of the perturbing body (Moon or Sun). Putting $\delta' = 0$, this amounts for the Moon to $5.4 \times 10^{-9}\, M R^2$. The largest change in l.o.d. is produced by Mf (cf. Lambeck 1980, p. 140), and amounts to 0.34 ms. By Eq. (5), with $\zeta = 0.3$, the corresponding change in C is, in absolute value, $[\delta C]_{fb} = 3.4\times10^{-12}\,MR^2$. Thus, the feedback effect amounts here to 0.063% of the direct effect. By Eq. (7), the corresponding feedback effect on J_2 has an amplitude of 2.6×10^{-12} for a period of 13.66 days, yielding a periodic $|dJ_2/dt|_{Mf}$ amounting to $1.4\times10^{-17}\,s^{-1}$. The latter value is about 16 times larger than the residual signature in the LAGEOS data, $|dJ_2/dt|_{LAGEOS} \cong 8.9\times10^{-19}\,s^{-1}$. Similarly, seasonal, decade and irregular changes in l.o.d., with magnitudes comprised between one and several milliseconds, will lead to values of $|dJ_2/dt|$ around $10^{-11}\,yr^{-1}$, and must be separated from the signal corresponding to post-glacial viscous rebound (Denis 1987).

6 Conclusion

In this paper, we examined a number of phenomena, acting at a variety of time scales, which contribute in producing changes in time of the polar moment of inertia, C. Some effects can presently be estimated with some confidence: tidal friction effects, which amount to $dC/dt = -(1.1\pm0.1)\times10^{26}\,km\cdot m^2\cdot yr^{-1}$; tidal straining effects for individual oceanic and body tides; feedback effects caused by observed changes of the Earth's rotation speed. The effects of other phenomena are presently much more difficult to assess and, no doubt, will give rise to many investigations in the coming years.

There is little doubt that post-glacial rebound and melting of present-day glaciers contribute to changes in the inertia tensor and in the gravity field which space-geodetic methods are capable to detect. However, very precise modelling has to be done before the observations can provide us with reliable information about the viscosity structure of the mantle. Also, several decades of observations will be necessary to decide how much of the variation is secular, and how much of the variation is due to long-period phenomena.

Acknowledgments. The authors wish to thank most heartily the organizers of this meeting, as well as the Zentrum für interdisziplinäre Forschung der Universität Bielefeld (ZiF) and its staff. The material conditions provided were ideal for having stimulating and fruitful discussions and exchanging useful information. P. V. is grateful to the Alexander von Humboldt Stiftung, and C. D. acknowledges with thanks financial support from the Fonds National de la Recherche Scientifique (Brussels). One of the editors (P.B.) reviewed our original manuscript quite thoroughly and suggested significant improvements.

References

Anderson DL (1974) Earthquakes and the rotation of the Earth. Science 186:49–50
Anderson DL, Hanks TC (1972) Formation of the Earth's core. Nature (Lond) 237:387–388
Anderson DL, Minster JB (1979) The frequency dependence of Q in the Earth and implications for mantle rheology and Chandler wobble. Geophys JR Astr Soc 58:431–440
Birch F (1965) Energetics of core formation. J Geophys Res 70:6217–6221
Birch F (1968) On the possibility of large changes in the Earth's volume. Phys Earth Planet Int 1:141–147
Bretterbauer K (1987) Geodesy and Cosmogony. Proc winter seminar on geodynamics '87 "Rotating Earth". Geod Geophys Res Inst Hungarian Acad Sci, Sopron, pp 257–278

Brosche P (1987) Tidal friction and the Earth's rotation. In: Holota P (ed) Proc Int Symp Figure and Dynamics of the Earth, Moon, and Planets. Prague, September 1986

Brosche P, Sündermann J (1985) The Antarctic circumpolar current and its influence on the Earth's rotation. Dtsch Hydrogr Z 38:1–6

Burša M (1987) Final report of IAG Special Study Group 5-99. Tidal friction and the Earth's rotation. Proc XIX Gen Ass IUGG, Vancouver, August 1987

Carey SW (1976) The expanding Earth. Elsevier, Amsterdam

Cazenave A (ed) (1986) Earth rotation: solved and unsolved problems. Reidel, Dordrecht

Christodoulis DC, Smith DE, Williamson RG, Klosko SM (1988) Observed tidal braking in the Earth/Moon/Sun system. J Geophys Res 93 (B6):6216–6236

Denis C (1985) Core formation and rotation of the Earth in the geological past. Proc Journées Luxbg Géodyn 60:12–17

Denis C (1986) On the change of kinetical parameters of the Earth during geological times. Geophys J R Astr Soc 87:559–568

Denis C (1987) On time changes of the gravitational quadrupole moment J_2. Terra Cognita 7:498

Denis C (1989) The hydrostatic figure of the Earth. In: Teisseyre R (ed) Physics and evolution of the Earth's interior. Vol 4, Gravity and Low-Frequency Geodynamics, Chap. 3. PWN, Warszawa and Elsevier, Amsterdam, pp 111–186

Dicke RH (1964) The theoretical significance of experimental relativity. Blackie, London

Dirac PAM (1937) The cosmological constants. Nature (Lond) 139:323

Djurovic D, Paquet P (1988) Short-term variation of the Earth's rotation and the solar activity. In: Babcock AK, Wilkins GA (eds) The Earth's rotation and Reference frames for geodesy and geodynamics. IAU, pp 353–358

Hide R (1984) Rotation of the atmospheres of the Earth and planets. Phil Trans R Soc Lond A313:107–121

Höpfner J (1983) Umfassende Analyse der Breitenbestimmungen des geodätisch-astronomischen Observatoriums Potsdam. Ph.D. thesis TU, Dresden

Jeffreys H (1970) The Earth: its origin, history and physical constitution (5th edn.) Univ Press, Cambridge

Jeffreys H (1972) Creep in the Earth and planets. Tectonophys 13:569–581

Jordan P (1966) Die Expansion der Erde. Vieweg, Braunschweig (Updated English translation: The expanding Earth. Pergamon, Oxford 1971

Lambeck K (1977) Tidal dissipation in the oceans: astronomical, geophysical and oceanographic consequences. Phil Trans R Soc Lond A287:545–594

Lambeck K (1980) The Earth's variable rotation: geophysical causes and consequences. Univ Press, Cambridge

Mazzarella A, Palumbo A (1988) Earth's rotation and solar activity. Geophys J 97:169–171

Mignard F (1986) Tidal and non-tidal acceleration of the Earth's rotation. In: Cazenave A (ed) Earth rotation: solved and unsolved problems. Reidel, Dordrecht, pp 93–110

Minster JB, Jordan TH (1978) Present-day plate motions. J Geophys Res 83:5331–5354

Moritz H, Mueller II (1988) Earth rotation – theory and observation. Ungar, New York

Munk WH, McDonald GJF (1960) The rotation of the Earth. Univ Press, Cambridge

Nakiboglu SM, Lambeck K (1980) Deglaciation effects on the rotation of the Earth. Geophys J R Astr Soc 62:49–58

Newhall XX, Williams JG, Dicky JO (1986) Earth rotation from lunar laser ranging. JPL Geod Geophys Preprint 153, Jet Propul Lab, Dec 1986

Oversby VM, Ringwood AE (1971) Time of formation of the Earth's core. Nature (Lond) 234:463–465

Peltier WR (1983) Constraint on deep mantle viscosity from LAGEOS acceleration data. Nature (Lond) 304:434–436

Peltier WR (1985) New constraints on transient lower mantle rheology and internal mantle buoyancy from glacial rebound data. Nature (Lond) 318:614–617

Peltier WR (1986) Slow changes in the Earth's shape and gravitational field: constraints on the glaciation history and internal viscoelastic stratification. In: Anderson AJ, Cazenave A (eds) Space geodesy and geodynamics. Academic Press, London, pp 75–109

Rochester MG (1970) Core-mantle interactions: geophysical and astronomical consequences. In: Mansinha L, Smylie DE, Beck AE (eds) Earthquake displacement fields and the rotation of the Earth. Reidel, Dordrecht, pp 136–148

Rochester MG (1984) Causes of fluctuations in the rotation of the Earth. Phil Trans R Soc Lond A313:95–105

Rochester MG, Smylie DE (1974) On changes in the trace of the Earth's inertia tensor. J Geophys Res 79:4948–4951

Runcorn SK (1964) Changes in the Earth's moment of inertia. Nature (Lond) 204:823–825

Runcorn SK (ed) (1969) The application of modern physics to the Earth and planetary interiors. Wiley, London

Sabadini R, Peltier WR (1981) Pleistocene deglaciation and the Earth's rotation: implications for mantle viscosity. Geophys J R Astron Soc 66:553–578

Scheidegger AE (1963) Principles of Geodynamics (2nd edn). Springer, Berlin Heidelberg

Schwiderski EW (1980) On charting global ocean tides. Rev Geophys Space Phys 18:243–268

Teisseyre R, Denis C, Majewski E, Zielinski J (1989) The Earth's gravity field. In: Teisseyre R (ed) Physics and evolution of the Earth's interior. Vol 4, Gravity and low-frequency geodynamics, Chap 3. PWN, Warszawa and Elsevier, Amsterdam, pp 1–77

Yabushita S (1986) Hypothesis of variable G and the secular accelerations of the Sun and the Moon. Earth Moon Planets 34:139–148

Yoder CF, Ivins ER (1985) Changes in the Earth's gravity field from Pleistocene deglaciation and present day glacial melting. AGU Spring Meeting, Baltimore 1985

Yoder CF, Williams JG, Parke ME (1981) Tidal variations of Earth rotation. J Geophys Res 86 (B2):881–891

Yoder CF, Williams JG, Dickey JO, Schutz BE, Eanes RJ, Tapley BD (1983) Secular variation of Earth's gravitational harmonic J_2 coefficient from Lageos and nontidal acceleration of Earth rotation. Nature (Lond) 303:752–762

Zschau J (1977) Phase shifts of tidal sea load deformations of the Earth's surface due to low viscosity layers in the interior. Proc 8th Int Symp Earth Tides, Bonn. Schweizerbart'sche Verlagsbuchhandlung, Stuttgart, pp 372–398

Zschau J (1978) Tidal friction in the solid Earth: loading tides versus body tides. In: Brosche P, Sündermann J (eds) Tidal friction and the Earth's Rotation. Springer, Berlin Heidelberg New York, pp 62–94

Zschau J (1983) Rheology of the Earth's mantle at tidal and Chandler wobble periods. Proc 9th Int Symp Earth Tides, Bonn. Schweizerbart'sche Verlagsbuchhandlung, Stuttgart, pp 605–630

Zschau J (1986) Tidal friction in the solid Earth: constraints from the Chandler wobble period. In: Anderson AJ, Cazenave A (eds) Space geodesy and geodynamics. Academic Press, London, pp 315–344

Global Consequences of the Tidal Secular Deceleration for the Solid Earth and its Fluid Core

J. Hinderer and H. Legros

1 Introduction

The problem of the secular deceleration of the Earth has been intensively studied in astronomy and geophysics in the frame of the evolution of the Earth/Moon/Sun system. In general, these studies are devoted to the different processes like mantle anelasticity or oceanic dissipation that can account for the braking of the Earth rotation; an important effort has also been undertaken in the observations to obtain the most accurate determination of this braking rate. Our contribution will mainly concern the various geodynamic consequences due to the secular deceleration. First we will consider the response of the fluid core when the mantle slows down. Then we investigate the radial and zonal deformation and related geometric and dynamic effects caused by changes in the rotational potential. The possibilities of resonant amplification in the Earth and core wobbles caused by forcing mechanisms related to the orbital motion of the Earth around the Sun supposed to be stable over geologic times are pointed out. Finally, possible implications of the secular change in the spin rate on the magnetic scaling laws, and especially on the history of the Earth magnetic field, are given.

2 Secular Deceleration of the Core-Mantle System

Usually, the problem of the tidal braking of the Earth is treated by considering the following axial component of the Euler equations for conservation of angular momentum (e.g., Lambeck 1980):

$$\frac{d(C\Omega)}{dt} = CF_3 \qquad (1)$$

where C is the polar moment of inertia of the whole Earth, Ω the axial angular velocity and $F_3 < 0$ [of dimension $(\text{time})^{-2}$] any given braking torque. If the changes in inertia are neglected in Eq. (1), specifying F_3 leads to the rate of secular deceleration $d\Omega/dt$. When there is a fluid core, one has to simultaneously consider one Euler equation for the core and one for the mantle (or one for the whole Earth, i.e., for the

Laboratoire de Géodynamique, Institut de Physique du Globe, 5, rue René Descartes, F-67084 Strasbourg Cedex, France.

system core+mantle) which are coupled by core-mantle interaction torques (e.g., Hinderer et al. 1982):

$$\dot{\omega}_3 + \frac{C^c}{C}\dot{\omega}_3^c = \frac{C^m}{C}F_3$$
$$\dot{\omega}_3 + \dot{\omega}_3^c + \beta\omega_3^c = 0 \qquad (2)$$

where ω_3 denotes now the angular velocity of the mantle and ω_3^c the relative core angular velocity with respect to the mantle; C^m is the mantle moment of inertia and β [of dimension $(\text{time})^{-1}$] stands for a (linear) friction process at the core – mantle boundary (CMB) of viscous or electromagnetic origin. F_3 can model either a tidally induced braking torque on the mantle caused by an anelastic rheology or torques exerted on the solid mantle from the oceans or the atmosphere (see Brosche and Sündermann 1978, 1982).

When the core is perfectly decoupled from the mantle ($\beta = 0$), it is easy to show that the core rotates uniformly at its initial value (no braking) while the mantle deceleration $d\omega_3/dt$ is proportional to F_3.

Assuming initial co-rotation, i.e., $\omega_3^c = 0$ for $t = 0$ and denoting Ω_0 this value, the integration of the system (2) is straightforward and leads to the following solutions (assuming F_3 = constant):

$$\omega_3 = \Omega_0 + \frac{C^m}{C}F_3 t - \frac{C^m C^c}{\beta C^2}F_3\left[\exp\left(-\frac{C\beta t}{C^m}\right) - 1\right] \qquad (3\,a)$$

$$\omega_3^c = \frac{C^m}{\beta C}F_3\left[\exp\left(-\frac{C\beta t}{C^m}\right) - 1\right] \qquad (3\,b)$$

We clearly see that the mantle rotation rate is linearly decreasing in time [second term in (3 a)]; because of the ratio $C^m/C \approx 0.9$, we see that this rate is slightly smaller (by about 10%) for a given braking mechanism F_3 than in the frictionless case. The main consequence of the friction between core and mantle is to decelerate the fluid core at practically the same rate as the mantle, the only difference being the small perturbation (3 b). By the way, it is interesting to notice that ω_3^c is always positive, meaning eastward drift of the core with respect to the mantle (for a similar conclusion, see Bondi and Lyttleton 1948). The upper bound of this eastward drift is:

$$\lim_{t\to\infty} \omega_3^c = -\frac{C^m}{C\beta}F_3 \qquad (4)$$

which is of the order of 10^{-10} rd. s^{-1} (hence comparable to the mean value of the westward drift of the Earth magnetic field) for times t large with respect to the relaxation time $C^m/C\beta$ which is about 30000 years from the usual estimate of (laminar) viscous friction (e.g., Rochester et al. 1975) and for a braking term $F_3 \approx -6\times10^{-22}$ rd. s^{-2}. Another consequence is that friction coupling leads to a small additional perturbation in the mantle rotation [third term in Eq. (3 a)]. If the braking torques can vary in time (see, e.g., Brosche and Hövel 1982), the differential velocity ω_3^c

would be also variable and could then play a role in the possible connection between geomagnetic reversals and tidal friction (Brosche 1981).

Although the core contributes for about 10% (ratio C^c/C) to the change in the kinetic energy of the whole system caused by the tidal braking, the energy really dissipated at the CMB by friction is $P = C^c\beta(\omega_3^c)^2 \approx 10^6$ W with the previous values of β and differential rotation rate of the core ω_3^c; for comparison, the power necessary to drive the geodynamo is about 10^9 W (Rochester et al. 1975) and the rate of dissipation related to mantle convection about 10^{11} W (Stacey 1969).

3 Geodynamic Consequences Due to the Change in the Rotational Potential

The volume potential Ψ caused by the rotational motion of the Earth is a spatial function that can be separated into a purely radial part and a degree n = 2 spherical harmonic part (see, e.g., Hinderer et al. 1982). The only contributions to the potential related to a change in the axial spin rate Ω are then:

$$\Psi(r, \theta, \lambda) = \Psi_{00}(r) + \Psi_{20}(r, \theta) = \frac{\Omega^2 r^2}{3} - \frac{\Omega^2 r^2}{3} \, \frac{3\cos^2\theta - 1}{2} \, . \tag{5}$$

The first term will cause an elastogravitational deformation of the Earth with radial symmetry, the second one a zonal deformation of degree n = 2 and order m = 0.

3.1 Radial Part

The system of equations (motion, gravitational potential) to be solved are the classical ones given by Alterman et al. (1959) with conditions of regularity at the origin and continuity of radial displacement, stress and perturbation in gravitational potential; more details can be found in Saito (1974), Mansinha and Shen (1976). In order to obtain an approximate solution for the radial deformation of the Earth, we have simply considered a spherical model supposed to be homogeneous in the equilibrium state, and of uniform compressibility and rigidity. After integration of the differential equation in $\mathrm{div}(\vec{u})$, the radial displacement $u_r(r)$ in the mantle can be written with the help of Bessel functions. We have then for the surface (r = a) and outer core (r = b) displacements:

$$u_r(a) = h_0 \frac{\Psi_{00}}{g} \qquad u_r(b) = n_0 \frac{\Psi_{00}}{g} \tag{6}$$

with the help of "Love numbers" of degree n = 0; g is the mean value of the surface gravity. Over geologic times, the Earth behaves like a fluid rather than an elastic body; therefore it is necessary to take the values of the Love numbers in (6) in the fluid limit (when the rigidity vanishes).

The change in time of the Earth radius caused by the tidal deceleration is:

$$\dot{u}_r(a) = \frac{2 h_{0f} a^2 \Omega}{3\,g} \dot{\Omega} \, . \tag{7}$$

For $d\Omega/dt<0$ (deceleration), (7) implies a reduction ($du_r/dt<0$) in the Earth radius, and similarly for the average core radius.

Numerically, setting $h_{0f}=0.218$, we found a reduction of about 470 m (meter) in the last 400 Ma, and 5.47 km if we extrapolate the deceleration rate $d\Omega/dt\approx-5.98$ 10^{-22} rd. s^{-2} (Christodoulidis et al. 1988) till 4 aeons. The change in the dimension of the core was found slightly smaller. This contraction of the Earth already mentioned by Stoneley (1924) may have several consequences:

a) Variation in the Inertia Tensor

Because of the radial symmetry of the deformation, the changes in inertia are isotropic (Rochester and Smylie 1974) and we have $\Delta A=\Delta B=\Delta C$. It must be noted that the contraction implies a secular decrease in the polar moment of inertia ($dC/dt<0$) leading to a small (second-order) additional accelerating feedback effect on the Earth rotation like the one caused by the zonal deformation (e.g., Denis 1986).

b) Variation in the Mean Absolute Gravity Value

Setting the radial dependence of the gravity $g(r)=GM/r^2$, where G is the constant of gravitation and M the mass, and assuming GM to be constant, any change in the Earth radius will cause a change in gravity (Lambeck 1973):

$$\frac{\Delta g(a)}{g(a)}=-\frac{2\Delta u_r(a)}{a}\,. \tag{8}$$

The contraction of the Earth previously mentioned would lead to an increase in gravity of about 150 milligal in about 500 Ma.

c) Generation of Stresses in the Lithosphere

As the radial part of the rotational potential is of the same order of magnitude than the zonal part, the induced stresses $\sigma_{ij}(\Psi_{00})$ will be comparable in amplitude but with a different spatial distribution than the ones caused by the harmonic part $\sigma_{ij}(\Psi_{20})$, which are sometimes invoked in the breakup of the original lithosphere (see Denis 1986; Denis et al. 1987).

3.2 Zonal Part

The problem of the deformation of the Earth caused by the zonal part of the potential relative to the tidal deceleration has been exposed in detail elsewhere (Denis 1986; Legros 1987). This harmonic deformation of the whole Earth leads to a radial displacement that can be written at the surface ($r=a$):

$$u_r(a,\theta)=h_f\frac{\Psi_{20}}{g}\,\frac{3\cos^2\theta-1}{2} \tag{9}$$

where h_f is a fluid volume Love number; a similar expression stands for the core-mantle boundary displacement $u_r(b)$ when exchanging h_f by another parameter n_f.

Let us briefly mention some related consequences. First there is a change in the flattening of the Earth; for instance, its expression at the surface is:

$$\varepsilon = \frac{u_r(a, \pi/2) - u_r(a, 0)}{a} = \frac{1}{2} h_f \frac{\Omega^2 a}{g} . \tag{10}$$

We immediately see that the tidal braking ($d\Omega/dt < 0$) leads to a decrease in the surface geometric flattening ε. For instance, setting $h_f = 1.937$, the value of ε has decreased in the last 500 Ma from 1/232 to the present value close to 1/299.8. A similar conclusion holds for the ellipticity of the core-mantle boundary ε^c.

Other important consequences of the deformation (9) are the decrease in time of the following parameters: dynamical ellipticities of the Earth and fluid core, precessional constant (see also Burša 1985) and gravitational quadrupole.

4 Resonance in the Paleowobbles

As previously for the axial spin rate components, the changes in direction (wobbles) of the rotational vectors of the Earth and its fluid core are described with the help of the Euler equations for conservation of angular momentum. When account is taken of the pressure coupling between core and mantle as well as of the elastogravitational deformation of the solid part (e.g., Sasao et al. 1980), the eigenfrequencies of this system are well known (Hinderer et al. 1982):

$$\sigma_{cw} = \frac{A}{A^m} \Omega \alpha \left(1 - \frac{k}{k_f}\right)$$
$$\sigma_{nd} = -\Omega \left[1 + \frac{A}{A^m}\left(\alpha^c - \frac{q_0 h_1^c}{2}\right)\right] \tag{11}$$

where k is the classical volume Love number and k_f its fluid limit, h_1^c an elastic quantity proportional to the displacement of the CMB caused by the fluid rotational pressure, and $q_0 = \Omega^2 a/g$ the ratio of centrifugal force to surface gravity. σ_{cw} and σ_{nd} are the Chandlerian and nearly diurnal eigenfrequencies relative to a purely elastic Earth model without ocean; the perturbations caused by anelasticity, CMB friction and ocean loading can be found elsewhere (e.g., Legros 1987).

The observed value of σ_{cw} is close to $\Omega/435.2$ (see Smith and Dahlen 1981; Lenhardt, this Vol.) while the theoretically predicted one is slightly larger when anelasticity is omitted; for instance, we would have $\Omega/433.6$ when adding a recent estimate of ocean loading (Dickman and Steinberg 1986) to the value relative to an elastic model (with a fluid core) in the hydrostatic equilibrium state (Sasao et al. 1980). For σ_{nd}, the discrepancy between the hydrostatic theoretical value $-\Omega(1+1/466.6)$ after Sasao et al. (1980) and the value $-\Omega(1+1/434.2)$ derived either from VLBI measurement (Gwinn et al. 1986) or gravity tidal observations (Neuberg et al. 1987) is as large as 30 days in the period of the associated spatial nutation (Free Core Nutation).

In order to investigate if resonant processes have occurred in the Earth and core wobbles in the geological past, it is necessary to consider excitation functions of frequency close to σ_{nd} and σ_{cw} which are stable over long time scales. One of the most stable astronomical motion over geologic time is the orbital motion of the Earth around the Sun (e.g., Toomre 1974). We therefore consider the two following excitation sources:

a) an annual (of frequency $\sigma_a = \Omega^*/366.25$, where Ω^* is the present value of the Earth rotation rate) atmospheric forcing (thermally driven mass redistribution) leading to a loading potential (or pressure change) of degree 2 and order 1 spherical harmonic repartition; this term being close to the Chandlerian period will enhance the Earth wobbling motion.
b) a nearly diurnal gravitational excitation due to the tidal potential associated with the solar wave ψ_1 of frequency $\sigma_{\psi_1} = -\Omega - \Omega^*/366.25$; this term will of course induce a resonant amplification in the core wobble.

The eigenfrequencies (11) are directly dependent on $\Omega(t)$; besides, $\Omega(t)$ is also involved in the expression of the dynamical ellipticities α and α^c if we assume that the planet is in hydrostatic equilibrium. Some authors have studied, often in the frame of simplified Earth models, either the Chandlerian resonance (Cannon 1974; Lambeck 1975) or the nearly diurnal one (Toomre 1974; Hinderer et al. 1987). We intend to investigate here simultaneously both resonances, starting from the hydrostatic theoretical values on one hand, and from the best observed ones on the other hand (we therefore neglect the influence of non-hydrostatic contributions to the eigenperiods and their role in the temporal evolution of the wobbles). The variable dynamical ellipticities are denoted:

$$\begin{aligned}\alpha(t) &= \frac{\Omega^2(t)}{(\Omega^*)^2}\alpha_0 \\ \alpha^c(t) &= \frac{\Omega^2(t)}{(\Omega^*)^2}\alpha_0^c \,.\end{aligned} \tag{12}$$

We have then for the expression of the variable eigenfrequencies:

$$\begin{aligned}\sigma_{cw}(t) &= \Omega^3(t)\frac{X}{(\Omega^*)^2} \\ \sigma_{nd}(t) &= -\Omega(t) - \Omega^3(t)\frac{Y}{(\Omega^*)^2}\end{aligned} \tag{13}$$

where X and Y are constant values.

The conditions for exact resonance become:

$$\begin{aligned}\sigma_a - \sigma_{cw} &= \frac{\Omega^*}{366.25} - \Omega^3(t)\frac{X}{(\Omega^*)^2} = 0 \\ \sigma_{\psi_1} - \sigma_{nd} &= -\frac{\Omega^*}{366.25} + \Omega^3(t)\frac{Y}{(\Omega^*)^2} = 0\end{aligned} \tag{14}$$

where σ_a is the annual frequency of atmospheric forcing and σ_{ψ_1} the nearly diurnal tidal one.

Using the hydrostatic value for the Chandler eigenperiod in (14) leads to:

$$\frac{\Omega^*}{366.25} - \frac{\Omega^3(t)}{(\Omega^*)^2 433.6} = 0 \quad \text{or} \quad \frac{\Omega(t)}{\Omega^*} = \left(\frac{433.6}{366.25}\right)^{1/3} \tag{15}$$

implying a spin rate $\Omega(t) = 7.71\ 10^{-5}$ rd. s^{-1} for exact resonance.

Similarly, the resonance condition for the core wobble becomes in the hydrostatic assumption:

$$\frac{\Omega(t)}{\Omega^*} = \left(\frac{466.6}{366.25}\right)^{1/3} \tag{16}$$

i.e, for $\Omega(t) = 7.90\ 10^{-5}$ rd. s^{-1}.

The epoch of exact resonance in the past depends on the value of tidal deceleration $d\Omega/dt$. Setting $d\Omega/dt = -5.98\ 10^{-22}$ rd. s^{-2} (Christodoulidis et al. 1988), we get from (15) and (16):

$$T_{cw} \approx 224\ \text{Ma} \qquad T_{nd} \approx 325\ \text{Ma} \tag{17}$$

for the epochs of Chandlerian and nearly diurnal resonance, respectively. In a very similar way, if we rather start from the observed values of the rotational eigenperiods, the resonance conditions are changed to:

$$\frac{\Omega(t)}{\Omega^*} = \left(\frac{435.2}{366.25}\right)^{1/3} \quad \text{for the Chandler wobble} \tag{18}$$

$$\frac{\Omega(t)}{\Omega^*} = \left(\frac{434.2}{366.25}\right)^{1/3} \quad \text{for the nearly diurnal free wobble .}$$

With the same value of tidal braking as before, the epochs of resonant enhancement are then:

$$T_{cw} \approx 229\ \text{Ma} \qquad T_{nd} \approx 225\ \text{Ma}\ . \tag{19}$$

Notice the extremely close proximity of these last values (less than 5 Ma discrepancy) which is obviously due to the nearly coincident observed values of the Chandlerian and Free Core Nutation periods.

For a perfectly elastic Earth model without dissipative coupling at the CMB, the resonance would lead to infinite amplification in the Earth and core wobbles. This is of course not the real case because both viscomagnetic friction and mantle anelasticity set upper bounds on the amplification. We do not discuss here the problem of the ratio of resonant versus present value of the rotational motions; some elements of discussion concerning the resonant tidal flow in the Earth's fluid core can be found elsewhere (Hinderer and Legros 1988). Let us just mention that these resonance processes (and especially in the case of the simultaneous double resonance) may have had important geodynamic consequences: increased velocity field in the core, enhanced pole tide in the oceans, stress accumulation and dissipated energy; the induced thermal effects and their relationship with the disruption of Pangaea (185 – 200 Ma) will be treated in a forthcoming paper.

5 Secular Deceleration and Magnetic Scaling Laws

One of the key problems in planetary magnetism is to find sources of motion which are capable to sustain a magnetic field against ohmic dissipation (for a review see, e.g., Stevenson 1983). The similarity properties of magnetic states existing between different planets have suggested that the same dynamo process is operating; in this case, one would expect a scaling law relating the amplitude of the magnetic field (or magnetic moment) to given physical, dynamical or geometrical parameters (Jacobs 1977; Russell 1980). Two major driving mechanisms for planets with a fluid core have been proposed (e.g., Stevenson 1974): convection (either thermally or compositionally driven) and precession. One can show that the general form for the scaling law in terms of the amplitude of the magnetic moment M can be written (Hinderer 1987; Legros and Hinderer 1987):

$$M \sim \mu^{-1/2} \varrho^{1/2} \Omega^{1-a-b} L^{4-2a} \eta^{a} \left[\frac{f \dot{\psi} \sin \varepsilon}{\alpha^{c}} \right]^{b} \tag{20}$$

where μ and η are the magnetic permeability and diffusivity, ϱ the mean density, L the mean core radius, Ω the axial rotation rate; for correlations between magnetic field and spin rate, see also Stix (1982), Newton (1985). The term in brackets expresses the dynamical response of the fluid core during the planetary precession (see Hinderer et al. 1987) and involves the precessional angular velocity $d\psi/dt$, the obliquity ε, the core ellipticity α^c and the dimensionless quantity f of elastic nature. The constants a and b depend on the regeneration model: $a = b = 0$ is the convection model used by Busse (1975, 1976); $a = -1$ and $b = 1$ correspond to the precessional model of Dolginov (1975, 1976), and $a = 1$ and $b = -1/2$ to another precessional law proposed by Stevenson (1974). Finally, $a = -1/2$ and $b = 1$ can model a turbulent precessional scaling law.

If we assume that μ, η, ϱ and L have not significantly changed in the geological past on the contrary to the axial rotation rate Ω and related parameters like the core ellipticity α^c (proportional to Ω^2 after Sect. 3.2) and precessional velocity $d\psi/dt$ which is proportional to $(C-A)/C\Omega$ (e.g., Amalvict et al. 1986), itself proportional to Ω, the ratio of ancient versus recent magnetic moments becomes:

$$\frac{M}{M^*} = \left[\frac{\dot{\psi} \alpha^{c*}}{\dot{\psi}^* \alpha^{c}} \right]^{b} \left[\frac{\Omega}{\Omega^*} \right]^{1-a-b} = \left[\frac{\Omega}{\Omega^*} \right]^{1-a-2b} \tag{21}$$

where all the starred (*) quantities are present-day values.

We immediately see that there must be an evolution (dependent on a and b) of the geomagnetic field with geological time as a function of the Earth secular deceleration. It is noticeable that the convective model ($a = b = 0$) which is independent on the precession leads to a ratio $M/M^* = \Omega/\Omega^* > 1$. This means that the magnetic moment must have been larger in earlier times than it is now; for instance, using again the previous value for the tidal despinning $d\Omega/dt$, we found that M has been about 1.5 times larger 2×10^9 years ago. It appears clearly that this scaling law would not be in agreement with the constancy of the Earth paleomagnetic moment over geologic time (see Merrill and McElhinny 1983). The best scaling laws from the point of view of the stability of the magnetic field would be the Dolginov's model ($a = -1$, $b = 1$) corrected for the resonant amplification in the fluid core (term in f/α^c in (20)) or the non-corrected ($f/\alpha^c = 1$) Stevenson's model ($a = 1$, $b = 0$).

A more careful investigation of the relationships between paleointensities, dynamo process and tidal deceleration would be of great interest but is beyond the scope of this study.

6 Conclusion

This study has shown different geodynamic implications related to the secular deceleration of the Earth. The existence of friction between core and mantle leads to a deceleration of the fluid core that is practically the same as the mantle one; however, there remains an eastward differential rotation rate of the order of 10^{-10} rd. s^{-1}. The fluid deformation of the Earth due to the variable rotational potential induces a contraction in the Earth radius of about 500 m in the last 500 Ma in addition to a decrease of the geometric flattening from 1/232 to 1/299.8 in the same time interval. There are correlated changes in the dynamic ellipticity, precessional constant and quadrupole coefficient J_2. The possibilities for exact resonances in the past in the Earth and core wobbles caused by excitations related to the stable orbital motion of the Earth around the Sun have been investigated. The epochs of resonance are dependent on the assumptions made on the dynamic flattenings involved in the nearly diurnal and Chandlerian eigenfrequencies. In the hydrostatic approximation, the epochs were found to be quite different, 225 Ma for the annual Chandlerian resonance and 325 Ma for the nearly diurnal one. But, if we start from the recent observed values of the rotational eigenperiods, there is an intriguing coincidence (less than 5 Ma difference) of both resonance epochs near 225 Ma which could be connected with the disruption of Pangaea via thermal effects. Finally, we have briefly considered the influence of the secular braking on the scaling laws expressing the amplitude of the magnetic field as a function of physical or dynamic parameters according to a given dynamo process. It appears that a more accurate knowledge of the paleointensity of the Earth's magnetic field would help to dismiss some regeneration mechanisms. These results point out the importance of the secular deceleration for the deformation, rate of differential rotation of the fluid core, temporal evolution of the wobbles and even for the amplitude of the magnetic field of the Earth.

Acknowledgment. Financial support of this work was provided by CNRS, INSU, France (DBT contribution No 81).

References

Alterman Z, Jarosch H, Pekeris CH (1959) Oscillations of the Earth. Proc R Soc A 252:80–95
Amalvict M, Hinderer J, Legros H (1986) Influence of elasticity on the precessions of the Earth and its fluid core. Proc 10th Int Symp Earth Tides, pp 321–326
Bondi H, Lyttleton RA (1948) On the dynamical theory of the rotation of the Earth I. The secular deceleration of the core. Proc Camb Phil Soc 44:345–359
Brosche P (1981) Geomagnetic reversals and tidal friction. Naturwissenschaften 68:139
Brosche P, Hövel W (1982) Tidal friction for times around the presence. In: Brosche P, Sündermann J (eds) Tidal friction and the Earth's rotation II. Springer, Berlin Heidelberg New York Tokyo, pp 175–189
Brosche P, Sündermann J (eds) (1978) Tidal friction and the Earth's rotation. Springer, Berlin Heidelberg New York
Brosche P, Sündermann J (eds) (1982) Tidal friction and the Earth's rotation II. Springer, Berlin Heidelberg New York
Burša M (1985) The Earth's precession and nutations in geological epochs. Bull Astron Inst Czech 36:139–141
Busse FH (1975) A model of the geodynamo. Geophys J R Astron Soc 42:437–459
Busse FH (1976) Generation of planetary magnetism by convection. Phys Earth Planet Int 12:350–358
Cannon WH (1974) The Chandler annual resonance and its possible geophysical significance. Phys Earth Planet Int 9:83–90
Christodoulidis DC, Smith DE, Williamson RG, Klosko SM (1988) Observed tidal braking in the Earth/Moon/Sun system. J Geophys Res 93:6216–6236
Denis C (1986) On the changes of kinetical parameters of the Earth during geological times. Geophys J R Astron Soc 87:559–568
Denis C, Karafistan A, Amalvict M, Legros H (1987) Changing curvature and tectonic activity of planetary bodies. Proc CNES Meeting on Comparative Planetology and Earth Sciences, La Londe-les-Maures, September 1987 (abstract in Geophys Rep Publ Inst Astrophys Liège, Belgium, 87/005)
Dickman SR, Steinberg DJ (1986) New aspects of the equilibrium pole tide. Geophys J R Astron Soc 86:515–529
Dolginov Sh (1975) The mechanism of planetary dynamos. Kosm Issled 13:367–374
Dolginov Sh (1976) On the question of the energy of the precessional dynamo. NASA Spec Publ SP-397:167–170
Gwinn CR, Herring TA, Shapiro II (1986) Geodesy by radiointerferometry: Studies of the forced nutations of the Earth, 2, Interpretation. J Geophys Res 91:4755–4765
Hinderer J (1987) Sur quelques effets en rotation et déformation d'une planète à noyau liquide, manteau élastique et couche fluide superficielle. Thesis, Univ Strasbourg, 362 p
Hinderer J, Legros H (1988) Tidal flow in the Earth's core: a search for the epoch and amplitude of exact resonance in the past, in "Structure and Dynamics of the Earth's deep Interior". In: Srrylie DE, Hide R (eds) AGU Monograph Series 46:79–82
Hinderer J, Legros H, Amalvict M (1982) A search for Chandler and nearly diurnal free wobbles using Liouville equations. Geophys J R Astron Soc 71:303–332
Hinderer J, Legros H, Amalvict M (1987) Tidal motions within the Earth's fluid core: resonance process and possible variations. Phys Earth Planet Int 49:213–221
Jacobs JA (1977) Planetary magnetism. J Geomag Geoelect 29:503–505
Lambeck K (1973) Temporal variations of rotational origin in the absolute value of gravity. Studia Geoph Geod 17:269–271
Lambeck K (1975) The Chandler annual resonance. Phys Earth Planet Int 11:166–168
Lambeck K (1980) The Earth's variable rotation. Univ Press, Cambridge
Legros H (1987) Sur quelques problèmes de dynamique planétaire. Thesis, Univ Strasbourg, 220 p
Legros H, Hinderer J (1987) Precession of planetary bodies and magnetic implications. Proc CNES Meeting on Comparative Planetology and Earth Sciences, La Londe-les-Maures, September 1987
Mansinha L, Shen PY (1976) Rotational deformation of the Earth. Phys Earth Planet Int 11: 200–206
Merrill RT, McElhinny MW (1983) The Earth's magnetic field. Academic Press, London

Neuberg J, Hinderer J, Zürn W (1987) Stacking gravity tide observations in Central Europe for the retrieval of the complex eigenfrequency of the nearly diurnal free wobble. Geophys J R Astron Soc 91:853–868

Newton RR (1985) The secular acceleration of the Earth's spin. Geophys J R Astron 80:313–328

Rochester MG, Smylie DE (1974) On changes in the trace of the Earth's inertia tensor. J Geophys Res 79:4948–4951

Rochester MG, Jacobs JA, Smylie DE, Chong KF (1975) Can precession power the geomagnetic dynamo? Geophys J R Astron Soc 43:661–678

Russell CT (1980) Planetary magnetism. Rev Geophys Space Phys 18:77–106

Saito M (1974) Some problems of static deformation of the Earth. J Geophys Res 22:123–140

Sasao T, Okubo S, Saito M (1980) A simple theory on dynamical effects of a stratified fluid core upon nutational motion of the Earth. In: Fedorov EP, Smith ML, Bender PL (eds) Proc IAU Symp No 78, Nutation and the Earth's rotation, Kiev, May 1977. Reidel, Dordrecht, pp 165–183

Smith ML, Dahlen FA (1981) The period and Q of the Chandler wobble. Geophys J R Astron Soc 64:223–281

Stacey FD (1969) Physics of the Earth. Wiley, New York

Stevenson DJ (1974) Planetary magnetism. Icarus 22:403–415

Stevenson DJ (1983) Planetary magnetic fields. Rep Prog Phys 46:555–620

Stix M (1982) The rotation and the magnetic field of the Earth. In: Brosche P, Sündermann J (eds) Tidal friction and the Earth's rotation II. Springer, Berlin Heidelberg New York Tokyo, pp 29–50

Stoneley R (1924) The shrinkage of the Earth's crust through diminishing rotations. R Astron Soc Geophys Suppl 1:149–155

Toomre A (1974) On the "nearly diurnal wobble" of the Earth. Geophys J R Astron Soc 38:335–348

Astronomical Frequencies in Paleoclimatic Data and the Dynamical Ellipticity of the Earth

M.F. Loutre[1], V. Dehant[1,2], and A. Berger[1]

1 Introduction

It is now recognized that climate is continually changing at different time scales. The astronomical theory seems to provide evidence for the link between Quaternary paleoclimates and eccentricity of the Earth's orbit, obliquity and the longitude of the perihelion ($\tilde{\omega}$) measured from the equinox of date which is mainly used through the climatic precession parameter ($e \sin \tilde{\omega}$) (Fig. 1). Moreover, slowly varying parameters, such as the Earth's moments of inertia, the Earth-Moon distance and the angular velocity of the Earth, can have an impact, at the geological time scale, on the frequencies of these astronomical parameters. On the other hand, the waxing and waning of the Quaternary ice sheets can also influence the values of the astro-climatic periods.

2 Climate Changes at the Astronomical Frequencies

2.1 Climatic Variations

Let us first assume, following Mitchell (1976), a record of the air surface temperature to be available over the whole history of the Earth. Its idealized variance spectrum illustrates the large variability in time of this parameter (Fig. 2). The analysis of the spectrum shows a background level of variability, deriving from internal stochastic mechanisms and corresponding to a low degree of predictability, which appears to increase in amplitude towards the longer time scales, superimposed on which is a band-limited variability due to external forcing processes and correspondingly to a much higher degree of predictability (spikes and broader peaks). The spikes are astronomically dictated; they are strictly periodic components of climate variation, such as the diurnal and annual variations and their harmonics, whereas the peaks represent variations that are, according to Mitchell, either quasi-periodic or aperiodic but, with a preferred time scale of energization. The peak at 3 – 7 days is associated with the synoptic disturbances mainly at mid-latitudes. The peak near 2500 years might be related to the characteristic time scale of the oceans. The next three peaks are related to the Quaternary ice ages. They are also related to deterministic

[1] Université Catholique de Louvain, Institut d'Astronomie et de Géophysique Georges Lemaître, 2 Chemin du Cyclotron, B-1348 Louvain-la-Neuve, Belgium.
[2] Present affiliation: Observatoire Royal de Belgique, 3, Avenue Circulaire, B-1180 Bruxelles, Belgium.

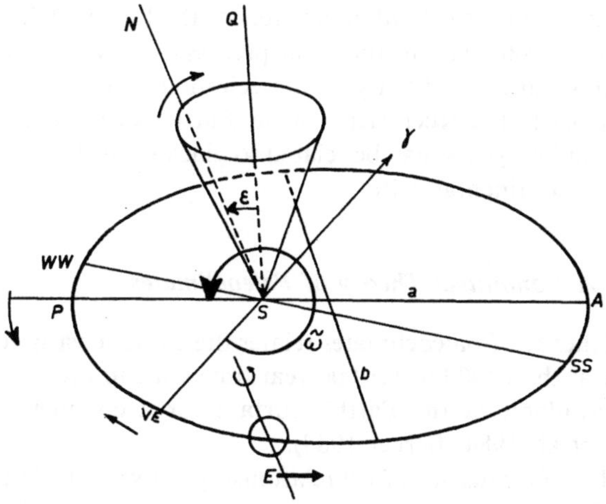

Fig. 1. Elements of the Earth's orbit. The orbit of the Earth, E, around the Sun, S, is represented by the ellipse $P\gamma EA$, P being the perihelion and A the aphelion. Its eccentricity, e, is given by $(a^2-b^2)^{1/2}/a$, a being the semi-major axis and b the semi-minor axis. WW and SS are respectively the winter and the summer solstice, VE is the vernal equinox; WW, SS and γ, which replaces VE in a geocentric system, are located where they are today. SQ is perpendicular to the ecliptic and the obliquity, ε, is the inclination of the equator upon the ecliptic – i.e., ε is equal to the angle between the Earth's axis of rotation SN and SQ. $\tilde{\omega}$ is the longitude of the perigee relative to the moving vernal equinox, and is equal to $\pi + \Psi$. The annual general precession in longitude, Ψ, describes the absolute motion of γ along the Earth's orbit relative to the fixed stars. π, the longitude of the perihelion, is measured from the reference vernal equinox of 1950 A.D. and describes the absolute motion of the perihelion relative to the fixed stars. For any numerical value of $\tilde{\omega}$, 180° is subtracted for a practical purpose: observations are made from the Earth, and the Sun is considered as revolving around the Earth (Berger 1980)

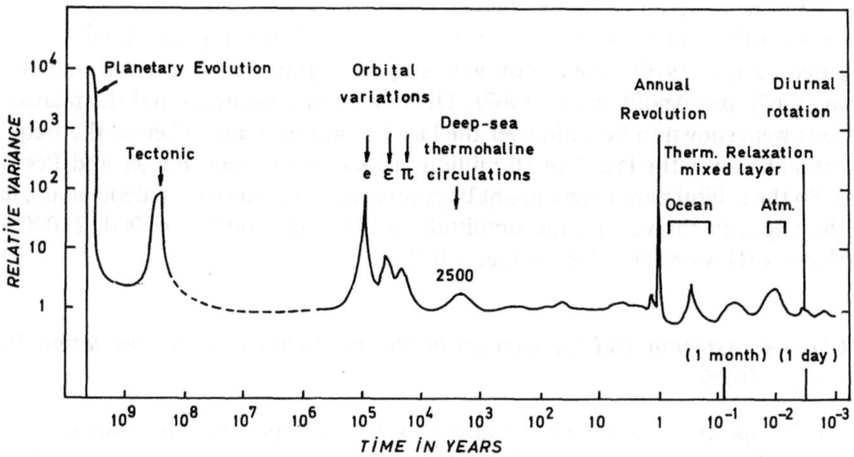

Fig. 2. Tentative spectrum of climatic variations with an estimate of the relative variance of climate over all periods of variation. A background level of variability, deriving from internal stochastic mechanisms and corresponding to a low degree of predictability, appears to increase in amplitude towards the longer time scales and to be overtopped by band-limited variability, due to external forcing processes and corresponding to a high degree of predictability (Adapted from Michell 1976; Berger 1979)

astronomical variations in the orbital parameters of the Earth (Milankovitch 1941) – the cycle at around 21 000 years to the axial precession, the period of about 41 000 years with the change in the obliquity of the ecliptic or the axial tilt, and the cycle at about 100 000 years to the eccentricity of the Earth's orbit (Berger 1988). Finally, the peak near 350 million years may be related to glaciations due to orogenic and tectonic effects and to continental drift.

2.2 Milankovitch Astronomical Theory of Paleoclimates

The astronomical theory of paleoclimates relates the climatic variations such as those of the Quaternary at the 10 000 to 100 000-year time scale to those in the solar energy which would be available at the Earth's surface for a completely transparent atmosphere (Berger et al. 1984; Berger 1988).

Although the idea of a cause-to-effect relationship between the long term variations of the Earth's orbital elements and those of the climate originates from the early part of the nineteenth century, it is only during the first decades of the twentieth century that mainly the Yugoslav astronomer Milankovitch (1941) indicated that, under the assumption of a perfectly transparent atmosphere, a minimum in the northern hemisphere summer insolation at high latitudes would generate a glacial period. In fact, this theory requires that the northern high-latitude summers must be cold to prevent the winter snow from melting in order to allow a positive value in the annual budget of snow and ice which will initiate a positive feedback: the subsequent increase of the surface albedo will amplify the initial radiative perturbation, resulting in a further extension of the snow cover.

For any latitude of the Earth, insolation is a single-valued function of the total radiative energy emitted by the sun S_0, of the semi-major axis a of the ecliptic, of its eccentricity e, of its obliquity ε and of the longitude $\tilde{\omega}$ of the perihelion measured from the moving vernal equinox (Berger 1978) (see Fig. 1).

Between 1970 and 1974, a new astronomical solution for paleoclimatology was computed (Berger 1976) based upon works by Bretagnon (1974), Sharaf and Boudnikova (1967) and Anolik et al. (1969). The new orbital elements and the related insolations were shown to be valid over the last 1.5 million years, whereas the frequencies are stable over the last 5 to 10 million years (Berger 1984; Berger and Pestiaux 1984). So the orbital parameters might be compared to Quaternary paleoclimatic data and their spectra showed a large similitude in the range 100 000, 41 000, 23 000 and 19 000 years (Hays et al. 1976; Berger 1977).

2.3 Classical Astronomical Frequencies in Diverse Geological and Insolation Data all over the World

Since 1976, spectral analysis of climatic records of the past 800 000 years or so, has confirmed these results and provided substantial evidence that, at least near the frequencies of variation in obliquity and precession, a considerable fraction of the climatic variance is driven in some way by insolation changes accompanying changes in the Earth's orbit (Imbrie and Imbrie 1980; Berger 1989). For example, if some long deep-sea cores are used and the uncertainty in the geological time scale and in the

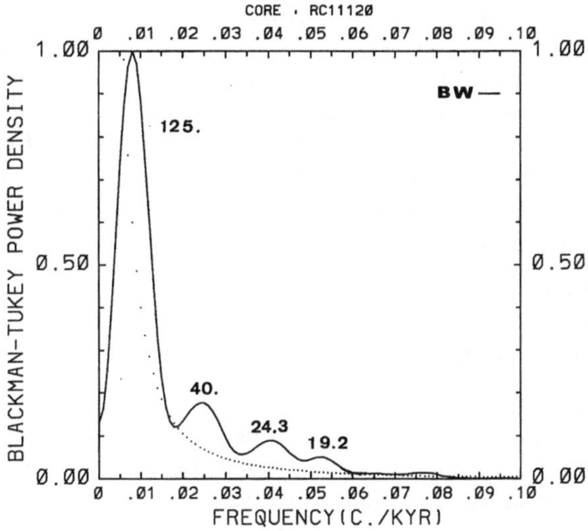

Fig. 3. Spectral analysis of the core RC11120. This graph – showing the relative importance of different climatic cycles in the isotopic record of an Indian Ocean core – confirmed predictions of the Milankovitch theory (data from Hays et al. 1976 and Shackleton 1977; Van der Mersch et al. 1988). The bandwidth (*BW*) gives a measure of the stability of the computed spectrum and thus of the reliability of the eventually detected peaks: a small bandwidth corresponds to a high value of the variance (i.e., decrease the stability of the spectrum but also decrease the bias or the resolution in frequency of the spectrum) (Jenkins and Watts 1968)

spectral analysis is accounted for, Berger and Pestiaux (1987) have shown that the following peaks are significantly present in the astronomical bands (Fig. 3):

103000 with a standard deviation of 24000
42000 with a standard deviation of 8000
23000 with a standard deviation of 4000

However, the interpretation of the results is not always as clear as that. The 100 kyr cycle, so dominant a feature of the late Pleistocene record, does not exhibit a constant amplitude over the past 2–3 million years. Clearly, this periodicity disappears before 10^6 years ago, at a time the ice sheets were much less developed over the Earth, reinforcing the idea that the major ice sheets may have played a role themselves in modulating this 100 kyr cycle.

The shape of the spectrum depends also upon the location of the core and the nature of the climatic parameters analyzed (Hays et al. 1976; Ruddiman and McIntyre 1981; Imbrie 1982). For example, the 41-kyr cycle is not seen and the 23-kyr cycle is dominant in Atlantic sea-surface temperatures of the last 250000 years in core V30-97. This is not too surprising as these spectra depend upon the way the climate system reacts to the insolation forcing and upon which type of insolation it is sensitive to. Indeed, contrary to the well-received Milankovitch idea that the high polar latitudes must record the obliquity signal, whereas low latitudes record only the precessional one, the latitudinal dependence of the insolation parameters is more complex: clearly, mid-month high-latitude insolation in summer displays a stronger signal in the precession band than in the obliquity one (Berger and Pestiaux 1984). The mid-month daily insolation is defined from a constant increment of the true longitude of the Sun starting at the spring equinox, the dates associated with the mid-month values being around the twentieth of each month.

In addition, cores with a high sedimentation rate and covering the last glacial-interglacial cycle have permitted to resolve also the higher frequency part of the spectrum (Pestiaux et al. 1988): besides the 19000-year precessional peak, three other periods were indeed significantly detected and related to the precessional ones:

10300 with a standard deviation of 2200
4700 with a standard deviation of 800
2500 with a standard deviation of 500

These preferential frequency bands of climatic variability outside the direct orbital forcing band are still too broad to allow for a definite physical interpretation. A tentative explanation, though, may be given in terms of the climatic system's non linear response to variations in the insolation available at the top of the atmosphere. The first characteristic of the 10.3, 4.7 and 2.5-kyr near-periodicities is, indeed, that they are roughly combination tones of the 41, 23 and 19-kyr peaks found in the insolation parameters.

3 Earth's Orbital Parameters

The perturbations of the orbital elements are due to the gravitational effects of the different planets on the Earth's orbit. The Lagrange differential equations of the planetary motion can be formulated as:

$$\frac{d\sigma_{k,n}}{dt} = \sum_{j=1;j\neq n}^{9} \sum_{i=1}^{6} f_{i,n}\,(\sigma_{1,n}, \ldots, \sigma_{6,n})\,\frac{\partial R_{j,n}}{\partial \sigma_{i,n}} \tag{1}$$

$$\sigma_{k,n} = \{h, k, p, q, a, \lambda\}_n \quad 1 \leq k \leq 6; \; 1 \leq n \leq 9 \; .$$

$$R_{j,n} = \sum_{r,s,t,u} A(a_j, a_n, m_j, m_n)\, e_j^r e_n^s \left(\sin\frac{i_j}{2}\right)^t \left(\sin\frac{i_n}{2}\right)^u \cos\varphi_{j,n}$$

$$\varphi_{j,n} = b_1\lambda_j + b_2\lambda_n + b_3\pi_j + b_4\pi_n + b_5\Omega_j + b_6\Omega_n$$

σ's represent the 6 orbital parameters, λ, π and Ω are the longitude of the Earth, the perihelion and the node respectively. R is the perturbing function which shows that the solutions will progress in powers of small parameters: the ratio, m, of the mass of the planets to that of the sun, the eccentricity, e, of the planets' orbit and their inclination, i, to the reference plane. The long-term solution of these equations have been expanded in trigonometrical series by Bretagnon (1974).

A similar system of equations governs the Earth's rotation and precession. These Poisson equations link the angle of luni-solar precession (ψ) to the angle ξ, between the equator of date and the ecliptic of epoch, which in turn permit to find the expression of the general precession in longitude (ψ_g) and of the obliquity (ε). The

Poisson equations take the following form (Sharaf and Boudnikova 1967; Dehant et al. (1987a):

$$\frac{d\psi}{dt} = \frac{1}{\sin\xi}\,\tilde{P}\,\frac{\partial}{\partial\xi}\left(\frac{Z_\odot}{R_\odot}\right)^2 ,$$

$$\frac{d\xi}{dt} = -\frac{1}{\sin\xi}\,\tilde{P}\,\frac{\partial}{\partial\psi}\left(\frac{Z_\odot}{R_\odot}\right)^2 , \tag{2}$$

with

$$\tilde{P} = \frac{3}{2}\,\frac{n_\odot^2}{\omega}\,\frac{C-A}{C}\left((1-e_\odot^2)^{-3/2} + \frac{m_☾}{m_\odot}\,\frac{a_\odot^3}{a_☾^3}\,(1-e_☾^2)^{-3/2}\left(1-\frac{3}{2}\sin^2 i_☾\right)\right) ,$$

where $n_\odot$ is the mean motion of the Sun in a geocentric reference frame, ω the rotational angular velocity of the Earth, A and C the Earth's moments of inertia around the polar and equatorial principal axis of inertia, $a_☾$ and $a_\odot$ the semi-major axis of the orbit of the Moon around the Earth and of the Earth around the Sun respectively, $e_\odot$ the eccentricity of the apparent orbit of the Sun around the Earth, $i_☾$ the inclination of the lunar orbit on the eliptic, $m_☾$ and $m_\odot$ the mass of the Moon and the Sun respectively, $R_\odot$ the distance from the center of mass of the Earth to that of the Sun and $Z_\odot$, the component along the polar principal axis of inertia of the position vector of the Sun measured from the center of mass of the Earth. After expanding $(Z_\odot/R_\odot)^2$ in trigonometrical series, the equations are solved order by order according to the amplitudes of the terms in the development of the (i, Ω) and (e, π) systems, so that the value of ψ and ξ can be obtained.

At the first order, we have

$$\xi = h , \tag{3}$$

and

$$\frac{d\psi}{dt} = \tilde{P}\cos h = k , \tag{4}$$

where h is a constant, whose present-day value, corresponding to $t_0 = 1950$ A.D., is 23°320556.

The frequency k, introduced through the Poisson's equations for precession, can then be written as follows:

$$k = \frac{3}{2}\,\frac{n_\odot^2}{\omega}\,\frac{C-A}{C}\left((1-e_\odot^2)^{-3/2} + \frac{m_☾}{m_\odot}\,\frac{a_\odot^3}{a_☾^3}\,(1-e_☾^2)^{-3/2}\left(1-\frac{3}{2}\sin^2 i_☾\right)\right)\cos h . \tag{5}$$

4 Astro-Climatic Elements

Provided the semi-major axis and the solar constant are considered as constant, three astronomical parameters are of primary importance in paleoclimatology and are still at the basis of modern versions of the Milankovitch theory (it means versions where numerical values have been greatly improved and where details of the explanation of why the astronomical cycles should affect climate have been deeply modified). First, the eccentricity, e, affects the total amount of energy received by the Earth over one year by changing the mean distance from the Sun to the Earth: $r_m = a(1-e^2)^{1/4}$. The other two parameters, $e \sin \tilde{\omega}$ and ε, mainly influence the geographical and the seasonal patterns of irradiation.

Taking care to include the most important high order terms, each of the classical astro-insolation parameters can be expressed as a quasi-periodic function of time (Berger 1978):

$$e = e_0 + \sum_i E_i \cos(\lambda_i t + \varphi_i)$$

$$e \sin \tilde{\omega} = \sum_i P_i \sin(\alpha_i t + \zeta_i)$$

$$\varepsilon = \varepsilon^* + \sum_i A_i \cos(\tilde{f}_i t + \tilde{\delta}_i)$$

where $\tilde{f}_i$ and α_i have the following form:

$$s_i + k,\ 2(s_i + k),\ s_i + k,\ s_i + s_j + 2k,\ s_i + g_j + 2k,\ g_i + k,\ s_i - s_j \text{ and } s_i + g_j\ ,$$

with s_i and g_j the frequencies of the trigonometrical expansion of the solution of the Lagrange equations for the elements h, k and p, q respectively. Consequently, the frequencies of $e \sin \tilde{\omega}$, ε and e are linear combinations of 17 frequencies, 16 which are the frequencies of the expansion of h, k, p and q and the last one, k, which present-day value is 50."439273/year (Berger and Loutre 1987).

As long as we can assume that the frequencies s_i and g_j of the orbital elements are stable (at least locally in time), the variations of the period of the astro-paleoclimatic parameters e, ε and $e \sin \tilde{\omega}$ come only from those of k. According to (5), the slow variations of ω, C, A and $a_☾$ will thus have to be taken into account for pre-Quaternary geological times.

5 Influence of the Variation of the Lunar Orbit

The lunar distance decreases back in the past; the rate of lunar recession varies between 10^{-7} cm s^{-1} and 2 10^{-7} cm s^{-1} over about the last 600 10^6 years. Neglecting the effect of the solar tide, the inclination and the eccentricity of the lunar orbit, the recession of the Moon due to the semidiurnal tide may be written after Lambeck (1980) as:

$$a_☾ = a_0 \left[1 - \frac{13}{2} T \frac{\dot{a}}{a_0}\right]^{2/13}, \qquad (6)$$

where T is time measured back from the present, $a_0 = 3.84\ 10^{10}$ cm is the present value of the semi-major axis of the lunar orbit and $\dot{a}$ is the present average velocity of the lunar recession; the value $\dot{a} = 10^{-7}$ cm s^{-1} has been selected among data taken from Cazenave (1982), Lambeck (1980) and Scrutton (1978).

Table 1. Estimated values of the semi-major axis of the lunar orbit, of the length of the day and of the dynamical ellipticity of the Earth

Geological epoch	Age (MYR)	a_C (km)	Number of days in a year (Stoyko 1970)	l.o.d. (s) (Stoyko 1970)	$C-A/C$ 10^{-6} (Denis 1986)
Holocene	0	384000	365	86164	3241
Upper Cretaceous	72	381690	370	85055	3326
Lower Permian	270	374908	384	81962	3581
Upper Carboniferous	298	373892	388	81119	3656
Middle Devonian	380	370828	399	78888	3866
Lower Silurian	440	368493	407	77342	4022

Tidal stresses due to the Moon and the Sun induce deformations of the Earth. A consequence of the delay in the Earth's non-elastic response to the attraction of these two bodies is that the tidal bulge will be displaced with respect to the Earth-Moon axis. The attraction on this bulge due to the gravitational potential creates a torque and, consequently, a modification of the spin of the Earth (Jeffreys 1976; Calame and Mulholland 1978; Lambeck 1980). Further, since the angular momentum of the Earth-Moon system is conserved, the sum of the Earth's spin momentum and the system's orbital momentum is constant. So the effect of the Moon, through the Earth-Moon distance, is related to the rotational angular velocity of the Earth (Munk 1966). But, observed variations of ω may also include other influences such as, for example, the variations in the angular momentum of the atmosphere and oceans, change in the sea level, the growth of the inner core, the transfer of the angular momentum at the liquid core – solid mantle boundary. Consequently, these past values of ω have been taken from Stoyko's data (1970) which are in agreement with the determinations reviewed by Scrutton (1978) and Lambeck (1980). Moreover, these values have been chosen to keep coherency between the different data, as Denis also used them for the computation of H. Values for the length of the day related to ω are given in Table 1 for different geological epochs.

The changing rotational speed also gives rise to a varying oblateness of the different level surfaces within the Earth and consequently to a modification of the geometrical flattening at its outer surface, so that the dynamical ellipticity of Earth ($H = (C-A)/C$) will also change with time. The values of H have been computed by Denis (1986). In his model the Earth is assumed to have remained in a nearly hydrostatic state along its whole geological history. Moreover the inner structure is considered to have not undergone important modifications. The density distribution is the PREM (Dziewonski and Anderson 1981) and the computation is based on the classical theory of the Earth's figure expanded to order 3 in the flattening. The equatorial, A, and the polar, C, principal moments of inertia are computed for each geological time by integration over the whole volume of the Earth, taking into account the time-variable rotation and flattening of the Earth's layers. The results are given in Table 1.

Table 2 gives, for the different epochs and for the four major astro-climatic periods, the relative rate of change of the related frequency ($\Delta f/f$), the modification of the period due to the decrease of the Earth-Moon distance only (partial effect) and the

Table 2. Estimated values of the semi-major axis of the lunar orbit, the relative rate of change of the frequencies of the paleoclimatic parameters, the absolute modification of the period and the new period due to the total effect of the Earth-Moon distance variations and the new period due to the partial effect of these variations of the Earth-Moon distance only (Berger et al. 1989)

Geological epoch	Age (MYR)	$\delta f/f$ [%]	total effect		partial effect
			Δp	p_{new}	p_{new}
Period = 19000					
Holocene	0	0.00	0	19000	19000
Upper Cretaceous	72	1.91	−355	18645	18826
Lower Permian	270	7.72	−1361	17638	18308
Upper Carboniferous	298	9.07	−1579	17421	18230
Middle Devonian	380	12.94	−2176	16824	17992
Lower Silurian	440	15.86	−2600	16399	17811
Period = 23000					
Holocene	0	0.00	0	23000	23000
Upper Cretaceous	72	2.31	−518	22481	22745
Lower Permian	270	9.35	−1965	21034	21993
Upper Carboniferous	298	10.98	−2274	20725	21881
Middle Devonian	380	15.66	−3113	19886	21540
Lower Silurian	440	19.20	−3704	19296	21280
Period = 41000					
Holocene	0	0.00	0	41000	41000
Upper Cretaceous	72	4.11	−1619	39381	40196
Lower Permian	270	16.66	−5854	35145	37908
Upper Carboniferous	298	19.57	−6709	34291	37574
Middle Devonian	380	27.91	−8947	32053	36580
Lower Silurian	440	34.22	−10453	30546	35837
Period = 54000					
Holocene	0	0.00	0	54000	54000
Upper Cretaceous	72	5.41	−2773	51226	52615
Lower Permian	270	21.94	−9716	44284	48761
Upper Carboniferous	298	25.77	−11064	42936	48210
Middle Devonian	380	36.76	−14516	39484	46586
Lower Silurian	440	45.07	−16777	37222	45387

new period where all the influences have been taken into account, i.e., a_{C}, ω, H as well as their mutual interactions (total effect) (Berger et al. 1989).

A first remark is that the variations are always negative, i.e., in the past the periods of the paleoclimatic parameters have always been smaller than today. Moreover, the changes are very small over the last 100 million years but become rather significant for earlier times. It is also easy to see that the smaller the periods are, the less they are affected.

6 Impact of the Ice Sheets During the Quaternary

During a glacial, the water vapor taken from the oceans forms the ice sheets which melt during an interglacial. These two geological stages are thus characterized by a different mass distribution at the Earth's surface and inside the Earth. These different mass distributions modify the Earth's moments of inertia and thus the Earth's dynamical ellipticity ($C-A/C$). Consequently, the fundamental frequency k and the principal frequencies of the elements used in the computation of the Earth's orbital parameters and insolation might change.

The model that is used (Dehant et al. 1987b) essentially considers a spherical Earth, a sea-level change, an isostatic rebound and an ice sheet of which the size is close to the 18000 YBP reconstructions obtained from geological data. The shifting of the crust under the ice sheet only has been taken into account. There is no mass redistribution neither under the crust nor under the oceans. Three continents were assumed to extend respectively

from 10°N to 80°N and from 0° to 110°E,
from 10°N to 80°N and from 65°W to 130°W,
and from 55°S to 90°S and from 180°E to 180°W.

So the total surface of the continents is roughly equal to its present-day value (25 10^6 km^2).

The maximal extent of the ice sheets during a glacial is given for two idealized ice sheets. The first one extends from 60°N to 80°N in latitude and 0° to 110°E in longitude (the Greenland-Eurasian ice sheet) and the second one extends from 40°N to 80°N in latitude and 130°W to 65°W in longitude (the North-American ice sheet).

The polar and equatorial principal moments of inertia of the Earth have been computed for a glacial and an interglacial. They are given as a sum of integrals, each corresponding to a particular sector of the Earth considered as having a constant density. For example, for a glacial and for an Earth's sector located between λ_1 and λ_2 in longitude, the contributions of the sea ice, the ice sheet, the northern and southern continents and the oceans have been taken into account. Accordingly, the inner part of the Earth extending up to the crust, the marine lithosphere, the continental lithosphere not covered by ice sheets, and the deep ocean not affected by eustatic sea level changes, do not change. The contribution of Antarctica has not been taken into account, but it does not affect the global accuracy of the model because this ice sheet has only slightly changed between a glacial and an interglacial.

An analysis of the sensitivity to a variation of the subsidence depth and to the height of the ice sheet has been performed. The two parameters have been chosen because of their variation in time; in particular, the height of the ice sheet varies during the building-up and the melting and so does the delay in the subsidence or in the rebound of the continent due to viscous dissipation in the mantle. The results of a sensitivity analysis of the overall results to the subsidence depth, the total ice volume, and the height of the ice sheets are given in Table 3. It can be seen that the variation of the dynamical ellipticity of the Earth is but slightly different from 0. More precisely, H remains more or less constant for the isostatic equilibrium case (i.e., a subsidence of about 3/10 of the height of the ice sheet). For larger subsidences, the variation of H becomes positive, and for smaller ones it becomes negative.

Table 3. Variations, in percents, of the dynamical ellipticity of the Earth in function of different subsidence depths and different ice volumes and heights (Dehant et al. 1987b)

h_{ice} km	V_{ice} km^3	(*) m	Subsidence in m							
			0	350	500	650	850	1000	1500	2000
1.0	25.0	62	−0.30	+0.02	+0.16	+0.29	+0.47	+0.61	+1.07	+1.52
1.5	37.5	92	−0.45	−0.13	+0.01	+0.14	+0.32	+0.46	+0.92	+1.37
2.0	50.0	124	−0.60	−0.28	−0.14	−0.01	+0.17	+0.31	+0.77	+1.22
2.5	62.5	155	−0.75	−0.43	−0.30	−0.16	+0.02	+0.16	+0.62	+1.07
3.0	75.0	186	−0.90	−0.58	−0.45	−0.31	−0.13	+0.01	+0.47	+0.92

* Sea level equivalent due to the melting of the ice sheets.

The impact on the main astronomical frequencies have then been computed for an increase of $(C-A/C)$ by 1%: this implies the same percentage of increase in k, i.e., about 0."5043927/yr. In terms of its impact on the periods, it is easy to see that the smaller the periods are, the less they are affected as indicated in Table 3. In any case, as for a subsidence of 1 000 m (at the last glacial maximum) the variation of $(C-A/C)$ is roughly 0.5%, the final impact will always be less than 100 years and 500 years for the main periods of precession (roughly 20000 years) and obliquity (41 000 years) respectively. Even for the longer obliquity period of 54000 years, the change will only be 1 000 years at the maximum. This means that, even if a glacial situation will remain all the time, the 54000-years wave will induce a phase shift of 25000 years (one half of the period) after 25 cycles, which corresponds to about 1300000 years.

In the model, ice loading on a continent leads to the subsidence of only the loaded part of the continent. In fact, this additional mass of ice also causes subsidence and rebound of the adjacent continental areas (Hyde and Peltier 1985). On the other hand, the variation of the sea level also leads to subsidence or rebound of the marine lithosphere. These effects have not been taken into account here.

The geographical location and the height of the ice sheets could be improved for a better approach to the reality. Consequently, a sensitivity analysis of the paleoclimatic periods to the ice sheets characteristics was performed using values within the range of physical plausibility. The results of the different tests were only slightly different and the differences considered as not being significant.

A model accounting for the transient response to the ice sheets loading, instead of the snapshot reconstruction used in the present study where we considered only full (maximum) glacial conditions and plain interglacials, would permit to simulate the time evolution of the global effects of the ice sheets formation and melting, taking into account lagging subsidence and rebound of the lithosphere. However, as the order of magnitude of our results is so small, even for extreme cases, there seem to be no reasons to expect significantly different conclusions from such a time-dependent simulation.

7 Conclusion

Following our simplified model, the waxing and waning of the Quaternary ice sheets do not seem to affect the main frequencies of the orbital parameters and consequently the astronomical time scale developed in Berger (1978). However, because of the assumption used in the model, the question remains open.

On the contrary, the impact of the varying parameters of the lunar orbit and of the length of the day for pre-Quaternary times affect significantly these astronomical frequencies mainly for periods older than 300 10^6 years. A better value of the Earth's dynamical ellipticity would be necessary to have a more realistic estimate of these impacts over the last 2 billion years.

Acknowledgments. This research was partly supported by grants ANDRA 7-0164-A-00-A and CEA BC-4561 of the Commissariat Français à l'Energie Atomique and also by grants KWA/86-3.1.2.c of the Geological Survey of Belgium. V. Dehant is supported by the National Fund for Scientific Research (Belgium).

References

Anolik MV, Krasinsky BA, Pius LJ (1969) Trigonometrical theory of secular perturbations of major planets. Trudy Inst; Theor Astron Leningrad, 14, pp 1–48 (in Russian)

Berger A (1976) Obliquity and precession for the last 5000000 years. Astron Astrophys 51:127–135

Berger A (1977) Support for the astronomical theory of climatic change. Nature (Lond) 268:44–45

Berger A (1978) Long term variations of daily insolation and Quaternary climatic change. J Atmos Sci 35(12):2362–2367

Berger A (1979) Spectrum of climatic variations and their causal mechanisms. Geophys Surv 3:351–402

Berger A (1980) Milankovitch astronomical theory of paleoclimates, a modern review. Vistas Astron 24:103–122

Berger A (1984) Accuracy and frequencies stability of the Earth's orbital elements during the Quaternary. In: Berger A, Imbrie J, Hays J, Kukla G, Saltzman B (eds) Milankovitch and Climate. Reidel, Dordrecht, pp 3–39

Berger A (1988) Milankovitch theory and climate. Rev Geophys 26(4):624–657

Berger A (1989) Pleistocene climatic variability at astronomical frequencies. Quaternary International 2:1–14

Berger A, Loutre MF (1987) Origine des fréquences des éléments astronomiques intervenant dans le calcul de l'insolation. Sci Rep 1987/13, Inst Astron Géophys G. Lemaître, Univ Catholique de Louvain, Louvain-la-Neuve

Berger A, Pestiaux P (1984) Accuracy and stability of the Quaternary terrestrial insolation. In: Berger A, Imbrie J, Hays J, Kukla G, Saltzman B (eds) Milankovitch and Climate. Reidel, Dordrecht, pp 83–112

Berger A, Pestiaux P (1987) Astronomical frequencies in Paleoclimatic data. In: Duzheng Ye, Congbin Fu, Jiping Chao, Yoshino M (eds) The Climate of China and Global Climate. China Ocean Press, Springer, Berlin Heidelberg New York Tokyo, pp 106–114

Berger A, Imbrie J, Hays J, Kukla G, Saltzman B (eds) (1984) Milankovitch and Climate, Understanding the Response to Orbital Forcing. NATO ASI Series C, 126, Reidel, Dordrecht, 895 p

Berger A, Loutre MF, Dehant V (1989) Influence of the changing lunar orbit on the astronomical frequencies of Pre-Quaternary insolation patterns. Paleoceanography 4(5):555–564

Bretagnon P (1974) Termes à longues périodes dans le système solaire. Astron Astrophys 30:141–154

Cazenave A (1982) Tidal friction parameters from satellites observations. In: Brosche P, Sündermann J (eds) Tidal Friction and Earth's Rotation II. Springer, Berlin Heidelberg New York Tokyo, pp 4–18

Calame O, Mulholland JD (1978) Effect of the tidal friction on the lunar orbit and the rotation of the Earth and its determination by laser ranging. In: Brosche J, Sündermann J (eds) Tidal Friction and Earth's Rotation. Springer, Berlin Heidelberg New York, pp 45–53

Dehant V, Loutre MF, Berger A (1987a) Les variations à court et à long terme de la rotation de la Terre et de la précession astronomique. Sci Rep 1987/12, Inst Astron Géophys G. Lemaître, Univ Catholique de Louvain, Louvain-la-Neuve

Dehant V, Loutre MF, Berger A (1987b) Potential impact of the northern hemisphere Quaternary ice sheet on the Earth moments of inertia and on the frequencies of the astroclimatic orbital parameters. Sci Rep 1987/14, Inst Astron Géophys G. Lemaître, Univ Catholique de Louvain, Louvain-la-Neuve

Denis C (1986) On the change of kinetical parameters of the Earth during geological times. Geophys J R Astron Soc 87:559–568

Dziewonski AM, Anderson DL (1981) Preliminary Reference Earth Model. Phys Earth Planet Int 25:297–356

Hays JD, Imbrie J, Shackleton NJ (1976) Variations in the Earth's orbit: Pacemaker of the Ice Ages. Science 194:1121–1132

Hyde WT, Peltier WR (1985) Sensitivity experiments with a model of the Ice Age Cycle: the response to harmonic forcing. J Atmos Sci 42(20):2172–2188

Imbrie J (1982) Astronomical theory of the Pleistocene ice ages: a brief historical review. Icarus 50:408–422

Imbrie J, Imbrie JZ (1980) Modelling the climatic response to the orbital variations. Science 207 (29):943–953

Jeffreys Sir H (1976) The Earth: its Origin History and Physical Constitution, 6th ed. Univ Press, Cambridge

Jenkins G, Watts D (1968) Spectral Analysis and its Applications. Holden-Day, San-Francisco

Lambeck K (1980) The Earth Variable Rotation: Geophysical Causes and Consequences. Univ Press, Cambridge

Milankovitch M (1941) Kanon der Erdbestrahlung und seine Anwendung auf das Eiszeitenproblem. Ed Spec Acad Royale Serbe, Belgrade, 633 pp (Canon of Insolation and the Ice Age Problem. English Translation by Israël Program for Scientific Translation and published for the US Department of Commerce and the National Science Foundation, 1969)

Mitchell JM (1976) An overview of climatic variability and its causal mechanisms. Quat Res 6:481–493

Munk WH (1966) Variation of the Earth's rotation in historical time. In: Marsden BG, Cameron AGW (eds) The Earth Moon System. Plenum Press, New York

Pestiaux P, Van der Mersch I, Berger A, Duplessy JCl (1988) Paleoclimatic variability at frequencies ranging from 1 cycle per 10000 years to 1 cycle per 1000 years: evidence for non linear behaviour of the climate system. Climate Change 12:9–37

Ruddiman WF, McIntyre A (1981) Oceanic mechanisms for amplification of the 23000-year ice-volume cycle. Science 212:617–627

Scrutton CT (1978) Periodic growth features in fossil organisms and the length of the day and the month. In: Brosche P, Sündermann J (eds) Tidal Friction and the Earth's Rotation. Springer, Berlin Heidelberg New York, pp 154–196

Shackleton NJ (1977) The oxygen isotope stratigraphic record of the past Pleistocene. Phil Trans R Soc Lond B280:169–182

Sharaf SG, Boudnikova NA (1967) Secular variations of elements of the Earth's orbit which influences the climates of the geological past. Trudy Inst Theor Astron Leningrad 11:233–262 (in Russian)

Stoyko A (1970) La variation séculaire de la rotation de la Terre et les problèmes connexes. Ann Guébhard (Neuchâtel) 46:293–316

Van der Mersch I, Pestiaux P, Mélice JL, Berger A (1988) A user guide to the Pacdata Bank. Sci Rep 1988/6, Inst Astron Géophys G.Lemaître, Univ Catholique de Louvain, Louvain-la-Neuve

Quaternary Sea-Level Changes in the North Sea, an Analysis of Amplitudes and Velocities

H. Streif

1 Introduction

Quaternary evolution of the North Sea Basin and the climatic changes are indicated in two ways. During the glacial periods the British and/or the Scandinavian ice sheets covered parts of the North Sea area and deposited moraine, glaciofluvial, and glaciolacustrine deposits. An indirect response to the climate is indicated by sea-level changes. Phases of cooling are characterized by regressions and low sea-level stands; phases of warming are indicated by marine transgressions and high sea levels during the Holsteinian (Hoxnian) and Eemian (Ipswichian) interglacial periods and the Holocene. Similar marine regressions and transgressions have been interpreted from the early Pleistocene deposits, however, no indications of ice advances in the North Sea have been demonstrated for this period.

2 Geological Evidence and Processes

The lower boundary of the Pleistocene sequence lies close to the transition between the paleomagnetic Matuyama and Gauss polarity epochs, dated at about 2.47 Ma. B.P. (Long et al. 1988). In the southernmost North Sea this boundary occurs at or near an unconformity coincident with a major change in the style of sedimentation offshore, and the sediments overlying the unconformity have mainly reversed magnetic polarity (Cameron et al. 1987).

The *Lower to early Middle Pleistocene* sediments are mainly shallow marine. Seismic data indicate that a sequence of delta-related and offshore sediments exists in the southern North Sea (Cameron et al. 1987; Long et al. 1988). This agrees with stratigraphic evidence in the Netherlands for a massive expansion of north European rivers into the North Sea Basin during Early Pleistocene times (Zagwijn 1974). On the basis of micropaleontological data a series of paleogeographical maps was constructed by Zagwijn (1979) to illustrate the offshore expansion of the delta complex from the Tiglian (2.1 to 1.7 Ma. B.P.) through the Waalian (1.3 to 0.95 Ma. B.P.) stages to the Cromerian Complex (approx. 0.7 to 0.4 Ma. B.P.).

A series of marine transgressions and regressions appears to have occurred during the approximately 2.1 million years of the Lower and early Middle Pleistocene. It perhaps records the effects of eustatic sea-level changes in the periphery of the North Sea Basin, with periods of marine regression corresponding to periods of maximum

Niedersächsisches Landesamt für Bodenforschung, Stilleweg 2, D-3000 Hannover 51, FRG.

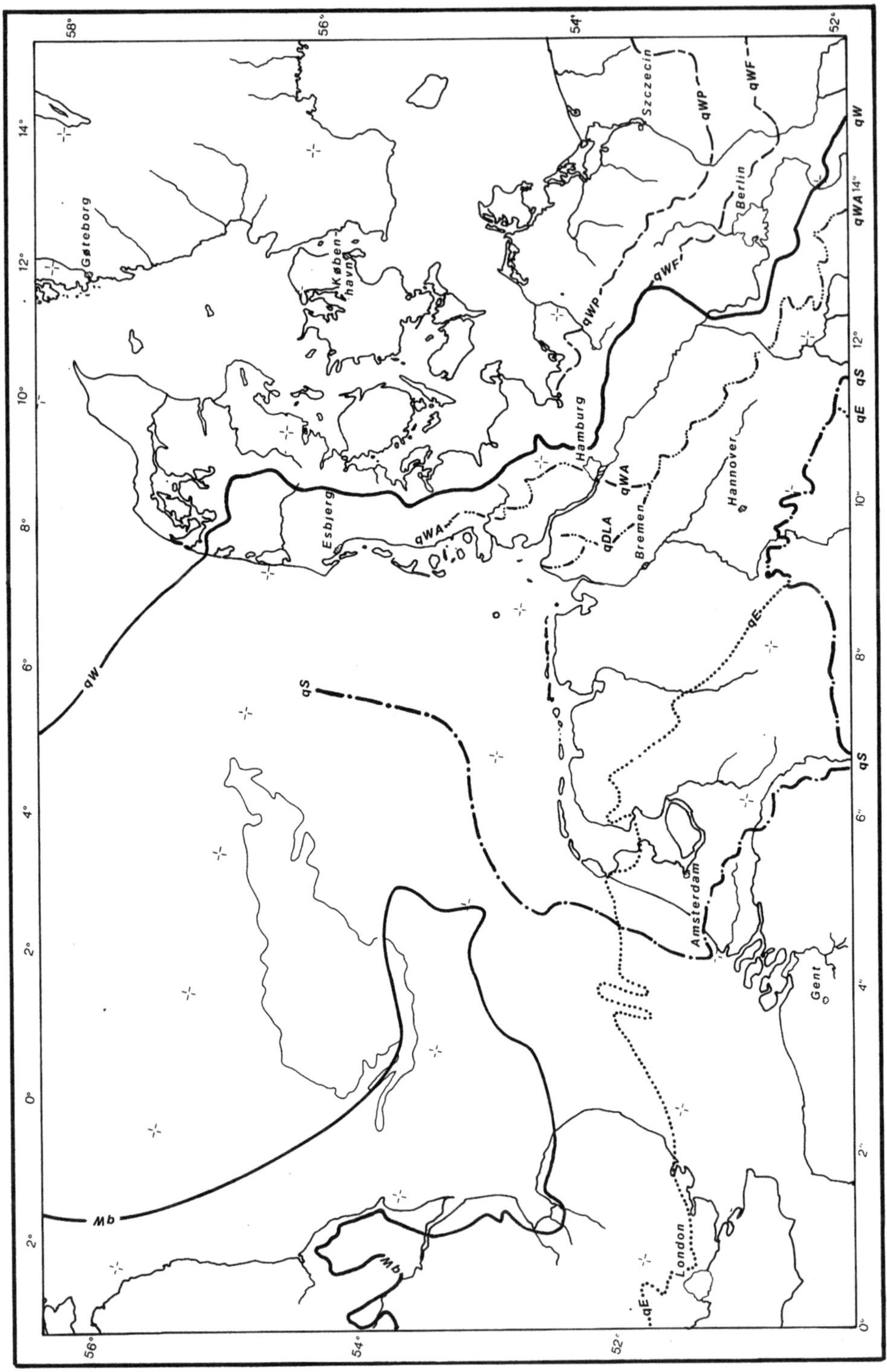
Gøteborg
København
Szczecin
Berlin
Hamburg
Hannover
Bremen
Esbjerg
Amsterdam
Gent
London
qW
qWP
qWF
qWA
qDLA
qS
qE
Mb

climatic deterioration. However, differential isostatic movements between the North Sea Basin and peripheral areas may also have caused relative sea-level changes, modification of the paleogeographic pattern, and quantitative variations in sediment discharge of the rivers in the drainage area. As no exact figures are available for the amplitude of the sea-level changes, and only rough estimates exist about the duration of the phases of warm climate; the dynamics of transgressions and regressions during the Lower and early Middle Pleistocene periods are practically unknown.

During the *late Middle and Upper Pleistocene* periods, the Scandinavian and British ice sheets temporarily covered parts of the North Sea area to various extents. The Lower and early Middle Pleistocene deposits mentioned above are truncated by a gently undulating regional unconformity. However, below the major seismic unconformity, there is sedimentological, and marine floral and faunal evidence in the central and northern North Sea for a dramatic deterioration in climatic conditions. According to Long et al. (1988) it occurs above the Middle Pleistocene Brunhes-Matuyama magnetic reversal.

Above the unconformity, which generally occurs less than 40 m below the seafloor (less than 80 m below present sea level), sedimentation in the North Sea Basin has been dominated by glacial erosion and deposition. Three regional glaciations affected NW Europe to different extents during the Elsterian (Anglian), Saalian (Wolstonian), and Weichselian (Devensian) glacial stages, when ice sheets covered parts of the North Sea, Britain, Scandinavia, Denmark, northern Germany, and the Netherlands (Fig. 1).

In the *Elsterian glacial period*, the earliest inland ice covered the area. A continuous ice sheet extended from the British Isles across the southern North Sea to the northern part of the Netherlands and northern Germany. The southern margin of this ice sheet, depicted in Figure 1, is demonstrated by local ice-push deformation of older sediments and by the southern limits of a complex system of anastomosing valleys, in part deeply eroded into early Middle and Lower Pleistocene and Tertiary deposits and filled during the Elsterian glaciation.

These valleys occur in the North Sea and in the adjacent land areas (Long et al. 1988). On the basis of the conspicuously undulating thalweg and the complex infill, it has been concluded that they were eroded subglacially under very high hydrostatic pressure by outbursts of meltwater beneath a continuous cover of land ice. The filling processes started under the ice cover with moraine slumping and the deposition of glaciofluvial sediments. After the ice had retreated from the area, thick accumulations of glaciolacustrine sediments, locally overlain by fluvio-deltaic deposits, filled most of the Elsterian valleys.

During the *Holsteinian interglacial period*, the remaining depressions were filled by marine sediments from the Holsteinian transgression, which extended across most of the North Sea and parts of the adjacent land areas (Fig. 2). The age of the Holsteinian interglacial period is still a subject of controversial discussions and ranges between

Fig. 1. Paleogeographic reconstruction of the ice margins of the Elsterian (*qE*), Saalian (*qS*; younger stages *qDLA* and *qWA*); and Weichselian (*qW*; younger stages *qWF* and *qWP*) glacial periods in the North Sea and in the adjacent onshore areas according to International Quaternary Map of Europe 1:2500000 (1970), Long et al. (1988), and Foged (1987)

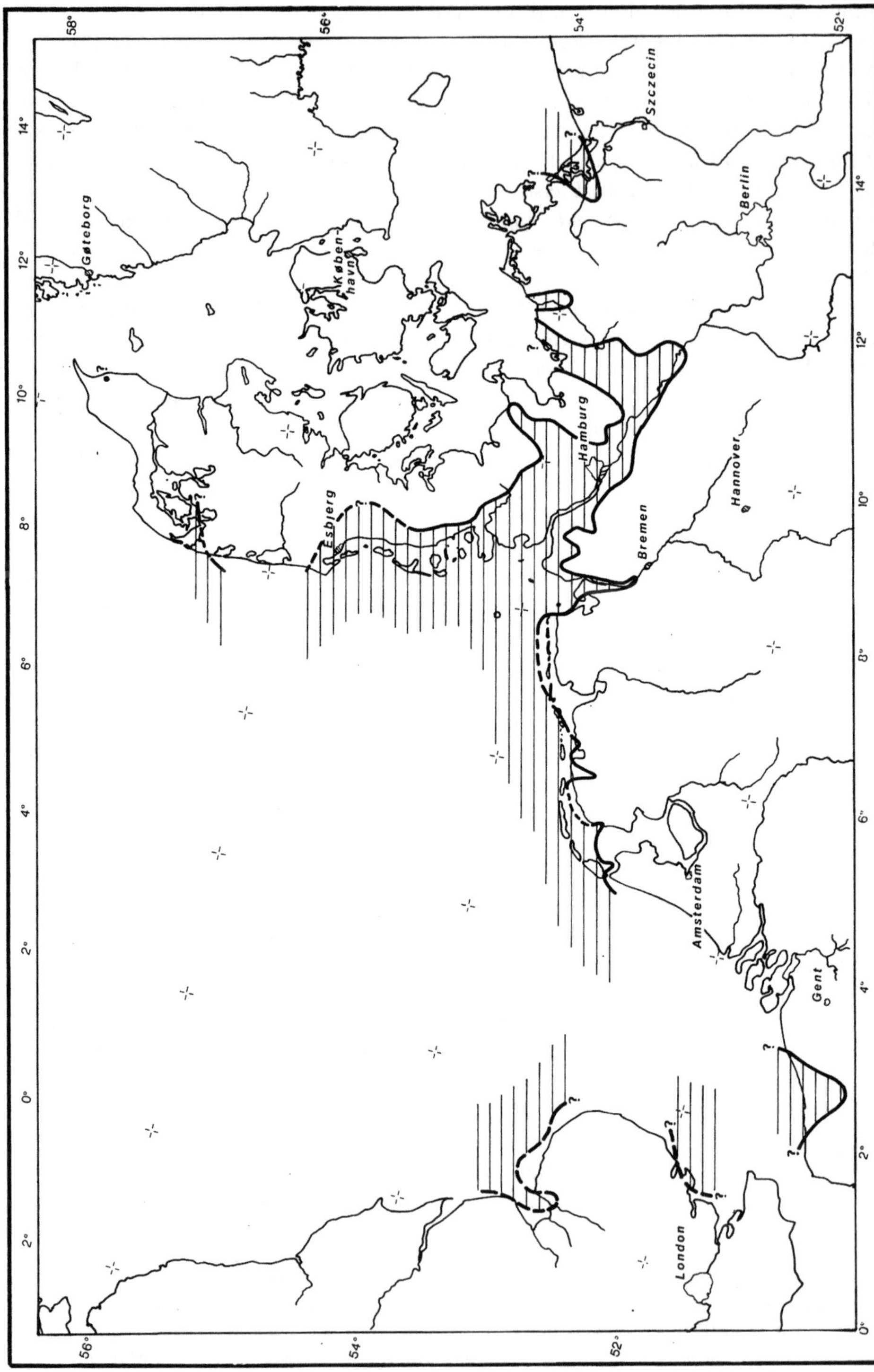
Gøteborg
København
Esbjerg
Szczecin
Berlin
Hamburg
Hannover
Bremen
Amsterdam
Gent
London
58°
56°
54°
52°
14°
12°
10°
8°
6°
4°
2°
0°
2°

195000 years B.P. and possibly 350000 to more than 370000 years B.P. (Linke et al. 1985; Sarnthein et al. 1986; Barabas et al. 1988). On the other hand, reliable dates are available for the duration of this period of warm climate. On the basis of the number of annual laminations in freshwater sediments Müller (1974a) demonstrated a length of 11000 to 12000 years for the Holsteinian interglacial period. Including the subsequent oscillations he assumed that the phase of temperate climate did not exceed 15000 to 16000 years.

The shallow marine sediments of the Holsteinian interglacial period occur in the Hamburg area in a 65-m-thick infill of a förde-like valley system between 10 and 75 m below present sea level. The transgression started with open water conditions and fauna indicating 20 to 30 m water depth. Hints for higher water temperature, shallower water depth and brackish conditions are found in the middle part; and more saline and slightly deeper water conditions recurred in the uppermost marine zone of the sequence (Knudsen 1986). However, as the thalweg of the preexisting valleys is not known in detail, knowledge about the Holsteinian transgression cycle remains very incomplete.

Additionally, the Holsteinian marine deposits occur on the margins of the southern North Sea between 25 m above sea level in East Anglia, −25 to −40 m in the Netherlands, and −20 to −30 m in northern Germany (Long et al. 1988). The difference in elevation is partly attributed to ice-push processes during the Saalian glaciation and partly to post-Holsteinian isostatic processes. All these facts make it difficult to draw reliable conclusions with regard to the amplitude and the dynamics of sea-level changes in the course of this interglacial period.

At the maximum extent of the *Saalian glaciation*, an extensive ice sheet covered the entire lowland area of Lower Saxony and parts of the Lower Rhine area, Germany, the northern and middle parts of the Netherlands (Fig. 1). In the North Sea, the ice covered an area up to 40 km off the Dutch coast (Long et al. 1988) and up to 104 km off the Danish coast (Foged 1987). The available evidence suggests that there was no connection between Scandinavian and possible Scottish ice sheets across the central or northern North Sea during the Saalian glaciation.

The *Eemian interglacial period* correlates with stage 5e of the deep-sea record (Shackleton and Opdyke 1973), which is dated between 125000 to 115000 years B.P. Marine sediments of this interglacial period have been identified at many sites in the southern North Sea and along the adjacent coastal zone (Fig. 3). They consist of shallow marine and intertidal deposits.

A length of 10000 years for the Eemian interglacial period has been determined by Müller (1974b) on the basis of the number of annual laminations in freshwater sediments. No marine beds have been recorded from the early Eemian period, but marine deposition has been demonstrated following Eemian zone IIIa. Fauna and flora have indicated climatic amelioration during the Eemian transgression from subarctic or high boreal conditions towards temperatures that were at least as warm as they are at present in the southern North Sea.

Although the record of the transgression cycle of the Eemian interglacial period is incomplete, rough estimates point at a general sea-level rise of about 100 m. Only the

Fig. 2. Paleogeographic map of the marine deposits of the Holsteinian interglacial period in the southern North Sea region

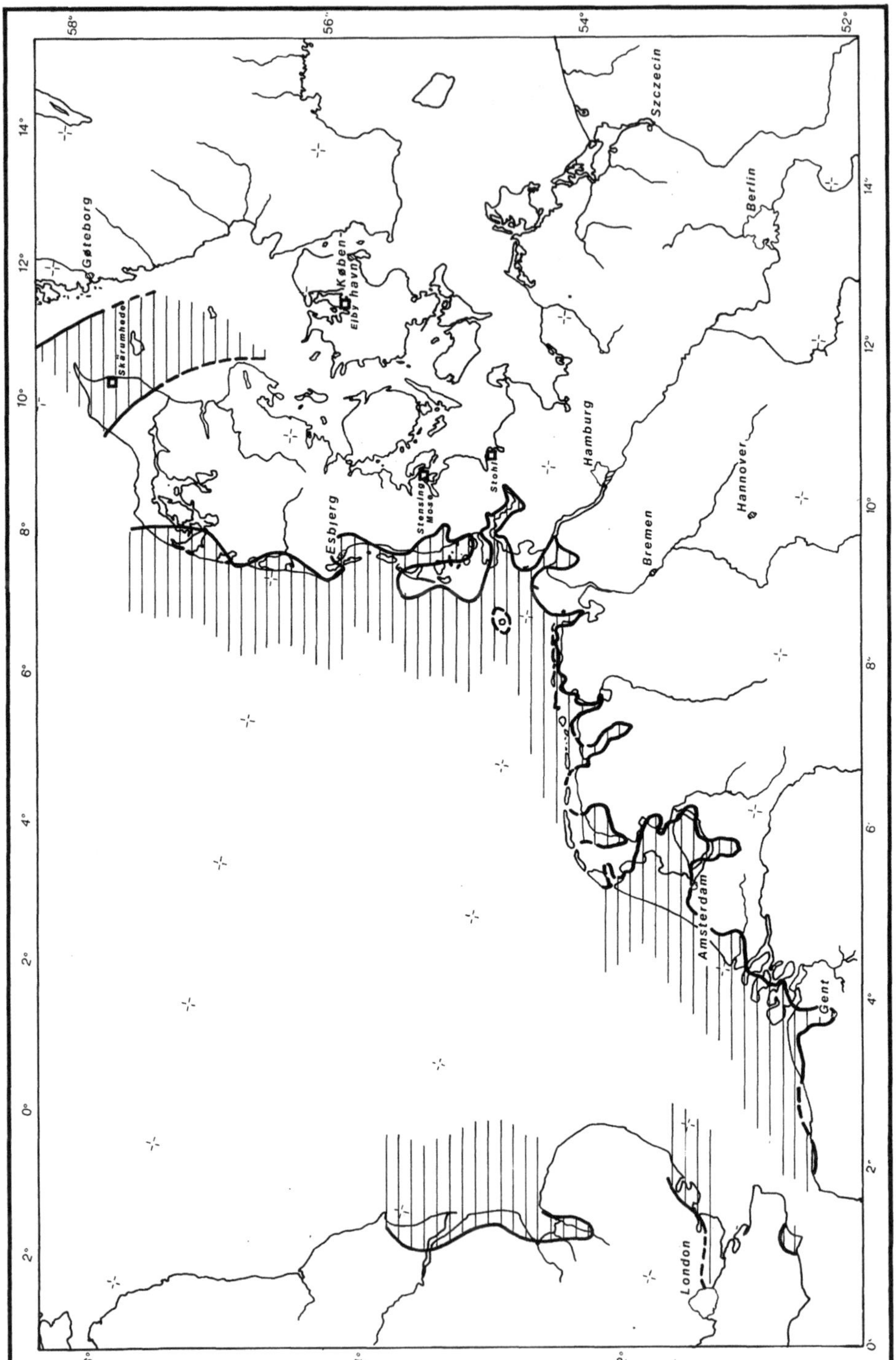
Göteborg
København
Skærumhede
Esbjerg
Stensing Mose
Stohl
Szczecin
Hamburg
Berlin
Bremen
Hannover
Amsterdam
Gent
London

uppermost 35 m of the transgression cycle can be reconstructed on the basis of palynologically dated transgression levels (Zagwijn 1983; Streif 1989b). The Eemian pollen zones IIIa and early IIIb indicate a very rapid sea-level rise. During this 450-years period, the average rate of sea-level rise was 4 m per century. It slowed down to about 0.6 to 0.7 cm per century during the time of pollen zone IIIc. Then a stagnation period of 4000 to 4800 years occurred during the period of the Eemian pollen zone IV and the beginning of zone Va. This was followed by a regression that lowered the sea level about 25 m during the time of the late Eemian pollen zones Vb and VI. The sea-level curve, for the Eemian interglacial period shows a uniform transgression – regression cycle with a steadily rising sea level, a period of stagnation, followed by a phase of steady lowering (Fig. 4). Indications for subordinate oscillations within this cycle or distinct phases of repeated retardation or acceleration of the sea-level rise have not been found.

In eastern Britain, marine terraces of the Eemian interglacial occur in their highest position at about 7 m above present sea level. The surface of marine Eemian sediments ranges at −5 m in Belgium and between −8 m in the western and −13 m in the northern part of the Netherlands. Along the German North Sea coast the surface lies at −9 to −7 m in Lower Saxony and at −5 m in western Schleswig-Holstein. This range of elevations shows a similar pattern to the Holsteinian units suggesting that both have been influenced by isostatic processes.

In the course of the Weichselian glacial period, sea level was always more than 35 m below its present elevation in the southern North Sea area. The phase of climatic deterioration to fully glacial conditions (25000 to 15000 years B.P.) caused regional sea level in the North Sea to fall to at least 100 m below its present height. This amount of lowering is in good accordance with dates from the Atlantic coasts of the USA and Africa (Emery 1969; McMaster et al. 1970) and gives strong hints on a worldwide eustatic lowering of the ocean level in lower and middle latitudes.

Ice sheets covered only smaller parts of the North Sea are during the maximum of the Weichselian glaciation, and there was no connection between the British and the Scandinavian ice sheets across the North Sea (see Fig. 1). In this phase, the Weichselian rivers discharged into the central North Sea. The most important drainage system of the north German hinterland was the „Elbe-Urstromtal“ (Figge 1980), which was the ice marginal valley during the Weichselian glacial maximum. The terrestrial parts of the present North Sea were exposed to periglacial conditions during the Weichselian glacial period.

During the Weichselian late glacial period (15000–10000 years B.P.), the central North Sea was a shallow brackish marine sea with a cover of sea ice (Stoker and Long 1984). As the temperature and sea level rose, the area of the sea ice became smaller and the sediments at the bottom of the shallow sea ceased to be reworked; Dinoflagellate cyst assemblages indicate that sea ice was absent from this are during the Bölling interstadial (Long et al. 1986), during which there was an influx of warmer Atlantic waters. Low salinity values and evidence of rapid sedimentation, possibly together with water stratification, suggest significant continued input from meltwater

Fig. 3. Paleogeographic map of the marine deposits of the Eemian interglacial period in the southern North Sea region

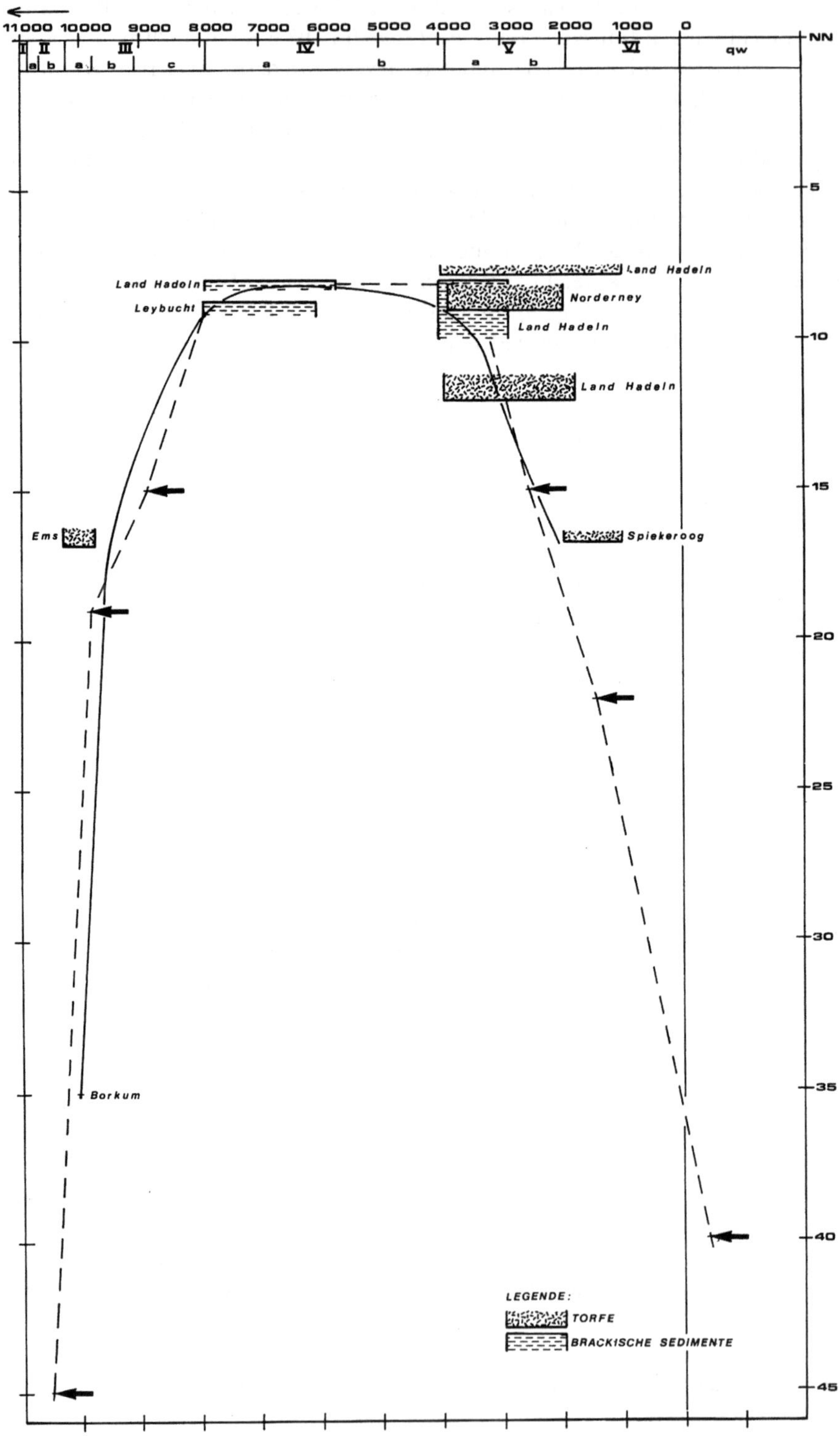
11000 10000 9000 8000 7000 6000 5000 4000 3000 2000 1000 0
I II III IV V VI qw
a b a b c a b a b
NN
5
10
15
20
25
30
35
40
45
Land Hadeln
Leybucht
Land Hadeln
Norderney
Land Hadeln
Land Hadeln
Ems
Spiekeroog
Borkum
LEGENDE:
TORFE
BRACKISCHE SEDIMENTE

sources. During the Younger Dryas, the connection with the North Atlantic remained open and there was only limited pack ice.

Fossil shorelines and heights of the sea level have been reconstructed for different stages of the Late Weichselian and early Holocene development of the southern North Sea by Jelgersma (1979: Fig. V-11, V-14, V-15) and Oele et al. (1979: color chart). However, these reconstructions are highly hypothetical because they are based on present-day seafloor morphology and water depth. In many cases, the thickness of Holocene deposits has not been taken into consideration and the effects of younger erosion, which in places has had a considerable impact, have been neglected.

During the *Holocene*, the earliest brackish-marine incursion into the southernmost North Sea may have occurred as early as 10000 years B.P. (Eisma et al. 1981). With continuing rise in sea level, tidal flat sedimentation became more widespread between 9000 and 8000 years B.P. and fully marine conditions spread out over most of the Southern Bight after 7000 years B.P. As datable transgression levels have not been detected in the southern North Sea region, the dynamics of sea-level rise during the phase of the Weichselian deglaciation and the early Holocene period (ca. 17000 to 8500 years B.P.) is unknown. From morphological finds it can be concluded that the sea level has risen from about −110 to −45 m within this time span.

Only the uppermost 46 m of the Holocene sea-level rise can be reconstructed on the basis of reliable radiocarbon dates. Ludwig et al. (1981) compiled sea-level dates and a series of new palynological and radiocarbon dates from the German sector of the North Sea. These were later supplemented by further dates (Behre et al. 1984). All dates available are compiled in Figure 5 and point to a rapid and uninterrupted rise of the North Sea from −46 m to −15 m in the time interval between 8600 and 7100 B.P. at an average rate of 2.1 m per century. This rise in sea level resulted in a rapid shift of the coastline landward and it occurred with transgressive overlaps of Holocene marine deposits on top of terrestrial and fluvial Pleistocene formations or Holocene peat. No examples of regressive overlaps are known from this phase of development of the North Sea.

The younger part of the Holocene transgression is recorded in the sequence of coastal deposits. This body of sediments reaches its maximum thickness of 35 m in the region of the barrier islands and on the seaward side of the open tidal-flat areas. It becomes thinner towards the landward part of the tidal flats and in the coastal marshes, and wedges out against the Pleistocene hinterland. Its seaward part consists of pure shallow marine to intertidal sand and silt. The middle part is built up by a cyclic sequence of clastic tidal and lagoonal deposits and intercalated peat layers. These intercalated layers merge together into continuous peat sequences close to the Pleistocene hinterland.

The cyclic alternation of transgressive overlaps (marine and brackish deposits overlying peat) and regressive overlaps (limnic sediments to semi-terrestrial peat overlying brackish and marine deposits) indicates repeated water-level fluctuations. On the basis of a great many radiocarbon dates, Streif (1989a) reconstructed these processes in a band-like sea-level curve of the last 8000 years. This curve is depicted in Figure 6.

◄

Fig. 4. Time/depth diagram of the transgression–regression cycle of the Eemian interglacial period. *Dashed line* after Zagwijn (1983), *solid line* after Streif (1989b)

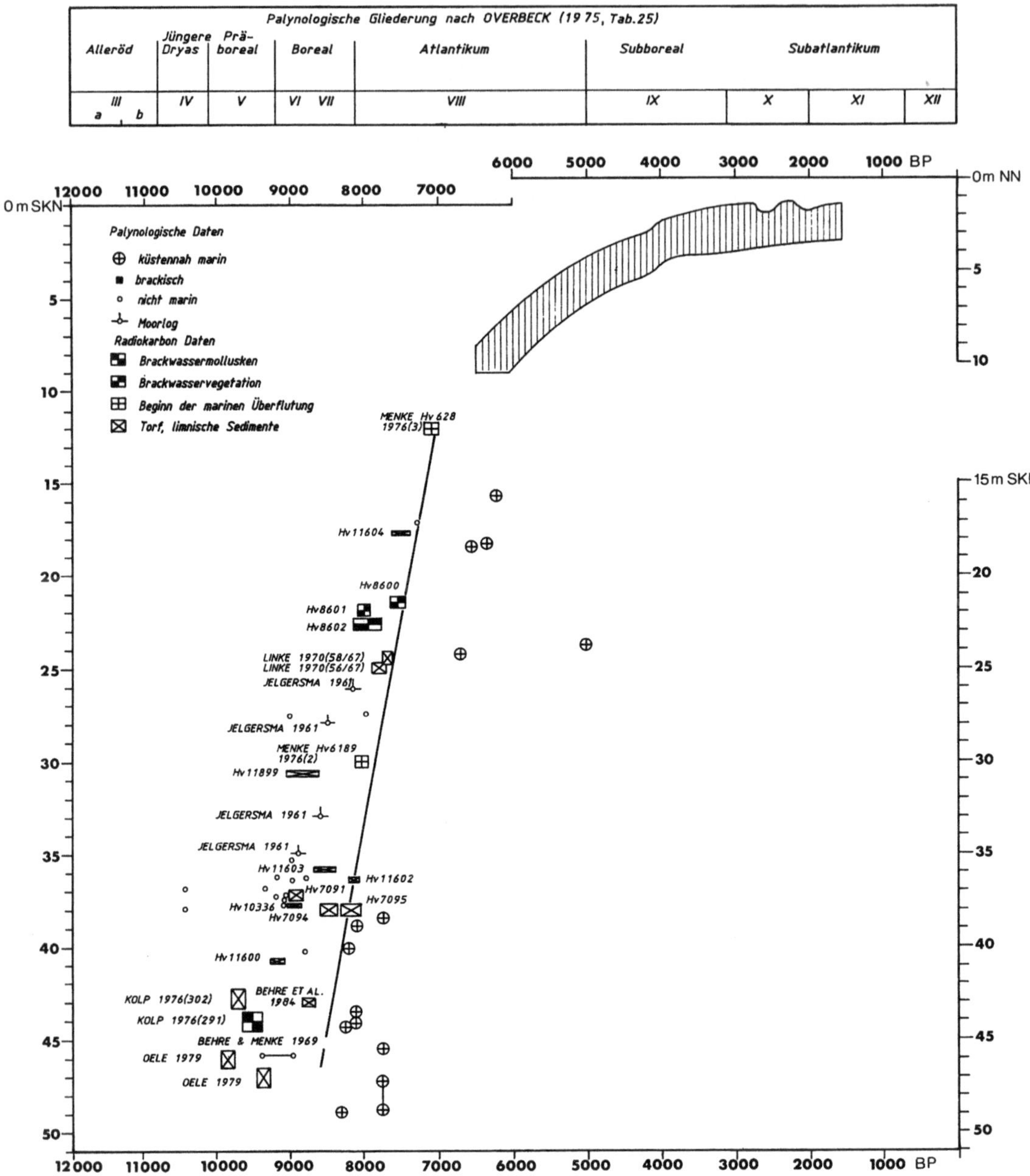

Fig. 5. Time/depth diagram of pollen- und radiocarbon dates from the Holocene transgression in the southern North Sea and the coastal zone (*hatched area*)

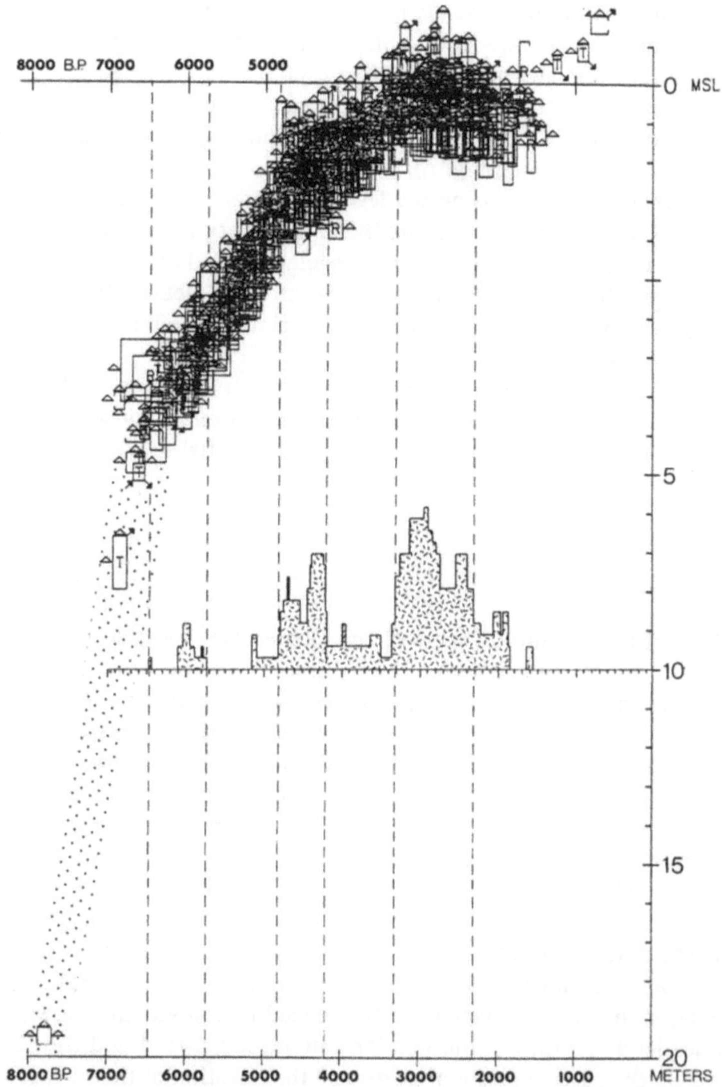

Fig. 6. Time/depth diagram of conventional radiocarbon dates from basal and intercalated peat layers from the coastal zone between the rivers Ems and Elbe. The graph shows a band like sea-level curve in the upper and the frequency distribution of samples from intercalated peat layers in its lower part

The older dates obtained in the coastal zone confirm the trend of sea-level rise demonstrated for the offshore area with a rapid and uninterrupted rise in sea level. However, from 6500 years B.P. onwards, temporarily opposite trends are indicated by repeated regressive overlaps of peat layers over clastic sediments of marine origin. Such regressive overlaps occur in phases of slowly rising water table, when the rate of bog growth fully compensates or even exceeds the contemporary rise of the sea level. The oldest intercalated peats dating from about 6500 and 6000 B.P. have a very limited distribution. They occur locally in valley-like depressions in the pre-Holocene surface. Ideal conditions for the formation of intercalated peat apparently prevailed between 4800 to 4200 B.P. and between 3300 to 2300 B.P. Intercalated peats from these periods occur throughout the coastal lowland area between the rivers Ems and Elbe. A small number of samples from locally occurring intercalated peat layers yield dates between 2000 to 1600 B.P. (see Fig. 6).

Within the general trend of rising sea level, temporary falls can be demonstrated at about 2800 B.P. and 2000 B.P. They are indicated by horizons of decomposed peat, by changes from fen peat to raised-bog peat vegetation, and by soil formation on top of clastic sediments (Preuss 1979; Behre and Streif 1980; Behre 1986). Additionally, there is strong evidence for a third phase of lowering of the sea level between 1125 and 1395 A.D. (Streif 1986).

Altogether, these finds reveal a complex transgression cycle during the middle and younger Holocene periods. It shows repeated phases of retardation and acceleration of the rise in sea level and, additionally, short phases of 100 to 300 years lowering of the sea level.

3 Conclusions

Characteristics of the 2.5 million years of the Quaternary are cyclic climatic changes which influenced the balance between the volumes of the oceans and the inland-ice sheets and which generated repeated eustatic sea-level changes. Numerous climatic cycles have been demonstrated for the Lower and early Middle Pleistocene. Although the age and the duration of these climatic oscillations are not known yet in detail, it has to be assumed that they coincided with variations of the sea level. Periods of maximum climatic deterioration may be expected to correspond with regional lowering and warm periods with high stands in sea level. Comparable regression – transgression cycles occurred in the late Middle and young Pleistocene and in the Holocene.

Phases of warm climate during the Holsteinian and Eemian interglacial periods have a similar length of 10000 to 12000 years, and the postglacial period of warm climate, the Holocene, has lasted 10000 years. The amplitude and the dynamics of sea-level changes during the Holsteinian interglacial period are not known. The total amount of sea-level rise during the Eemian interglacial period and the Holocene is about 100 m. On the basis of regional studies on the elevation of the surface of marine Eemian and Holocene sediments it can be demonstrated that at least 80% of the Eemian and 95% of the Holocene sea-level rise have to be attributed to eustatic sea-level changes and only 20% or 5% are caused by tectonic subsidence, isostatic movements or other influences. Consequently, the amplitudes of the transgression – regression cycles have to be considered as representative for worldwide eustatic processes in non glaciated tectonically relatively stable areas.

However, from the total amount of 100-m sea-level rise, only the uppermost 35 m or 45 m, respectively, can be reconstructed in detail. Although the amplitudes compare quite well, the dynamics of sea-level rise show clear differences which most probably mirror differences in the climatic development. The Eemian interglacial is characterized by a simple cycle with a very rapid rise in sea level (up to 4 m per century), a phase of stagnation, and a regression at the end of the warm period. The Holocene started with a lower rate of sea-level rise (2.1 m per century), followed by repeated short phases of retardation, stagnation, and even lowering of the sea level. From a paleoclimatical point of view, the thermal optimum of the Holocene occurred about 5000 to 5500 years ago. Trusting in the cyclicity of the natural Quaternary processes, we have to assume that we are approaching the end of the present warm period and cooling has to be expected within 1000 to 2000 years. However, unknown factors are human impacts on natural processes and the stability of the natural system.

In general, it must be stated that the present-day distribution of water and ice on the earth's surface, as well as comparable interglacial conditions, existed only during relatively short periods of the Quaternary of about 5000 years. During much longer time intervals, the ice-water balance differed distinctly from the present day and interglacial conditions. The sea level was more than 35 m and temporarily up to 100 m below its present elevation. From these finds, it also has to be concluded that present day tidal conditions are significantly different from those prevailing during most of the Quaternary.

References

Barabas M, Sarnthein M, Stremme H (1988) The age of the Holstein interglaciation: a reply. Quat Res 29:80–84, 2 Fig, 1 Tab

Behre K-E (1986) Meeresspiegelverhalten und Besiedlung während der Zeit um Christi Geburt in den Nordseemarschen. Offa 43:45–51, 2 Fig

Behre K-E, Streif H (1980) Kriterien zu Meeresspiegel- und darauf bezogene Grundwasserabsenkungen. Eiszeitalter Ggw 30:153–160

Behre K-E, Dörjes J, Irion G (1984) Ein datierter Sedimentkern aus dem Holozän der südlichen Nordsee. Probl Küstenforsch im südlichen Nordseegebiet 15:135–148, 6 Fig, 1 Tab

Cameron TDJ, Stoker MS, Long D (1987) The history of Quaternary sedimentation in the UK sector of the North Sea Basin. J Geol Soc Lond 144:43–58, 9 Fig

Eisma D, Mook WG, Laban C (1981) An early Holocene tidal flat in the Southern Bight. Spec Publ Int Assoc Sediment 5:229–237, 6 Fig, 1 Tab

Emery K (1969) The continental shelves. Sci Am 221(3):107–122, 11 Fig

Figge K (1980) Das Elbe-Urstromtal im Bereich der Deutschen Bucht (Nordsee). Eiszeitalter Ggw 30:203–211, 6 Fig

Foged N (1987) The need for Quaternary geological knowledge in geotechnical engineering. Boreas 16(4):419–424, 4 Fig, 1 Tab

International Quaternary Map of Europe (1970) Sheet 6 Kopenhagen; Hannover

Jelgersma S (1979) Sea-level changes in the North Sea basin. In: Oele E, Schüttenhelm RTE, Wiggers AJ (eds) The Quaternary history of the North Sea. Acta Univ Ups Symp Univ Ups Ann Quing Cel 2:234–248, 7 Fig

Knudsen KL (1986) Foraminiferal stratigraphy of the late Elsterian-Holsteinian sequence in borings at Eggestedt and the Hamburg area. Abstr INQUA Symp Holstein-Interglazial, Hamburg 1986:14–15

Linke G, Katzenberger O, Grün R (1985) Description and ESR dating of the Holsteinian Interglaciation. Quat Sci Rev 4:319–331, 6 Fig, 1 Tab

Long D, Bent A, Harland R, Gregory DM, Graham DM, Morton AC (1986) Late Quaternary palaeobotany, sedimentology, and geochemistry of a vibrocore from the Witch Ground Basin, central North Sea. Mar Geol 73:109–123, 6 Fig, 1 Tab

Long D, Laban C, Streif H, Cameron TDJ, Schüttenhelm RTE (1988) The sedimentary record of climatic variation in the southern North Sea. In: Shackleton FRS, West FRS, Bowen DQ (eds) The past three million years: Evolution of climatic variability in the North Atlantic region. Phil Trans R Soc Lond B 318:523–537, 4 Fig

Ludwig G, Müller H, Streif H (1981) New dates on Holocene sea-level changes in the German Bight. Spec Publ Int Assoc Sediment 5:211–219, 2 Fig, 1 Tab

McMaster RL, Lachance TP, Ashraf A (1970) Continental shelf geomorphology features of Portuguese Guinea, Guinea and Sierra Leone, West Africa. Mar Geol 9(3):203–213, 5 Fig

Müller H (1974a) Pollenanalytische Untersuchungen und Jahresschichtenzählungen an der holsteinzeitlichen Kieselgur von Munster-Breloh. Geol Jb A 21:107–140, 10 Abb, 2 Tab

Müller H (1974b) Pollenanalytische Untersuchungen und Jahresschichtenzählungen an der eemzeitlichen Kieselgur von Bispingen/Luhe. Geol Jb A 21:149–169, 5 Abb, 3 Tab

Oele E, Schüttenhelm RTE, Wiggers AJ (eds) (1979) The Quaternary history of the North Sea. Acta Univ Ups Symp Univ Ups Ann Quing Cel 2:248

Preuss H (1979) Die holozäne Entwicklung der Nordseeküste im Gebiet der östlichen Wesermarsch. Geol Jb A 53:3–84, 25 Fig, 6 Tab

Sarnthein M, Stremme HE, Mangini A (1986) The Holstein Interglaciation: time-stratigraphic position and correlation to stable-isotope stratigraphy of deep-sea sediments. Quat Res 26:283–298, 5 Fig, 3 Tab

Shackleton NJ, Opdyke ND (1973) Oxygen-isotope and paleomagnetic stratigraphy of equatorial Pacific core V28–238: Oxygen-isotope temperatures and ice volumes on a 10^5 Years Scale. Quat Res 3(1):39–55, 9 Fig, 4 Tab

Stoker MS, Long D (1984) A relict ice-scoured erosion surface in the central North Sea. Mar Geol 61:85–93

Streif H (1986) Zur Altersstellung und Entwicklung der Ostfriesischen Inseln. Offa 43:29–44, 3 Fig, 6 Tab

Streif H (1989a) Barrier islands, tidal flats, and coastal marshes resulting from a relative rise of sea level in East Friesia on the German North Sea coast. Proc KNGMG Symp Coastal Lowlands, Geology and Geotechnology 1987. Kluver, Dordrecht, pp 213–223, 3 Fig, 2 Tab

Streif H (1989b) Zum Ausmaß und Ablauf eustatischer Meeresspiegelschwankungen im südlichen Nordseegebiet seit Beginn des letzten Interglazials. In: Frenzel B (ed) Klimageschichte des Holozäns und des letzten Interglazials. Paläoklimaforschung 1:12, 2 Fig, 1 Tab

Zagwijn WH (1974) The palaeogeographic evolution of the Netherlands during the Quaternary. Geol Mijnbouw 53(6):369–385, 8 Fig, 12 Maps

Zagwijn WH (1979) Early and Middle Pleistocene coastlines in the southern North Sea basin. In: Oele E, Schüttenhelm RTE, Wiggers AJ (eds) The Quaternary history of the North Sea. Acta Univ Ups Symp Univ Ups Ann Quing Cel 2:31–42, 8 Fig

Zagwijn WH (1983) Sea-level changes in the Netherlands during the Eemian. Geol Mijnbouw 62(3):437–450, 7 Fig, 2 Tab

Tidal Rhythms in the Shell Secretion of Living Bivalves

C. A. Richardson

1 Introduction

The periodicity of microgrowth patterns in the shells of living bivalve molluscs has been the subject of considerable research during the past two decades. The early literature relating to research during the 1960's and 1970's has been extensively reviewed by among others Clark (1974), Scrutton (1978) and Lutz and Rhoads (1980). I do not therefore propose to go over in detail this by now familiar work, other than to draw briefly upon examples to illustrate specific points. For the first meeting in 1977 Scrutton (1978) reviewed current knowledge of how growth increments form in corals, bivalves and stromatolites and how faithfully they record elapsed time. At the second meeting Ohno (1983) presented a paper on the variability of the formation of growth increments in cockles *Cerastoderma* (= *Cardium*) *edule* shells. It is my intention to present to this meeting a review of the current knowledge of tidal microgrowth patterns in living bivalve molluscs, using illustrations from my own work as well as other researches carried out during the past decade. The key to understanding the periodicity of the microgrowth patterns in acetate peels of polished and etched shell sections has been to study the way in which the organism reacts to environmental fluctuations. The only way to do this is to work with living material. Once the patterns are fully understood, then perhaps the results of the work can be extrapolated to fossil relatives from the geological past.

Until the late 1970's the concept of daily microgrowth patterns in bivalve shells had been widely accepted as the status quo. This conclusion was based on two independant studies. In the United States of America Pannella and MacClintock (1968) counted the number of growth increments in *Mercenaria mercenaria* that had been marked and transplanted into the natural environment for 368 and 723 days respectively and concluded that daily bands were produced. On the other side of the Atlantic in the United Kingdom House and Farrow (1968), Farrow (1971, 1972) and later Whyte (1975) similarly claimed that the results of analysis of cockle shells, *C. edule* collected on different occasions showed that increments were deposited with a solar periodicity. This view was shared by other workers, among them Kennish (1978, 1980), and more recently has received support from Fritz and Lutz (1986) working on the freshwater bivalve *Corbicula fluminea*. Evans (1972, 1975) was the first to question whether the accepted view of daily bands held for all bivalve species. Evans noticed that when he matched the patterns of bands in the shell of the basket cockle *Clinocardium nuttalli* collected from the natural environment with the tidal prediction for the

School of Ocean Sciences, Marine Science Laboratories, Menai Bridge, Anglesey, LL59 5EH, GB.

area in which they were collected, the patterns had a tidal periodicity. Due to the mixed semi-diurnal tidal patterns, cockles produced two bands a day when they were emersed twice a day at neaps but produced only one clear band a day when they were emersed during spring tides. In 1975 a series of papers reporting the results of studies carried out in the early part of the 1970's appeared in *Growth Rhythms and the History of the Earth's Rotation* (eds. Rosenberg and Runcorn). The general view at that time was that the bands in bivalve shells had a daily periodicity, although Evans (1975) demonstrated otherwise. Dolman's (1975) contribution on the periodicity of patterns in cockle *C. edule* shells from the North East of England further strengthened the idea that some shells might contain tidal bands. However, it was in the late 1970's and early 1980's that an extensive series of experiments demonstrated unequivocally that the banding patterns in the intertidal cockle *C. edule* were tidally produced (Richardson et al. 1979, 1980a, 1981). Richardson et al. (1979 et seq.) date-marked the shells of *C. edule* and grew them experimentally in the laboratory and under natural conditions. Such experiments showed that the commonly held view that all bivalve shell banding patterns were of daily origin was an incorrect one. Irrefutable evidence was presented which demonstrated that the banding in intertidal cockles was of a tidal periodicity, one band being produced every time the animals were emersed.

In this paper I will review the experiments relating to the formation of tidal bands in bivalve shells and show the importance of the environment in influencing the periodicity of the patterns, using examples from several different species.

2 Experimental Approach

One way to clearly understand the periodicity of the microgrowth patterns in bivalve shells is to date-mark the shell at a given point during the life of the animal to which all subsequent growth can be related. Several methods have been tried including 1) cold shock marking: animals are maintained out of water in a moist atmosphere in a refrigerator at 4 °C for periods from 24–72 h (Richardson et al. 1979; Richardson 1987a, b, 1988b and 1989); 2) abrasion, in which the growing margin of the shell is either gently abraded with a file (Pannella and MacClintock 1968; Ekaratne and Crisp 1982; Richardson 1987b and 1989) or notched with a small drill (Fritz and Lutz 1986); and 3) chemical treatment; immersion in 2.25×10^{-4} M acetazolamide has been used successfully by Ekaratne and Crisp (1982). The techniques rely on the fact that the normal pattern of shell banding is temporarily interrupted by the marking procedure. The severity of the procedure is reflected in the disruption of the growth increments in the post-shock period of growth. Thus, Richardson (1987b and 1989) found that file marking disrupted shell deposition and the increments in the post-shock region were narrow and compressed for several days after marking. Similarly, Fritz and Lutz (1986) suggested that injury caused during notching of the ventral shell margin of *C. fluminea* resulted in the poorer growth of specimens after transfer to experimental cages. However Richardson (1987a and 1989) found that cold shock marking, exposure to 24 h periods at 4 °C, was superior to file marking. The experimental bivalves responded by depositing a strongly defined line which could nearly always be identified in acetate peel replicas of shell sections.

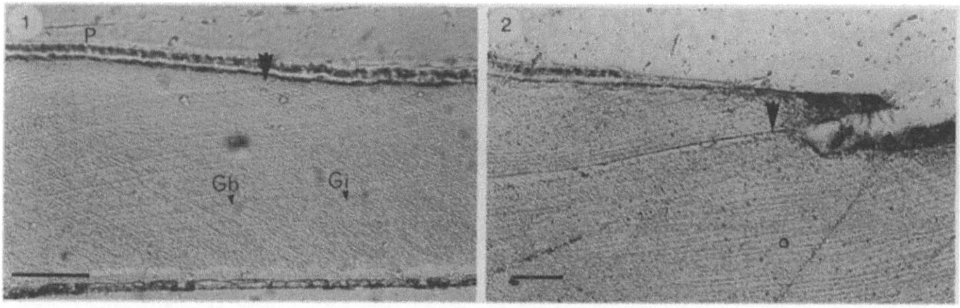

Fig. 1. Acetate peel replica of a shell section through *Mytilus edulis* in the vicinity of the cold shock mark. The mussel responds by depositing a strong band (*arrow*) with widely spaced tidal growth increments in the post-shock region. *Gb* growth band; *Gi* growth increment; *P* periostracum. Bar line = 50 μm

Fig. 2. Acetate peel replica of a section through *Mytilus edulis* from the intertidal shore in which the mussel was file marked (*arrow*). The normal pattern of growth is disrupted for several days in the post-shock region of the shell. Bar line = 100 μm

Figure 1 shows the appearance of the bands in an acetate peel of the shell of *Mytilus edulis* in the vicinity of the cold shock mark. The mussel responds by depositing a strong clearly defined band. Compare this with Figure 2 in which the outer shell margin has been abraded with a file. The subsequent deposition of the bands has been disrupted by the filing which interrupts the normal pattern of bands and narrow increments result. Marking internal bands has been a successful approach used by several workers. The organisms are marked, transplanted into the natural environment or transferred to simulated conditions and then recovered after various periods of time. The length of time the animals have been left in their respective environments has ranged from 1 to 2 years (Pannella and MacClintock 1968), 5–6 months (Fritz and Lutz 1986) and 1–2 months (Richardson et al. 1979; Richardson 1987a, 1988b, 1989), sufficient periods of time to study the effect of long-term seasonal fluctuations in water temperature, the short-term influence of the neap – spring tidal cycle and the effects of emersion and submersion on the formation of the patterns.

3 Effects of Tidal Emersion

The effect of tidal emersion on the formation of growth lines in bivalve shells is typified by the common European cockle *Cerastoderma* (= *Cardium*) *edule*. An acetate peel of a section through a cockle shell reveals a series of alternating dark and light lines. Figure 3 shows the appearance of the pattern. The darker narrower growth bands are fomed during tidal emersion, while the lighter wider growth increments are formed when the animal is immersed and actively feeding. Cockles transplanted into the natural environment at different tidal levels have been shown to produce one band each time they are emersed at low tide (Richardson et al. 1979, 1980a; Ohno 1983, 1985). Richardson et al. (1981) clearly showed from samples of cockles collected approximately every hour throughout a 27-h period that shell deposition occurred when they were actively feeding during immersion. Once the animals were emersed

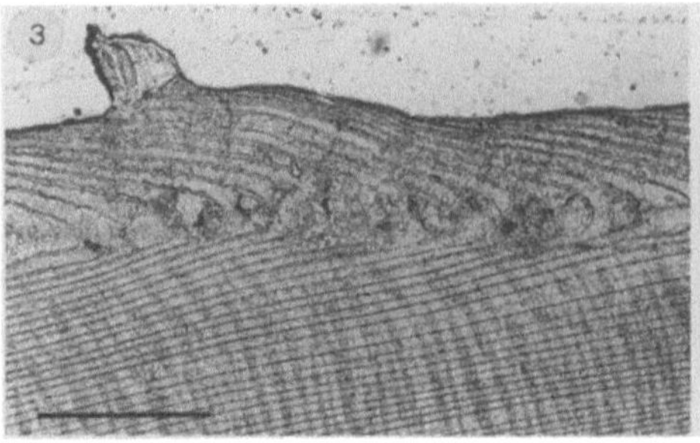

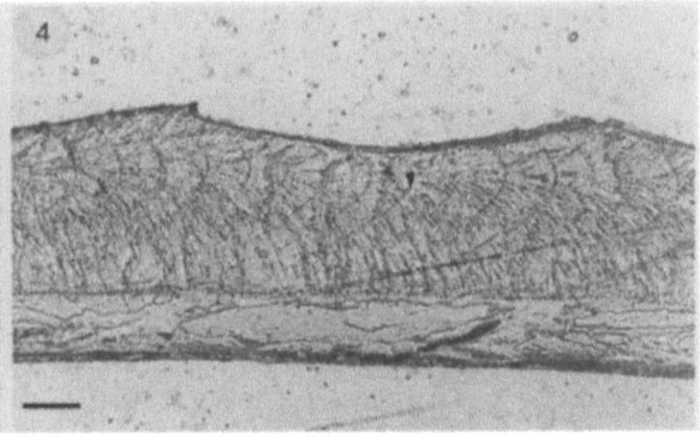

Fig. 3. Peel from a cockle shell *Cerastoderma edule* collected from below mean tide level. The dark narrow growth bands are formed during tidal emersion while the lighter growth increments are formed when the animal is immersed and actively feeding. Bar line = 100 μm

Fig. 4. Peel of a section through a clam *Tapes philippinarum* shell grown in continuously immersed conditions in the laboratory. The weak patterns (*arrow*) are related to the growth of the shell, not to the tides or days. Bar line = 100 μm

the rate of shell deposition became slower resulting in the formation of a dark band. Since emersion is responsible for the production of bands it seems reasonable to expect that the position on the shore of the animals and the precise form of the neap – spring tidal cycle will be important in influencing the appearance of the banding patterns.

As early as 1972, Evans (1972, 1975) observed the importance of the neap – spring tidal cycle in his study of *Clinocardium nuttalli* on the coast of Oregon, USA, in which the effect of bidaily tidal emersion at neaps was clearly enhanced. More recently Richardson et al. (1980b) carried out detailed short term measurements from shell sections of marked cockles *C. edule* from North Wales, U.K., in which each tidal increment could be related to a particular tidal immersion. They found there were small, 5%–10%, but significant differences in growth rate during the neap – spring lunar cycle. At mid- and high-tide levels and when continuously immersed on a raft the animals grew fastest at springs and slowest during neaps. However cockles from low water of spring tides showed narrower increments during spring tides than at neaps at this level. This anomaly was ascribed to the reduction in the period of immersion during springs and continuous immersion during neaps, although the effect represented quite a small change in total growth and the significance could only be demonstrated by repeated analysis over several tidal cycles. The converse was found in cockles near high-water neaps which experienced shorter periods for feeding during neap tides and longer periods during springs. Richardson et al. (1980b) suggested water flow at spring tides could be the factor responsible for the increased growth of fully immersed cockles on the raft, or those near mean tide level on the shore since the proportion of time immersed did not vary between springs and neaps. Richardson et al. (1980b) concluded that the thickness of the growth increments would not therefore be a good guide to determine the timing and position of springs and neaps along the shell banding pattern although alternations in the distinctness of the bands may be (Richardson et al. 1979, 1980a, 1981). Ohno (1983, 1985) similarly noted the importance of the position of the cockle in the intertidal zone in determining the width and clarity of the growth bands, and more recently Lønne and Gray (1988) confirmed these findings when they studied the periodicity of the banding patterns in cockles collected from two tidal levels on the shore in Norway.

Recent work has concentrated on studying how the effects of different forms of the neap – spring cycle affect the appearance of the shell banding patterns in several different bivalve species. Thus Ohno (1985) studied the appearance of the banding patterns in marked specimens of *Fragum unedo* grown along the coast of Ishigaki Island, Japan. He showed that the nature of the tides at Ishigaki Island, a coastline with a mixed semi-diurnal tide, had a profound influence on the banding patterns. Animals on the lower part of the shore were emersed only once a day during spring tides and remained immersed at neap tides. The pattern of bands shown in Figure 4 shows clearly daily emersion lines during spring tides, while at neaps when the animals remained immersed a weak pattern of apparently daily bands continued to be produced. Evans (1972) in his study of *Clinocardium nuttalli* observed clearly defined bands produced at emersion, with weaker bands between the emersion bands even though the animals remained immersed. From Figure 5 (Ohno 1985) no such weak bands could be seen between the daily emersion bands suggesting that *F. unedo* may behave in a different way to *C. nuttalli*. In contrast, Richardson (1987b) working on the Malaysian cockle *Anadara granosa* could find no influence of the mixed semi-diurnal tidal regime on the appearance of the banding patterns at Penang Island, Malaysia. Like Ishigaki Island, the tidal pattern at Penang Island has two highs and

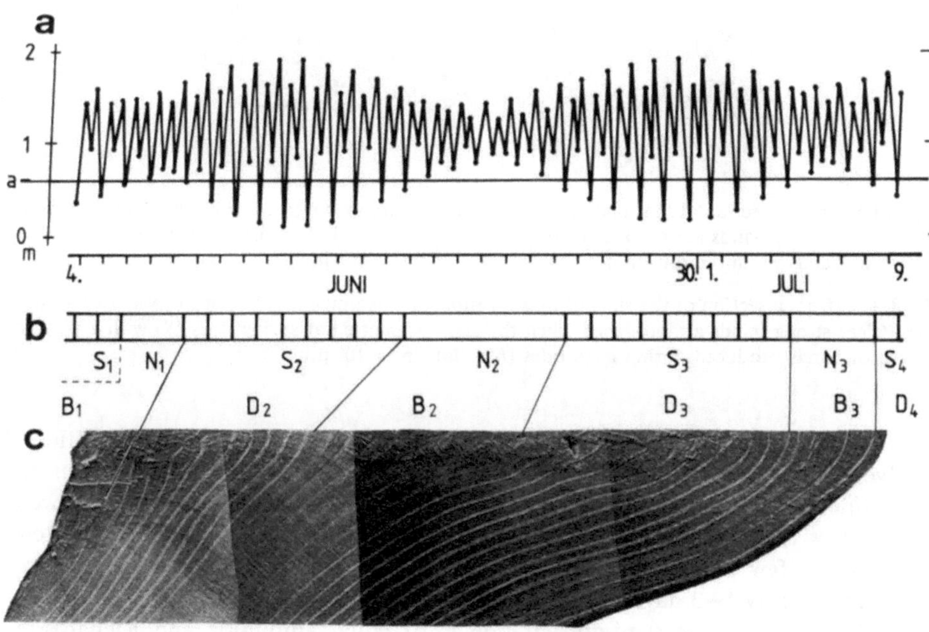

Fig. 5a–c. Alternation pattern of *Fragum unedo* and change of spring and neap tides. **a** The predicted tidal curve along the coast of Ishigaki Island; *a* the level of the cage; **b** Diagram showing the exposure of the cage at each low tide; and **c** Scanning electron micrograph of a section through the shell of *Fragum unedo* to show the alternation in strong and weak bands during spring and neap tides. (Reproduced by kind permission of Dr. Terufumi Ohno from Paleontographia, Abt A, 189 text, figure 25)

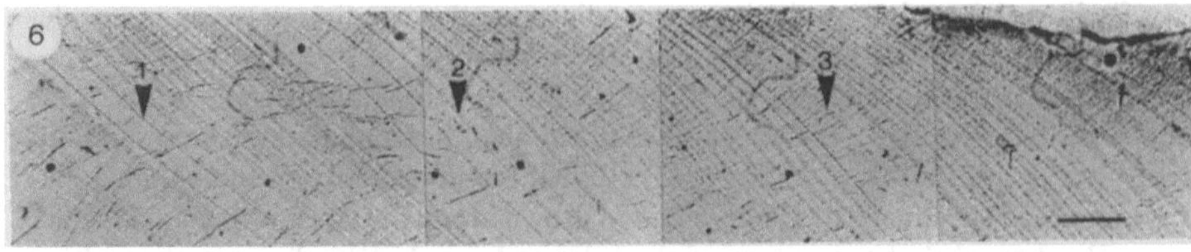

Fig. 6. Peel from a shell of the Malaysian cockle *Anadara granosa* to show the lunar spring-neap cycle of banding. The banding shows a pattern of narrow increments during springs alternating with a few wider increments during neap tides. *From right of figure between consecutive arrows* 26 bands counted in first group, 28 bands in second group and 25 bands in third group. Bar line = 200 μm. (From Richardson 1987b, J Exp Mar Biol Ecol 111, figure 8)

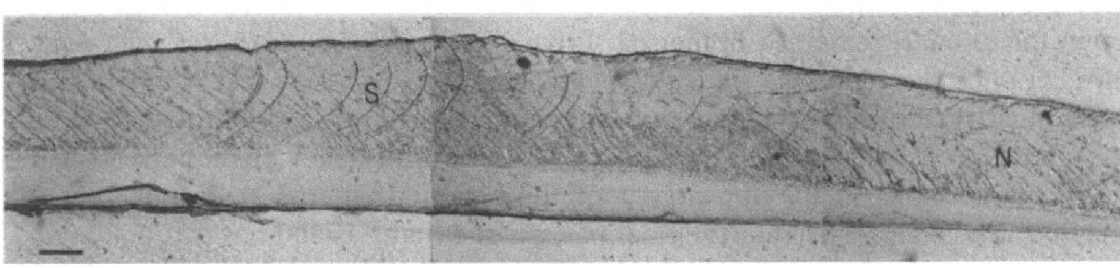

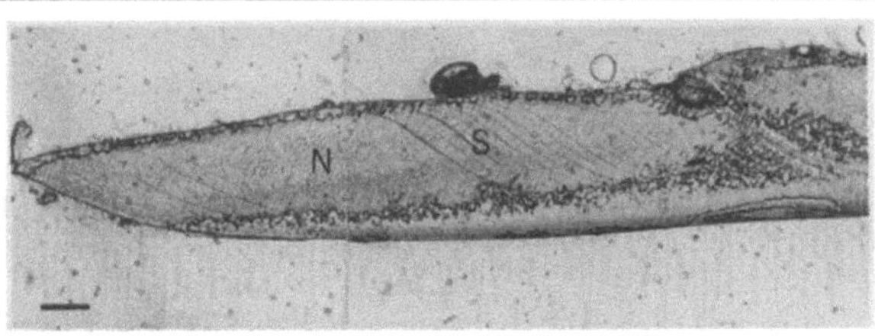

Fig. 7. Peel from a section of the shell of *Tapes philippinarum* grown in a box at low water of spring tides. Clear strong bands are produced when the clam is emersed at spring tides (*S*) with a weaker pattern of bands produced during neap tides (*N*). Bar line = 100 μm

Fig. 8. Peel from a section of the shell of *Spisula subtruncata* grown in a box at low water of spring tides. Clear strong bands are produced when the clam is emersed at spring tides (*S*) with a weaker pattern of bands produced during neap tides (*N*). Bar line = 100 μm

two lows of unequal amplitude each day with the greatest difference at spring tides. Bivalves on the lower part of the shore would be emersed only once a day during spring tides and remain emersed at neap tides. The banding in marked cockles showed a pattern of clear bands separated by narrow increments during spring tides alternating with a few wider increments during neap tides. These wider increments were produced for only 1–3 days and occasionally had weaker bands between (Fig. 6). In North Wales, U.K., the semi-diurnal tide is of equal amplitude and Richardson (1987a, 1988a) has studied the appearance of the band patterns in the shells of the clam *Tapes philippinarum* and the subtidal bivalve *Spisula subtruncata* maintained at extreme low water of spring tides. The appearance of the bands (Figs. 7 and 8) is essentially similar to those seen in shells of the cockle *C. edule* growing at extreme low water of neap tides (Richardson et al. 1979). The bivalves produced strong clear

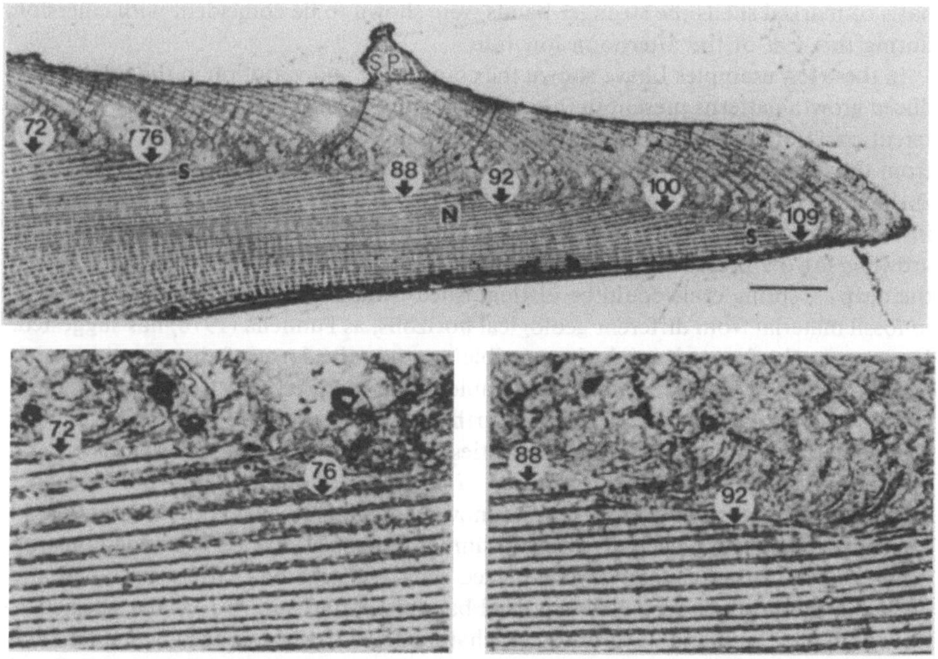

Fig. 9. Peel from *Cerastoderma edule* grown above low water, Abermenai Point, Traeth Melynog, UK between April and July 1977. Alternations in the distinctness of the bands formed during spring tides (*S*) are visible around bands *76* and *104* but barely distinguishable during neap tides (*N*) between bands *90* and *95*. The *enlargements* show these differences more clearly. *SP* spine. Bar line = 100 μm. (From Richardson, Crisp and Runham 1981, J Mar Biol Assoc UK 61, Fig. 4)

bands when they were emersed at spring tides, with a weaker pattern of bands produced during neap tides.

One consequence of emersion, which was studied in detail by Richardson et al. (1979 and 1981), was the phenomenon of alternating bands in the shell of the cockle *C. edule* (Fig. 9). Early studies of cockle shells (Dolman 1975) had noticed that periodically groups of strong and weak bands were laid down, separated by regions in which the bands were more or less uniform in definition. Dolman (1975) suggested, without evidence, that the thin and thick bands could be related to tidal emersion in light and dark conditions, respectively. Since the low-tide period advances progressively during the lunar cycle, there would be periods when both low waters would be at twilight and others a week later when low water was around midnight and midday, thus producing a lunar periodicity in alternating patterns. Richardson et al. (1979, 1981) put forward a different explanation. They showed, on the basis of marked shells in which each band could be assigned to a particular low tide, that the stronger bands were laid down on the afternoon tide while the less distinct bands were formed during the cool of the morning. A similar conclusion was reached by Ohno (1985) working on *C. edule* and *M. mercenaria*. Alternating bands at spring low tides have been observed in the subtidal bivalve *Spisula subtruncata* growing at extreme low water of spring tides in the Menai Strait (Richardson 1988 a) (see Fig. 8) and on the basis of marked shells the stronger bands were shown to be coincident with emersion during the heat of the afternoon low tide.

In these few examples I have shown the complexity and variation in the tidally produced growth patterns present in the shells of living bivalves using examples from different parts of the world. Shells from mixed semi-diurnal tidal regimes, *C. nuttalli* from the Oregon coast, USA (Evans 1972), *F. unedo* from Japan (Ohno 1985), and *A. granosa* from Malaysia (Richardson 1987b), all show differences in the patterns in response to fluctuations in tidal conditions in the local vicinity where they were growing. At the present time it seems unlikely that subtle variations in the form of the neap – spring cycle could be distinguished from the appearance of the patterns in fossil material from different geological horizons, as Pannella (1976) has suggested. However in the future it might be possible to study the form of the neap – spring cycle from the growth patterns from suitable extant material, but from the outset the studies will be complex and involve further work particularly on modern living representatives from widely spaced localities. At the present time, for example, we do not know to what extent the influence of local variations in shore profile, shore aspect, sediment type and water current movements might have on the appearance of the bands of the same species growing at similar tidal positions. As I have demonstrated previously, an apparently similar mixed semi-diurnal tidal pattern in Japan and Malaysia can produce very different shell banding patterns in two bivalve species, *F. unedo* (Ohno 1985) and *A. granosa* (Richardson 1987b) which occupy similar tidal levels and habitats. Therefore, before any attempt is made to examine fossil shells we must be sure beyond any reasonable doubt that any potential sources of error in understanding the patterns in modern living representatives are fully known.

4 Effects of Continuous Immersion

It is appropriate at this point to consider the effects of the subtidal environment on the periodicity of microgrowth patterns in living bivalves, since it is from this area that most of the fossilizable material is found (Jones 1981). However, in order to understand the patterns in fossil shells we must first study living material.

There is little clear evidence for daily microgrowth bands in the shells of subtidal bivalves. Pannella and MacClintock (1968) and Rhoads and Pannella (1970) sought to explain the presence of the bands in the shells of intertidal and subtidal *Mercenaria mercenaria* as the result of changes during day and night. Later however, Pannella (1975) was forced to concede that semi-diurnal tides could be a more likely explanation for the presence of bands in intertidal shells. Thompson (1975) similarly shared Pannella's early view that bands in *Mercenaria* represented changes in growth rate during day and night and claimed that most shell growth took place during the night when the shells were recorded as gaping and the animals were filtering. Cunliffe and Kennish (1974) and Kennish (1978, 1980) have also studied shell microgrowth patterns in *M. mercenaria* in detail and have recognised subdaily, daily and bidaily patterns. Their conclusions on the periodicity appear to have been based on the earlier work of Pannella and MacClintock (1968) and from their own experiments in which several thousand notched clams were transplanted into the intertidal and subtidal zones (Kennish 1980). However, as reported in Richardson (1987a), photographs of sections of

Mercenaria reproduced in Cunliffe and Kennish (1974) and Kennish (1980) show more bands than can be simply accounted for by a daily periodicity.

Jones (1981) counted the number of bands in the shells of recent *Spisula solidissima* and attempted to quantify the periodicities in Barker's (1964) five orders of periodic layers. Barker (1964) was one of the first to identify and describe a hierarchy of five cyclic groupings of growth layers in the shells of several marine bivalves. He assumed these growth layers reflected the annual changes of temperature and salinity (first order), equinoctial storms and tides (second order), the fortnightly tidal cycle (third order), day and night cycles (fourth order) and daily tidal rhythms (fifth order). When Jones (1981) counted the number of fourth order layers formed per year for several years of growth in surf clams, he found the numbers counted per year were very variable both within and between specimens. The number of layers counted never approached 365 and usually the number per annual growth layer decreased with increasing age (Jones 1981, Table 1). Jones (1981) also found a similar difficulty in identifying the fortnightly cycles in the bivalve shell (Barker's 3rd order layering), casting doubts on the reliability that periodicities in subtidal shells represent periodic environmental fluctuations.

Several recent studies have examined the effects of continuous immersion on marked bivalves transplanted into different subtidal environments. Richardson et al. (1979) and Richardson (1987a) showed that the bands in cockle and clam shells held under subtidal conditions were weaker than those laid down under tidally emersed conditions. In the earlier work of Richardson et al. (1979, 1980a) evidence was presented to show that the bands in continuously immersed *C. edule* were of a tidal origin. Total counts of the bands in the shell corresponded to an approximate tidal rhythm of formation and Richardson et al. (1980a) concluded that an endogenous rhythm of shell deposition occurred in *C. edule* when specimens were maintained in constant conditions of continuous darkness, temperature, food supply and water flow. The formation of endogenous tidal bands was later questioned by Ohno (1983) who suggested from his experiments with *C. edule* that they produce bands without any kind of rhythm. Nevertheless, further support for an endogenous tidal rhythm of shell deposition was presented by Richardson (1987a) who showed that in the absence of tidal emersion, clams, *T. philippinarum* produced a pattern of bands with an approximate tidal periodicity. However, although the average number of bands laid down in the shells coincided with the number of expected tidal periods, several of the shells showed deviation from the expected number. Richardson (1988b) put forward an alternative explanation for the variation in band numbers in continuously immersed clams held in constant conditions. He suggested the rhythm of weak banding in subtidal clams (see Fig. 4) was not related to the tides or days but to the growth of the shell. In a series of experiments in which continuously immersed clams were fed a range of concentrations of microalgae Richardson (1988b) showed that the total number of bands in the shell correlated with the growth of the shell (Fig. 10a). Further support for the relationship between shell bands and growth has come from work on the mussel *M. edulis.* Richardson (1989) showed there was a positive and significant correlation between shell growth and the number of bands deposited in shells grown in subtidal conditions in the natural environment and in continuously immersed conditions in the laboratory (Fig. 10b). The results of this recent work cast further doubt on the use of microgrowth patterns in subtidal shells as possibly recording the number of days/tides in the year.

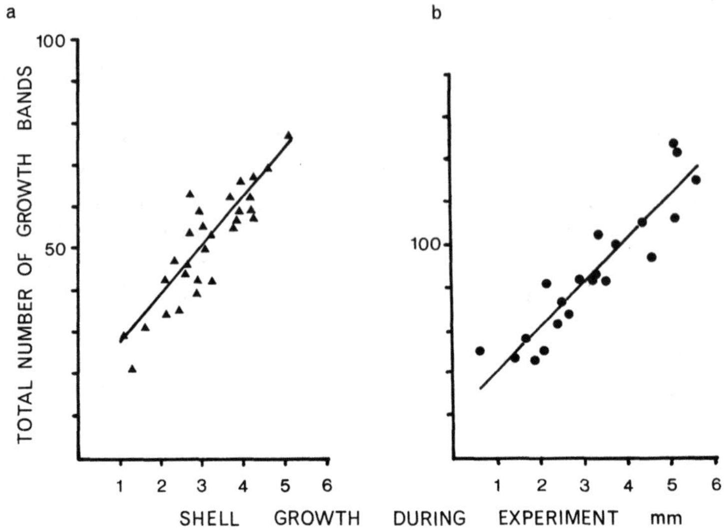

Fig. 10 a, b. Relationship between the total number of growth bands and shell growth in continuously immersed bivalves. **a** Clams *Tapes philippinarum*, regression equation is $Y = 11.89X + 13.81$ ($r = 0.881$) (from Richardson 1988b, J Exp Mar Biol Ecol 122, figure 9), and **b** mussels *Mytilus edulis*, regression equation is $Y = 20.68X + 20.36$ ($r = 0.920$). (From Richardson 1989, J Mar Biol Assoc UK 69, figure 4A)

It is clear that the patterns reflected in the shells of living bivalves are varied and complex. Intertidal species contain clear microgrowth patterns and appear to be the most promising material for study, although they are poorly represented in the fossil record. In subtidal species the occurrence of endogenous rhythms of shell growth, whether associated with innate tidal rhythms or with the production of shell material, must throw doubt on the use of such patterns for predicting the number of tides/days in the year. However, the tidal banding patterns in cockle shells show a seasonal variation in width with crowding of lines during the winter months when the sea water temperatures are at their lowest, and with wide growth increments during the warmer summer months (Richardson et al. 1980c). The seasonal variation in the patterns could be used for studying the environmental conditions during the growth of the animal. For example, the tidal bands in cockle shells from Norway (Richardson et al. 1980c) are severely compressed indicating reduced shell growth during the 5–6 months of the winter when ice covers the seashore. In contrast, the width of tidal bands in bivalve shells from tropical regions, e.g. *A. granosa*, do not show pronounced variations in increment width owing to the absence of seasonal water temperature fluctuations. These observations when extrapolated to fossil material could provide valuable evidence for changes in climatic conditions.

The use of statistical methods, such as Time Series Analysis to analyse periodicities in the shells of intertidal cockles (Dolman 1975; Lønne and Gray 1988) along with a better understanding of the patterns in living species might in time lead to an interpretation of patterns in fossil material. At this stage it would be premature to use the patterns in fossil shells to calculate the number of days/tides in the year while so much variability is present in the patterns of living material.

References

Barker RM (1964) Microtextural variation in pelecypod shells. Malacologia 2:69–86

Clark GR (1974) II: Growth lines in invertebrate skeletons. A Rev Earth Planet Sci 2:77–99

Cunliffe JE, Kennish MJ (1974) Shell growth patterns in the hard-shelled clam. Underwater Nat 8:20–24

Dolman J (1975) A technique for the extraction of environmental and geophysical information from growth records in invertebrates and stromatolites. In: Rosenberg GD, Runcorn SK (eds) Growth Rhythms and the History of the Earth's Rotation. Wiley, London, pp 191–222

Ekaratne SUK, Crisp DJ (1982) Tidal micro-growth bands in intertidal gastropod shells, with an evaluation of band dating techniques. Proc R Soc Lond B214:305–323

Evans JW (1972) Tidal growth increments in the cockle *Clinocardium nuttalli*. Science 176:416–417

Evans JW (1975) Growth and micromorphology of two bivalves exhibiting nondaily growth lines. In: Rosenberg GD, Runcorn SK (eds) Growth Rhythms and the History of the Earth's rotation. Wiley, London, pp 119–134

Farrow GE (1971) Periodicity structures in the bivalve shell: experiments to establish growth controls in *Cerastoderma edule* from the Thames estuary. Palaeontology 14:571–588

Farrow GE (1972) Periodicity structures in the bivalve shell: analysis of stunting in *Cerastoderma edule* from the Burry Inlet (South Wales). Palaeontology 15:61–72

Fritz LW, Lutz RA (1986) Environmental perturbations reflected in internal shell growth patterns of *Corbicula fluminea* (Mollusca: Bivalvia). Veliger 28:401–417

House MR, Farrow GE (1968) Daily growth banding in the shell of the cockle, *Cardium edule*. Nature (Lond) 219:1384–1386

Jones DS (1981) Repeating layers in the molluscan shell are not always periodic. J Paleontol 55:1076–1082

Kennish MJ (1978) Effects of thermal discharges on mortality of *Mercenaria mercenaria* in Barnegat Bay, New Jersey. Environ Geol 2:223–254

Kennish MJ (1980) Shell microgrowth analysis: *Mercenaria mercenaria* as a type example for research in population dynamics. In: Rhoads DC, Lutz RA (eds) Skeletal Growth of Aquatic Organisms. Plenum Press, New York, pp 255–292

Lønne OJ, Gray JS (1988) Influence of tides on microgrowth bands in *Cerastoderma edule* from Norway. Mar Ecol Prog Ser 42:1–7

Lutz RA, Rhoads DC (1980) Growth patterns within the molluscan shell: an overview. In: Rhoads DC, Lutz RA (eds) Skeletal Growth of Aquatic Organisms. Plenum Press, New York, pp 203–248

Ohno T (1983) A note on the variability of growth increment formation in the shell of the common cockle *Cerastoderma edule*. In: Brosche P, Sündermann J (eds) Tidal Friction and the Earth's Rotation II. Springer, Berlin Heidelberg New York Tokyo, pp 222–228

Ohno T (1985) Experimentale Analysen zur Rhythmik des Schalenwachstums einiger Bivalven und ihre paläobiologische Bedeutung. Palaeontographica Abt A 189:63–123

Pannella G (1975) Palaeontological clocks and the history of the earth's rotation. In: Rosenberg GD, Runcorn SK (eds) Growth Rhythms and the History of the Earth's Rotation. Wiley, London, pp 253–283

Pannella G (1976) Tidal growth patterns in recent and fossil mollusc bivalve shells: a tool for the reconstruction of paleotides. Naturwissenschaften 63:539–543

Pannella G, MacClintock C (1968) Biological and environmental rhythms reflected in molluscan shell growth. J Paleontol 42:64–80

Rhoads DC, Pannella G (1970) The use of molluscan growth patterns in ecology and paleoecology. Lethaia 3:143–161

Richardson CA (1987a) Tidal bands in the shell of the clam *Tapes philippinarum* (Adam and Reeve, 1850). Proc R Soc Lond B230:367–387

Richardson CA (1987b) Microgrowth patterns in the shell of the Malaysian cockle *Anadara granosa* and their use in age determination. J Exp Mar Biol Ecol 111:77–98

Richardson CA (1988a) Tidally produced growth bands in the subtidal bivalve *Spisula subtruncata* (Da Costa). J Moll Stud 54:71–82

Richardson CA (1988b) Exogenous and endogenous rhythms of band formation in the shell of the clam *Tapes philippinarum* (Adams and Reeve, 1850). J Exp Mar Biol Ecol 122:105–126

Richardson CA (1989) An analysis of the microgrowth bands in the shell of the mussel *Myilus edulis* (L). J Mar Biol Assoc UK, 69:477–491

Richardson CA, Crisp DJ, Runham NW (1979) Tidally deposited growth bands in the shell of the common cockle *Cerastoderma edule* (L). Malacologia. Proc 6th Eur Malac Congr 18:277–290

Richardson CA, Crisp DJ, Runham NW (1980a) An endogenous rhythm in shell deposition in *Cerastoderma edule*. J Mar Biol Assoc UK 60:991–1004

Richardson CA, Crisp DJ, Runham NW (1980b) Factors influencing shell growth in *Cerastoderma edule* Proc R Soc Lond B210:513–531

Richardson CA, Crisp DJ, Runham NW, Gruffydd LLD (1980c) The use of tidal growth bands in the shell of *Cerastoderma edule* to measure seasonal growth rates under cool temperate and subarctic conditions. J Mar Biol Assoc UK 60:977–989

Richardson CA, Crisp DJ, Runham NW (1981) Factors influencing shell deposition during a tidal cycle in the intertidal bivalve *Cerastoderma edule*. J Mar Biol Assoc UK 61:465–476

Scrutton CT (1978) Periodic growth features in fossil organisms and the length of the day and month. In: Brosche P, Sündermann J (eds) Tidal friction and the Earth's Rotation. Springer, Berlin Heidelberg New York, pp 154–196

Thompson I (1975) Biological clocks and shell growth in bivalves. In: Rosenberg GD, Runcorn SK (eds) Growth Rhythms and the History of the Earth's Rotation. Wiley, London, pp 149–161

Whyte MA (1975) Time, tide and the cockle. In: Rosenberg GD, Runcorn SK (eds) Growth Rhythms and the History of the Earth's Rotation. Wiley, London, pp 177–189

Implications of Some Recent Sedimentological Studies to the History of the Earth-Moon System

J. D. A. Piper

1 Introduction

Animals and plants are influenced by environmental rhythms in their tissues which may secrete material reflecting these rhythms; if this material is preserved in fossil remains it can provide a record of ancient tidal cycles. Since Wells (1963) noted the interrelationship of two growth cycles in Palaeozoic corals and provided the first data on the number of days per year in the remote past, many attempts (ably summarized by Scrutton 1978) have been made to determine the history of the Earth's rotation from the study of skeletal growth increments in fossils.

There are several impediments to expanding this database and improving on its time coverage:

1. The metazoans arose and diversified in late Precambrian times and began to secrete hard parts in their bodies near the beginning of Cambrian times ca. 590 Ma ago. The history of the Earth's rotation during the long period of Precambrian times cannot be resolved in this way. Whilst stromatolites from the Proterozoic eon, formed by the sediment-trapping activities of blue-green algae and bacteria, have been studied (e.g. Pannella 1975), the results are conflicting and of uncertain reliability: there is no doubt that cyclic groupings of laminae are present but it is unclear whether the solar day, diurnal tides or semi-diurnal tides are represented, and the data from modern stromatolites suggest that they are unlikely to record any single event precisely. It is possible that subtidal Precambrian stromatolites (which had developed before the appearance of destructive burrowing and feeding organisms) are effective recorders of ancient tidal parameters, although in the absence of a modern analogue it is difficult to be certain of this (Scrutton 1978).
2. The significance of the growth rhythms in many organisms is not well understood. The work of Richardson (this Vol.) shows that modern bivalves rarely show a complete record of the mixed (semi-diurnal and diurnal) fortnightly tidal growth patterns: only those animals which are critically juxtaposed in the inter-tidal zone are favourably disposed for recording the full cycle in their calcification.
3. Animals with well-developed periodicities likely to reflect the lunar day are bottom dwellers, have growth potentials strongly dependent on local environments, and are restricted to the intertidal and immediate subtidal zones. This is a transitional geological environment of high local variability requiring prevailing transgressive conditions for its preservation. The wider offshore environments with better preservation

Department of Earth Sciences, University of Liverpool, Liverpool L69 3BX, England.

potential and widespread uniformity are less suitable: the fossils illustrating prominent growth rhythms may be derived or have cycles in which the lunar and solar elements are difficult to separate.

Although these problems have restricted progress in this field over the last decade, new information of the periods of ancient tides has recently been forthcoming from the study of cycles recorded in some rhythmically bedded fine-grained sediments referred to as "rhythmites". The study and interpretation of these successions present new prospects both for extending the investigation of palaeotides into the remote past, and for recognising seasonal variations to isolate the number of days in the year. Since the time periods represented are much longer than the life span of a fossil, the credibility of the identified cycles is likely to be much enhanced.

2 Sedimentary Rhythmites

These rocks comprise multiple cycles of sedimentary laminae, usually of fine sand to silt. Individual units are graded (fining upwards) and the laminae are often conspicuous by a colour contrast between the coarser and finer layers. Individual laminae are traceable for hundreds of metres, and sometimes much more, although synsedimentary erosion may sometimes preclude the full development of cycles. While known for some years, these deposits had originally been interpreted in the context of an annual spring or summer melt influx in a periglacial environment (varves); deposits of this origin are well described from areas adjoining Pleistocene ice sheets. Although a tidal control is an alternative explanation for these complexly-cycled rhythmic deposits, no modern sediments comprising extensive graded laminae produced in offshore marine environments below the levels of wave agitation appear to have been reported. Nevertheless, it does appear that tidal ebb flow can produce powerful current jets capable of transporting fine detrital material into deep waters up to many kilometres offshore (Özsoy 1986).

Williams (1981, 1985, 1989a) has made a detailed study of a laminated member of the Elatina Formation ; this comprises siltstones and fine sandstones deposited in the Adelaide Geosyncline offshore from a periglacial environment during the Marinoan Glaciation (Late Precambrian, ca 650 Ma). Rhythmic laminations resulted from deposition from turbidity currents or settling from suspension in quiet waters below wave base, and their thickness variations have recorded a complexity of periodic signals. By detailed measurements of drill core material at Pichi Richi Pass in South Australia, Williams (1989a) identifies lamina-cycles containing on average 12 laminae, a dominant long-term oscillation with an average period of 26.2 lamina-cycles, and spectral analysis reveals additional harmonics of 13.1, 8.8, 6.6 and 5.3 lamina-cycles. Williams (op. cit.) also investigated rhythmites of the correlative Reynella Siltstone exposed on the coast 300 km to the south. This section shows thicker laminae with more abrupt changes in thickness suggestive of fluctuations in water depth; study of a thick undisturbed section was not therefore possible although lamina-cycles of 14–15 laminae are clearly present. This figure is marginally greater than the mean count at Pichi Richi Pass suggesting that the much thinner lamina-cycles at the latter locality may be truncated by non-deposition at the more clay-rich dark bands between cycles. Rhythmites, possibly of somewhat older (Sturtian, ca. 800 Ma) age, were also studied from the Chambers Bluff Tillite of northern South Australia. Syn-depositional scouring has limited the value of these rhythmites but lamina-cycles with up to 25 laminae are present. Their structure suggests the same origin as the Elatina and Reynella cycles.

Since no modern analogues of these deposits have yet been described, their interpretation must remain somewhat uncertain. Evidence for regional permafrost at the time of deposition suggests a strongly seasonal, arid climate (Williams 1986) and it was logical to interpret the laminae as varves. Williams (1981, 1985) originally postulated that the Elatina laminae were varves and noted the close similarity of the Elatina periods to the Sunspot cycle (ca. 11 years) and the double sunspot (Hale Magnetic) Cycle (ca. 22 years); the variation of lamina thickness in the Elatina rhythmites in fact shows a close comparison with the mean annual sunpot numbers in the sample of 15 cycles recorded continuously since 1830 (Williams 1981). However, no mechanism predicting such a powerful control of late Precambrian climates by solar cyclicity has yet been proposed. In addition, the occurrence of 14–15 laminae per lamina-cycle in the Reynella Siltstone, the presence of cycles with twice as many laminae in the Chambers Bluff Tillite, and the occurrence of distinct pairs of semi (diurnal?) laminae in the Reynella, all indicate a complex interaction of periods unrelated to the sunspot cycle. By considering an alternative tidal model, Williams (1989a, b) equates the prominent long period cycle of 26.2 (± 0.9) lamina-cycles in the Elatina data with an annual cycle. Since these deposits originated in a strongly seasonal periglacial environment (Williams 1986), the thickest laminae are presumed to record the summer peaks of rapid deposition, while low sedimentation rates (thin laminae) are correlated with the glacial winter. The corollary that the lamina-cycle records the lunar fortnightly cycle permits the range of periodicities recorded in the Elatina to be correlated with a spectrum of ancient tidal cycles from semi-diurnal to the lunar nodal cycle. The recognition of successive high and low spring tides leads Williams to estimate that there were 13.1 ± 0.5 lunar months during the late Precambrian year comprising 400 ± 20 days. The computed Earth-Moon distance at ca. 650 Ma from an inferred 19.5 year lunar nodal cycle is 96.9% of the present distance (Williams 1989a).

Williams' study illustrates the full range of potentialities and problems in the study of rhythmites. It demonstrates that a regional study is properly required to place these deposits in their environmental setting: the diurnal signature of the Elatina Formation at Pichi Richi Pass, for example, gives way to a mixed diurnal-semidiurnal signature in the Reynella Siltstone at Hallett Cove implying that the latter area was more proximal to the open sea at the time of deposition. A comparable study by Archer and Kvale (1988) has determined a first order periodicity of ca 30 laminae in Lower Carboniferous siltstones of Indiana. They suggest that laminae formation is linked to the semidiurnal rise and fall of palaeotides and that the periodicity matches a half-month tidal cycle of 15 days. Larger cycles comprising 24 to 25 of these smaller cycles are also suggested and appear to represent 12 to 13 lunar months; these authors also note periodicities indicative of other tidal cycles. The ratios of the tidal periods determined by Archer and Kvale may have an application to the determination of Carboniferous tidal and rotational parameters, and thereby provide a critical test of the data already derived from contemporaneous fossils.

These recent developments suggest that a concerted study of pre-Phanerozoic rhythmites could yield critical information about the dynamics of the Earth-Moon system (Williams 1987), but it is clear that borecore material will generally be required for a proper analysis. The microbanding in Proterozoic sediments has traditionally been interpreted as an annual record, although it is now evident that some occurrences, at least, merit a tidal interpretation. Early Proterozoic iron formations are extensively banded; they comprise multiple alternations of cryptocrystalline silica and iron oxides (hematite and magnetite) and are locally cyclic. In the early Proterozoic Weeli Wolli Formation (ca. 2500 Ma) of the Hamersley Basin in Western Australia, Trendall (1972, 1973), recognised a cycle of ca 23–25 laminae with a maximum count of 28 laminae per cycle. Williams (1989c) has counted up to 28–30 microband couplets per rhythmite cycle in this formation; he suggests that samples containing fewer couplets are underestimates because some adjacent bands have amalgamated. These deposits appear to have been formed by chemical evaporitic (Garrells 1987) or volcanic-exhalative (Simonson 1985) mechanisms; a tidal modulation of either of these processes is possible (see discussion in Williams 1989c). In the McArthur Basin of Northern Australia Jackson (1985) describes dolomitic rhythmites in a black shale facies of the ca. 1750 Ma Wollogorang Formation; the best preserved rhythmites contained 22±4 internal subdivisions. The origin of these rhythmites is evidently different from the examples in South Australia; they are seldom graded and Jackson (op. cit.) regards them as the distal part of a lacustrine (lake) deposit. It is possible (Williams 1989a) that the 14.8 and 28.8 laminae per cycle identified by Sonett and Trebisky (1986) from spectral analysis of a short section in the McArthur deposits record fortnightly and monthly grouping of diurnal increments, while the maximum value of ca. 28 microbands per cycle in the Hamersley Basin may suggest fortnightly sets of semidiurnal laminae (Williams 1989a) among other possible interpretations (Williams 1989c). Unfortunately, unless a seasonal signature is also present, it is not possible to determine the number of synodic months per year.

These rhythmite results merit comparison with the determination by Mohr (1975) of a 25.6 days/synodic month from the ca 2400 Ma Biwalik stromatolites of Minnesota, and the determinations by Pannella (1972) of 32 days/month and 448 days/year from the (probably correlative) Gunflint Formation. The difference between these two estimates is largely the difference between mean and maximum counts (Scrutton 1978), but Mohr's estimate may also be on the low side because of disturbances and non-deposition in the Biwalik succession. In Figure 1 the results of Williams (1989a, b) are summarised with data for days per year and days per month determined from palaeontological data and compiled by Scrutton (1978).

As discussed in detail by Scrutton (op. cit.) and others, the Phanerozoic palaeontological data define a gradual increase in the length of the day and record a slowing of the Earth's axial spin rate. This is commensurate with the hypothesis of lunar tidal friction. The increase in the length of the day over this last interval of geological time is appreciably less than the contemporaneous value computed from tidal eclipse data (Müller and Stephenson 1975); it has probably been variable (Sündermann and Brosche 1978) although this cannot positively be confirmed by the present palaeontological data (Scrutton 1978). The sedimentological data appear to support the sparse stromatolite evidence by showing that the progressive increase in number of days per synodic month and number of days per year does not continue back into

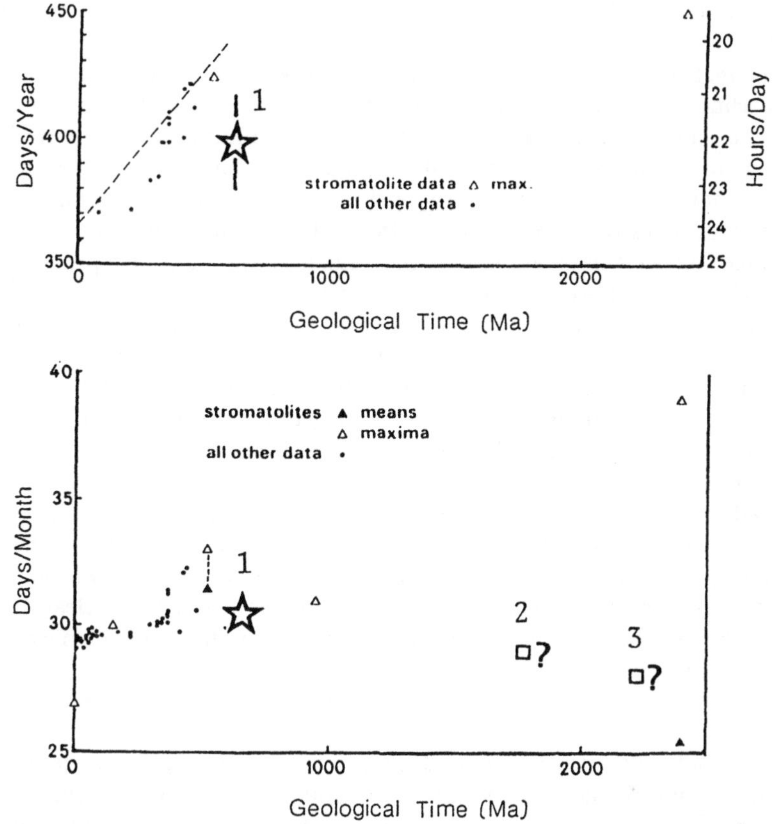

Fig. 1. Graphical representation of the days per presumed synodic month and the days per year derived from palaeontological data after Scrutton (1978). Rhythmic results noted in this paper have been added. The large symbol (*1*) refers to the data of Williams (1989a); note that the value for days per month from the Elatina data are not as well constrained as the months per year. The small symbols (*2* and *3*) refer to the McArthur and Hamersley results. Note that there are several possible interpretations of the cycles in these latter two formations. In addition, the ages of the ironstone formations from which the Lower Proterozoic stromatolite data are derived are also uncertain; most Superior-type banded ironstones were probably deposited within a relatively short interval at ca 2400–2200 Ma (Hutchinson 1981)

Proterozoic times. Instead there may have been a small decrease in the number of days per month near the Proterozoic-Phanerozoic boundary, following perhaps little change over the long interval of Proterozoic times.

A variety of palaeographic and tectonic evidence suggests that tidal friction was appreciably lower during Proterozoic times (Piper 1978), and a lower rate of deceleration of the Earth is therefore to be anticipated. The tentative picture to emerge from Figure 1 is that tidal friction was indeed small and/or its effects were masked by an accelerative couple which has not operated during Phanerozoic times. Creer (1975) discusses mechanisms which may have produced changes in the moment of inertia of the Earth. Factors leading to a decrease in this parameter with a consequent increase in the rate of rotation are: 1. The decline in mantle temperatures and consequent global contraction. This effect is likely to have produced only a very small decrease in the length of the day over geological times (Creer, op. cit.). 2. The elevation of phase changes within the mantle; this appears to be a more important consequence of the decline in radiogenic heat production. 3. The differentiation of the Earth to form mantle and core; this is unlikely to have had an important effect during Proterozoic and Phanerozoic times, firstly because the bulk of the core formation appears to have taken place very early in the Earth's history, and secondly because the core contributes only about 10% to the total moment of inertia.

These mechanisms, in common with the weak accelerative couple produced by the thermal tide in the atmosphere, cause long-term secular change and would not be able to explain the discontinuous change implied by Figure 1. A single short-term cause however, is implicit in the contrasting styles of lithosphere behaviour in Proterozoic and Phanerozoic times: the polar configuration and symmetry of the continental lithosphere prior to 800 Ma contrasts with the geoid-constrained configuration and equatorial symmetry illustrated by this lithosphere after 650 Ma and by the supercontinent of Pangaea (Piper, this Vol.). Furthermore, Phanerozoic tectonics have been dominated by the effects of uprising mantle plumes generated by instabilities in thermal and/or chemical boundary layers within the D″ layer at the base of the mantle; tectonic and magmatic events linked to plumes are rarely recorded within the older Proterozoic configuration. The changing style of convective heat release from the mantle probably resulted in profound changes in the thermal and dynamic conditions in the Earth's interior with important but uncertain consequences to the moment of inertia. This would be a likely cause of changes near the Proterozoic-Phanerozoic boundary.

Further studies are clearly required on two fronts. Firstly, more rhythmite and stromatolite data are needed to clarify the comparison between Proterozoic and Phanerozoic changes presently suggested in no more than a tentative way by Figure 1. Secondly, model calculations of the tidal retarding couple in Proterozoic times should identify the necessity for considering possible secular or short-term accelerations in the Earth's rotation.

Acknowledgment. I am grateful to Dr. G.E. Williams for providing reprints of his work on the rhythmites of South Australia and for critical review of this paper.

References

Archer AW, Kvale EP (1988) Tidal deposits as indicators of seasonality and palaeolatitudes: an example from the Carboniferous of Indiana, USA (abstract). Palaeozoic Biogeogr Palaeogeogr Symp, Oxford, August 1988, 78

Creer KM (1975) On a tentative correlation between changes in the geomagnetic polarity bias and reversal frequency and the Earth's rotation through Phanerozoic time. In: Rosenberg GD, Runcorn SK (eds) Growth Rhythms and the History of the Earth's Rotation. Wiley, London, pp 293–318

Garrels RM (1987) A model for the deposition of the microbanded Precambrian iron formations. Am J Sci 287:81–106

Hutchinson RW (1981) Mineral deposits as guides to supracrustal evolution: In: O'Connell RJ, Fyfe WS (eds) Evolution of the Earth. Geodyn Ser, Am Geophys Un 5:120–140

Jackson MJ (1985) Mid-Proterozoic dolomitic varves and microcycles from the McArthur Basin, northern Australia. Sediment Geol 44:301–326

Mohr RE (1975) Measured periodicities of the Biwabik (Precambrian) stromatolites and their geophysical significance. In: Rosenberg GD, Runcorn SK (eds) Growth Rhythms and the History of the Earth's Rotation. Wiley, London, pp 43–56

Müller PM, Stephenson FR (1975) The acceleration of the Earth and Moon from early astronomical observations. In: Rosenberg GD, Runcorn SK (eds) Growth Rhythms and the History of the Earth's Rotation. Wiley, London, pp 459–534

Özsoy E (1986) Ebb-tidal jets: a model of suspended sediment and mass transport at tidal inlets. Estuar Coastal Shelf Sci 22:45–62

Pannella G (1972) Palaeontological evidence on the Earth's rotational history since the early Precambrian. Astrophys Space Sci 16:212–237

Pannella G (1975) Palaeontological clocks and the history of the Earth's rotation. In: Rosenberg GD, Runcorn SK (eds) Growth Rhythms and the History of the Earth's Rotation. Wiley, London, pp 253–284

Piper JDA (1978) Geological and geophysical evidence relating to continental growth and dynamics and the hydrosphere in Precambrian times: a review and analysis. In: Brosche P, Sündermann J (eds) Tidal Friction and the Earth's Rotation. Springer, Berlin Heidelberg New York, pp 197–241

Piper JDA (1987) Palaeomagnetism and the Continental Crust. Open Univ Press/Halstead Press, 434 pp

Scrutton CT (1978) Periodic growth features in fossil organisms and the length of the day and month. In: Brosche P, Sündermann J (eds) Tidal Friction and the Earth's Rotation. Springer, Berlin Heidelberg New York, pp 154–196

Simonson BM (1985) Sedimentological constraints on the origins of Precambrian iron-formations. Bull Geol Soc Am 96:244–252

Sonett CP, Trebisky TJ (1986) Secular change in Solar activity derived from ancient varves and the sunspot index. Nature (Lond) 322:615–617

Sündermann J, Brosche P (1978) Numerical computations of tidal friction for present and ancient oceans. In: Sündermann J, Brosche P (eds) Tidal Friction and the Earth's Rotation. Springer, Berlin Heidelberg New York, pp 125–144

Trendall AF (1972) Revolution in Earth History. J Geol Soc Austral 19:287–311

Trendall AF (1973) Varve cycles in the Weeli Wolli Formation of the Precambrian Hamersley Group, Western Aust. Econ Geol 68:1089–1097

Wells JW (1963) Coral growth and geochronometry. Nature (Lond) 197:948–950

Williams GE (1981) Sunspot periods in the late Precambrian glacial climate and solar-planetary relations. Nature (Lond) 291:624–628

Williams GE (1985) Solar affinity of sedimentary cycles in the late Precambrian Elatina Formation. Aust J Phys 38:1027–1043

Williams GE (1986) Precambrian permafrost horizons as indicators of palaeoclimate. Precambrian Res 32:233–242

Williams GE (1987) Cosmic signals laid down in stone. New Sci 114(1566):63–66

Williams GE (1989a) Late Precambrian tidal rhythmites in South Australia and the history of the Earth's rotation. J Geol Soc Lond 146:97–111

Williams GE (1989b) Precambrian tidal sedimentary cycles and Earth's palaeorotation. Trans Am Geophys Un (EOS) 70(33):40–41

Williams GE (1989c) Precambrian cyclic rhythmites: solar or tidal signatures? Phil Trans R Soc Lond A (in press)

Continental Configurations and Mantle Reference Frames over Geological Time

J. D. A. Piper

1 The Palaeomagnetic Constraint and the Dipole Assumption

The mapping of magnetic anomalies generated by ocean crustal spreading and the sampling of this crust by deep drilling have produced a detailed kinematic model for growth, destruction and relative movements of the outer quasi-rigid shell of the Earth (the Lithosphere) during the last part of geological times. The refinement of this model, and its extension into more ancient epoch is limited by the inexorable consumption of the ancient and dense oceanic lithosphere back into the asthenosphere at the subduction zones. Earth scientists are now actively extending the concepts of global mobility derived from the surface layer of the globe to the interior comprising the silicate mantle and iron-nickel core. Much interest centres on the nature of the mantle layer adjoining the core at the core-mantle boundary (CMB), the topography of this zone, and the implications of these parameters for the processes of heat release from the Earth's interior. The key observations come from seismic tomography, material properties at high temperatures and pressures, and satellite geodesy. Collectively, this evidence has important consequences for our understanding of mantle convection and the mobility of the Earth's interior, but it is limited by lack of the time dimension. The latter is provided by the study of the magnetic field incorporating geomagnetism (over historical times) and palaeomagnetism (over geological times) with which this review is largely concerned.

The record of the ancient magnetic field of the Earth preserved in rocks provides the only means of constraining the former locations of the crustal plates with a high degree of precision. Directions of the ancient field are converted into equivalent pole positions on the assumption that the field observed at the surface approximates to the one produced by a geocentric axial dipole source. Since this model is axially symmetric, palaeomagnetism can define the ancient latitude of a continental fragment and identify its orientation with respect to lines of longitude, but it cannot locate that longitude absolutely: any position around the small circle defined by the line of latitude is permissible.

The application of the geocentric dipole assumption to the interpretation of time-averaged palaeomagnetic data can be tested in several ways. Over the last part of geological time (<10 Ma) the palaeomagnetic poles plot close to the rotation axis. The average poles determined from this youngest part of the palaeomagnetic record are statistically identical to the geographic pole (Tarling 1983). They do not coincide precisely with this axis and the geomagnetic pole has, in any case, been observed to

Department of Earth Sciences, University of Liverpool, PO Box 147, Liverpool L69 3BX, England.

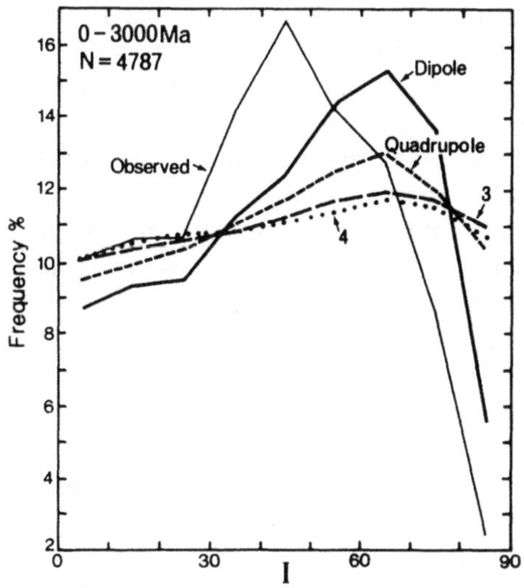

Fig. 1. The cumulative distribution of magnetic inclination |I| in the global palaeomagnetic data covering geological time (0–3000 Ma) compared with the distributions predicted from axial dipole, quadrupole and octopole sources. (After Piper and Grant 1989)

migrate partially around the geographic pole during the time interval covered by observatory records. An average axial geocentric source makes specific predictions about the distribution of the magnetic inclination, I, observed at the surface (Evans 1976); the distribution for a dipole source is distinct from the distributions produced by an axial quadrupole or more complex multipole sources (Fig. 1). On the assumption that continental drift will have caused the crust to randomly sample the field source over long periods of geological time, the average distribution of |I| can be used to test the validity of the assumption (Fig. 1). Piper and Grant (1989) find that the time-averaged distributions of |I| over Phanerozoic (0–590 Ma) and Precambrian (590–3700 Ma) times are not significantly different from the distribution expected from an axial dipole; a source modelled in this way is always found to be preferred over a more complex multipolar one when the field is averaged over time intervals of more than 100 Ma. It will be noted from Figure 1 that the distribution of |I| is skewed towards a preference for low values (and therefore lower latitudes). This is not predicted by any of the axial sources and could be caused by a long-term failure of the axial assumption and/or by a tendency for mantle motions to constrain the continental crust towards lower latitude regions of the Earth.

Using the kinematic model derived from seafloor spreading over the last 200 Ma Courtillot and Besse (1987) also show that the large continental areas of Africa, Eurasia and North America define the same polar wander paths when their relative movements are subtracted. This, like the accordance of poles over large continental areas for extended periods of geological time (Piper 1987), is another confirmation of the dipole assumption.

This "first order" conclusion can be substantiated by palaeoclimatic indicators during Phanerozoic times (< 600 Ma) which provide an independent measure of past geographic latitudes. Since climatic belts are subject to seasonal variations and are constrained by oceanic and continental distributions, their boundaries are never precisely parallel to lines of latitude and they cannot therefore individually provide critical evidence. However, the observation that such a wide range of indicators ranging from equatorial to polar in their characteristics all generally conform to the latitudes predicted by palaeomagnetic evidence (Nairn 1964; Tarling 1983) is a very strong indication that the model is a valid one.

Whilst many important conclusions about the gross form and distribution of the continental crust can be drawn from making this simplified assumption about the source of the geomagnetic field, there is now little doubt that there are short, and probably longer, term second-order complexities with an important geodynamic significance. Wilson (1970, 1971) observed that mean palaeomagnetic directions from each region of the globe do not yield pole positions coinciding precisely with the present geographic pole; instead they tend to plot on the far side of the geographic pole and to the right hand with respect to an observer at the sample location. Wilson and McElhinny (1974) extended this observation back to mid-Cenozoic times (25 Ma), while Marton (1986) found that the time-averaged geomagnetic field since Permian times (250 Ma) has approximated most closely to a dipole offset from the geocentre.

It is generally accepted that the magnetic field results largely from motions in the liquid outer core of the Earth (Rikitake 1966). New analyses of the short-term data from observatories and satellites can be inverted using certain assumptions, and can thereby be used to constrain fluid motions in the core and clarify the nature of the dynamo mechanism. A few implications of this work and of second order complexities in the field source are discussed in Sect. 4 below.

2 Configurations and Movements of the Continental Crust

The time sequence of palaeomagnetic poles from a crustal unit defines an *apparent polar wander* (APW) path. It is by definition and comparison of these paths that we identify the nature of large-scale tectonic processes in the geological past. The comparison may also permit the matching of segments of crust and enable the continents to be reconstructed in their former configurations. The differing tectonic models make contrasting predictions about comparisons between these paths (Piper 1987, 1990): a long period of common movement with continental plates in continuity will be recorded by a similar segment of APW in all of them; divergent movements however, will be reflected in differing directions and rates of motion and will be recorded as contrasting segments of APW. The exact nature of such relative movements cannot readily be resolved by comparing the APW paths both because palaeolongitude is indeterminant and because the APW paths are inevitably imperfect. It is never possible to define them completely because the rock record within any continental division is an incomplete one: the incidence of igneous and metamorphic events and the sequence of marine transgressions and regressions in the sedimentary column means that only a small number of relatively short events are generally represented. Furthermore this presupposes that all of these preserved rocks are effective recorders of the ancient field; this is never in practice the case. In addition to the fact that some rock facies do not possess a suitable mineralogy to acquire a stable remanence, many others will have been subject to later thermal and tectonic events responsible for obliterating and/or replacing the older magnetic remanence.

Present palaeomagnetic results comprise a heterogeneous data set with experimental results of very variable quality; the acquisition of these data commenced nearly 40 years ago and the areal and temporal coverage has been improving continuously since then. The inherent deficiencies noted above will always be present because they are constrained by geological history. They limit the versatility of palaeomagnetism when applied to crustal tectonics. Nevertheless, if we make critical assumptions, then it is possible to use the data to test limiting tectonic models. For example, if it is assumed that the continental crust was constrained to a single perfectly rigid supercontinent over a time interval in the past, then the model requires that all the pole positions spanning this interval conform to a single unified APW path. This conclusion must apply in this instance regardless of the quality of the data: if a conformity to a single path is not evident through all the imperfections of the data then the case for such a model disappears. It is in fact only this quasi-rigid solution which can be effectively tested over the earlier part of geological time; the evidence is then thinly spread over an immense period of time (~2000 Ma) and to be testable the model must make very stringent demands on the data.

In the event, the quasi-rigid solution can definitively be shown to be the correct one over the long interval of geological time referred to the Proterozoic eon (2500–590 Ma). Commencing with a reconstruction which has developed piecemeal over a decade, and was the logical outcome of the finding that Proterozoic palaeomagnetic poles illustrated a general conformity over large areas of crust (e.g., Piper et al. 1973; Irving and McGlynn 1976), it is possible to select any interval of these times and demonstrate a conformity of the data to a single APW path. It is then further possible to move forward by an arbitrary interval of time and find that the same unique reconstruction again brings the poles into conformity with a single path (Figs. 2 and 3). No other tectonic premise can explain this observation. Of course the poles are very old; some will be poorly dated and others will be poorly defined. However, the quasi-rigid solution requires that if a magnetisation is secondary rather than primary then it must conform to a later segment of the single path. The conformity to the single path evident in Figures 2 and 3 can therefore accommodate some uncertainties in the ages of magnetisation, and magnetisations which are demonstrably secondary can only be used to test this model. Since the APW path moves through markedly different sectors of the globe over the long interval of Proterozoic time (Figs. 2 and 3), no single overprinting event is evident and the efficacy of the test is enhanced.

This solution both predicts that the continental fragments were in direct continuity over an enormous interval of geological time, and that the magmatic and tectonic elements formed during these times will illustrate a continuity between the fragments. I have explored the latter prediction in the two previous volumes in this series (Piper 1978, 1982) and shown that there is a striking continuity of geological elements between the shields; in addition the unique isotopic and geochemical signature of Proterozoic times and the distinctive spectra of rock facies which characterised these times, are all explained by the premise that the continental crust was then an integral unit (Piper 1982, 1987). The chances of these two consecutive predictions being satisfied for the range of continental fragments (Fig. 4) by accident must be very small indeed.

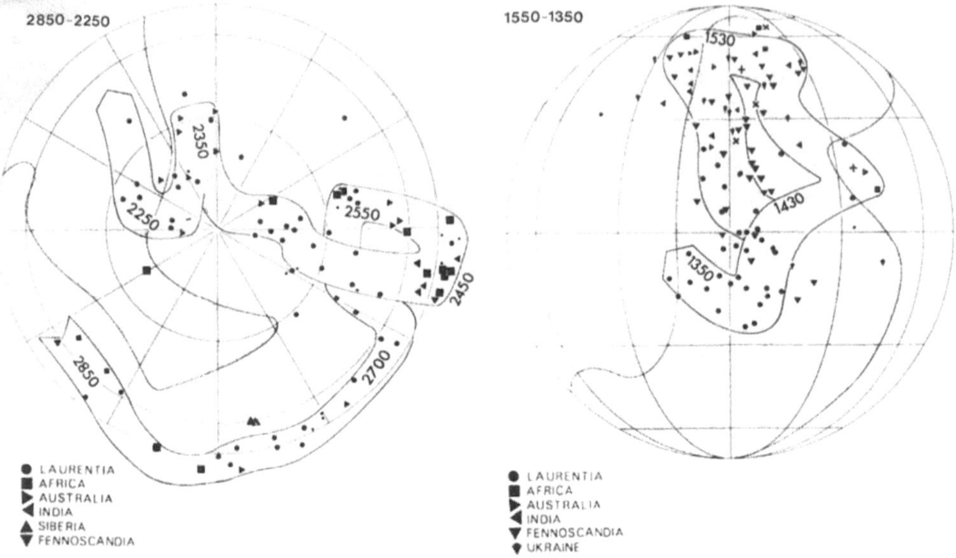

Fig. 2

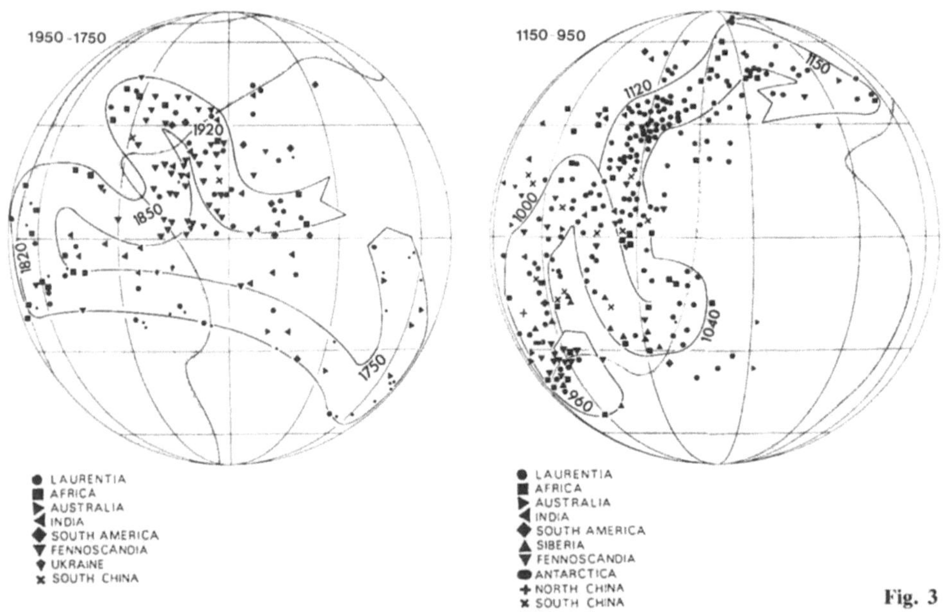

Fig. 3

Fig. 2 and 3. The definitive test of the quasi-rigid supercontinent configuration for Proterozoic times (see Fig. 4). In each case the palaeomagnetic poles for an arbitrary time interval early in the Proterozoic eon are plotted *on the left* and then again *on the right* for a much later time interval using the same single and rigid reconstruction. No other tectonic premise can be applicable to the interval of geological time covered by these tests because the poles from the second interval would then plot all over the projection when the reconstruction from the first interval is employed. (After Piper 1989)

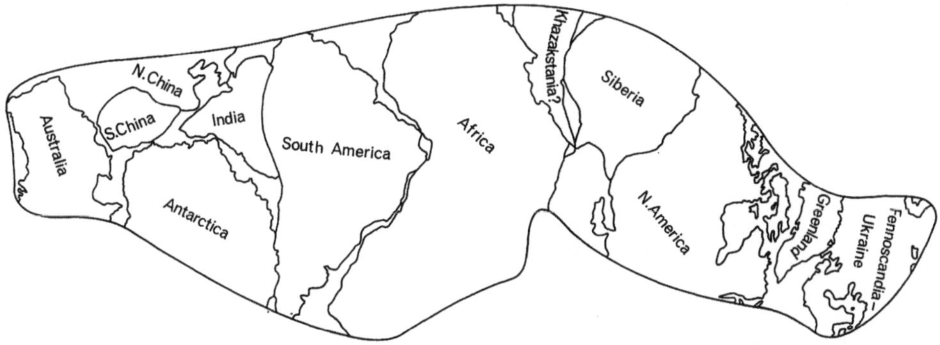

Fig. 4. The Proterozoic Supercontinent representing the configuration of the continental shields prior to 1000 Ma. There are little or no palaeomagnetic data from the Antarctica and Khazakhstania Shields, and these blocks are inserted on the basis of their shape and the trend of geological lineaments. Recent palaeomagnetic evidence (see Piper 1990) from the North and South China Shields supports the configuration shown here and previously suggested from the trend of geological structures and the general Palaeozoic association with the divisions of Gondwana (Piper 1982)

The improving aereal and temporal coverage of the Proterozoic interval is continuously enhancing the definition of the quasi-rigid model and the single APW path. The initial comparisons between Africa and North America (Piper 1975, 1976), were extended to incorporate Siberia (Sears and Price 1978; Morris et al. 1979), Fennoscandia (Patchett et al. 1978) and several constituents of Gondwana (Piper 1982). Some blocks have of necessity to be inserted like the residual pieces of a jigsaw to accommodate their shapes and the tectonic grain of the Precambrian outcrop (Fig. 4); the general locations of the China Shields in this reconstruction have since been confirmed by the emerging record of Precambrian palaeomagnetism in China (Zhang Huimin and Zhang Wenghi 1985; Piper 1990), although the position of Khazakhstania block remains speculation.

The distinctive characteristic of the Proterozoic APW path is the repeated tendency for the path to migrate outwards from a location close to the geometrical centre of the crust, execute a sharp bend or "hairpin", and then return to the starting point (Morris et al. 1979; Piper 1987). At least nine of these loops are recognised between 2500 and 600 Ma, each one executed over intervals of between 100 and 250 Ma. The radiating pattern from the centre of the reconstruction (Fig. 5) implies that the supercontinent possessed a pole-centred symmetry. This mean position is compatible with the $|I|$ distributions of the Proterozoic palaeomagnetic record (Piper and Grant 1989). The crust was presumably constrained to this position by the prevailing mantle convection system and it contrasts with the configuration of the later supercontinent of Pangaea (Fig. 6b). The crust was repeatedly displaced from this position, perhaps by a thermal perturbation in the Earth's interior, leading to a change in the figure of the Earth, and then restored to the stable pole-centred position over periods of tens of millions of years. A gradual change in the dynamic interaction between this crustal body and the mantle constraining system is illustrated by temporal changes in the directions of APW looping: at 2000–1600 Ma the loops were directed to the east (right-hand side) of the reconstruction, as shown in Figure 5; they became focussed within the continental area in mid-Proterozoic times (1500–1000 Ma), and then

Fig. 5. The cumulative APW path for the continental crust between 2100 Ma and 850 Ma plotted on a Lambert Equal Area projection of the whole globe and centred on the equator with respect to Laurentia (North America-Greenland) in present day coordinates; after Piper (1987). Note that this path defines a pole-centred symmetry with a repeated pattern of loops radiating from the geometrical centre of the continental crust over time intervals of 100–300 Ma

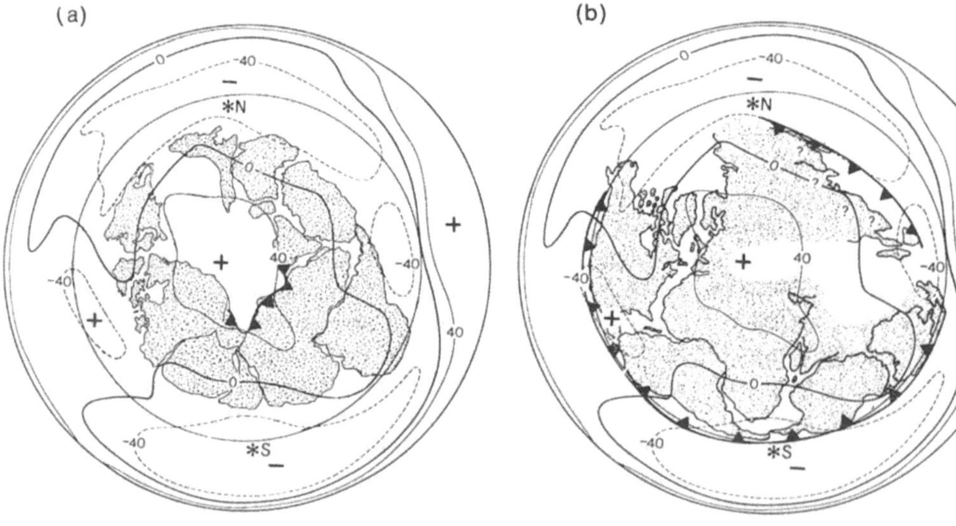

Fig. 6a, b. The distribution of the continental crust in late Proterozoic times (ca. 1000–600 Ma) (**a**) compared with the supercontinent of Pangaea (**b**) plotted on a Lambert Equal Area projections of the whole globe. The reconstructions are superimposed on the present residual geoid by aligning the mean palaeomagnetic poles at ca. 600 Ma and ca. 250 Ma respectively, and the peripheries of the supercontinents with the plane of symmetry defined by the polar negative zone. *S* and *N* refer to the positions of the geographic poles and the *sawtooth symbol* refers to the probable locations of destructive plate margins during the lifetimes of these supercontinents

moved to the west (left-hand side) of the reconstruction, as shown in Figure 5. These changes in the migratory pattern of the supercontinent were accompanied by the development of orogenic processes, including crustal accretion, along much of the eastern margin of the continent after 1800 Ma (Piper 1982, 1987).

The first unambiguous evidence for relative movements between divisions of the continental crust is observed at about 1100 Ma. Divergent APW paths from the Fennoscandian and Laurentian Shields at this time (Patchett et al. 1978) define a rotation of the former shield with a peripheral location into a new configuration (cf. Figs. 4 and 6a); magnetisations from both shields dated 1050–850 Ma accord with the secondary reconstruction. Breakup of the opposite extremity is identified by the absence of conformity of late Precambrian palaeomagnetic results from Australia, India and the China Shields with the primitive reconstruction (Piper 1987); however, the timing of this event cannot be well dated in this instance because of the poor age control of the data. This periphery was reorganised by these movements into a configuration close to the familiar Gondwana reconstruction (Fig. 6a).

The late Precambrian poles continue to show a tight conformity to a single APW path and imply that it was essentially rigid until early in Cambrian times (ca. 550 Ma). At this point there is a radial divergence of APW paths, and Middle Cambrian to Middle Ordovician poles from Gondwana, Laurentia, Fennoscandia and Siberia each plot discretely in quite different areas of the globe (Piper 1987, Fig. 9.4). This unambiguous separation of the APW paths illustrates the palaeomagnetic signature of continental dispersal. The event is recognised by a wide range of evidence indicative of continental rifting and associated magmatism, and it was followed by the global development of passive marine margins continuing into Middle and Upper-Ordovician times (Bond et al. 1984).

The reorganisation of the peripheral parts of the Proterozoic supercontinent in late Precambrian times produced a secondary configuration with an outer perimeter, typically characterised by subsiding margins (Bond et al., op. cit.), following a global small circle (Fig. 6a and Piper 1987). The two symmetrical wings of the reconstruction were separated by a narrow ocean where the instep in the Afro-Arabian sector was the location of active orogenesis and subduction (Fleck et al. 1980). The reconstruction is remarkably similar in shape to the later supercontinent of Pangaea (Fig. 6b) but the position of the tectonic features was exactly reversed: the outer perimeter of Pangaea was the site of active orogenesis and subduction during the lifetime of the supercontinent, while the narrow Tethyian seaway was characterised by passive margins during much of these times.

Le Pichon and Huchon (1984) have linked the symmetrical shape of Pangaea to the symmetry of the geoid, the sea-level equipotential surface of the Earth's gravity field. They observe two planes of symmetry in the geoid reflecting a dominant link to the Earth's rotation. One is equatorial and passes through the equatorial positive belt of gravitational potential, and the other is longitudinal and passes through the negative belt which has an axis passing through the rotation poles; the two planes intersect at the equator close to 90° and 270°E in present coordinates. A comparable symmetry is present in the form of Pangaea and the palaeomagnetic poles relevant to the lifetime of this supercontinent (ca. 400–200 Ma) confirm that the plane defined by the great circle perimeter was approximately constrained to the poles of rotation. Superimposing these symmetry elements using the mean palaeomagnetic pole for these times yields the relationship shown in Figure 6b. It implies that the peripheral zone of oceanic consumption around Pangaea was largely rimmed by the polar negative belt. The narrow Tethyian Ocean had an equatorial location. The complementary positions of subduction zones and passive margins in Pangaea imply that the oceanic hemisphere was a site of asthenosphere upwelling with complementary subduction focussed around the great circle perimeter. Since the largest long wavelength anomalies of the residual geoid originate in the lower mantle and are linked to mass variations here, this relationship between the gravitational potential and the continental crust implies that deep (whole?) mantle convection has at some intervals of Phanerozoic times constrained the distribution of crust at the surface.

This inter-relationship with whole mantle convection is implicit in the dynamics of the crust throughout Proterozoic times (Fig. 5), but efficient coupling has only been intermittently evident since late Proterozoic times (< 1000 Ma) during the lifetime of the circular Late Precambrian and Pangaean (Devonian-Cretaceous) supercontinents (Fig. 6). One possible explanation for this is the nature of heat release from the Earth's interior during the latter part of geological times. This has taken the form of thermal plumes originating in the vicinity of the core-mantle interface and rising episodically to be expressed as tectonic and magmatic events ("hotspots") in the lithosphere (Morgan 1981 and Sect. 3). Le Pichon and Huchon (1984) plot the breakup of the great circular Pangaean configuration. Although the Central Atlantic and Indian Oceans began to develop after 200 Ma, this was achieved at the expense of the Tethyian area and the circle was not breached until between 125 and 80 Ma ago. This interval includes the establishment of the oldest hotspots which are still active at the present time. After these times the dismembered continents abandoned the great circle configuration to move outside of the geoid maximum and towards the peripheral geoid low. The exception is Africa which has been blocked in its northward motion by the Eurasian continent and has stayed in the geoid positive area while continuing to "nudge" Eurasia and form the Alpine orogenic belt. It has been suggested that the maintenance of a supercontinent cap over a large surface area of the globe for 100–200 Ma would create a highly unstable thermal situation ultimately responsible for the breakup event (Anderson 1982). This is supported by the concentration of continental breakup episodes within the interval of mantle and core "stillstand" at 170–110 Ma (Sect. 4 and Fig. 12), and it might account for the inability of the Late Precambrian and Pangaean supercontinents to survive for more than a few hundreds of millions of years. Calculations suggest that shallow convection will in any case result (Christensen 1983) and be responsible for largely uncoupling the lower and upper mantle circulations.

Whilst there are major similarities between the Pangaean and Late Precambrian supercontinents indicative of a cyclic process in continental growth and destruction during the last part of geological times (Le Pichon and Gaulier 1986), there are some significant differences. Firstly, there is the reversal of tectonic features noted above and presumably reflecting the form of deep mantle convection responsible for constraining the crust in these ways. Secondly, the Late Precambrian supercontinent does not appear to have attained the hemispheric perimeter of Pangaea. However, because the Eurasian margin of Pangaea, as conventionally illustrated, has been expanded by the impingement of India into Asia leading to the lateral displacement of blocks comprising the eastern part of this continent during Cenozoic times (Molnar and Tapponnier 1977), the Pangaean shape may also have fallen short of a full hemisphere. Thirdly, the narrow ocean within the Late Proterozoic supercontinent did not attain an equatorial location like the Tethys, and indeed the APW path shows that the shape symmetry of this supercontinent did not become constrained to the contemporary geoid symmetry (as shown in Fig. 6a) until ca. 650–600 Ma and shortly before it was dismembered in Lower Cambrian times. At the present time we can only speculate that these differences reflect the evolving nature of geodynamic processes responsible for the major tectonic contrasts between the long interval of Lower-Middle Proterozoic times (ca. 2700–1000 Ma and Figs. 2–5), Late Proterozoic times (ca. 1000–600 Ma, Fig. 6a) and Phanerozoic times (ca. 600 Ma to the present, and including Fig. 6b).

3 True Polar Wander and the Hotspot Frame

The temporal migration of APW paths can, in principle, be produced by drift of the segment of lithosphere relative to a stationary spin axis, or by migration of the spin axis with respect to the tectosphere (i.e. the lithosphere and the segments of the underlying asthenosphere defined by a thermal boundary layer with which it moves laterally as a coherent unit). The latter motion is referred to as true polar wander (TPW). It can only be isolated from the effects of continental drift if contemporaneous APW paths are defined from more than one continent. Identical segments of APW recorded by continents which were not physically sutured during these time intervals may then be attributed to TPW. In principle TPW could be produced in at least two ways. Firstly, since the net torque acting on the plates due to push at the ridges and pull at the trenches is balanced by shear tractions at the base of the lithosphere, any change in these driving forces produces a change in the counteracting forces and could lead to TPW. Secondly, changes in the inertial tensor of the Earth can lead to movement of entire outer layers provided that the equatorial bulge is sufficiently mobile to adjust continuously to changes in the rotational axis (Goldreich and Toomre 1969). The distributions of large-scale density anomalies in the lithosphere (most notably the subduction zones) and the plumes collectively control the position of the spin axis. The principal axis of their combined effects is presently within a degree of the Earth's rotational axis (Jurdy 1981).

There are several possible models for TPW (Jurdy 1981). Either the tectosphere, or the mesosphere, or both could move relative to the rotational axis of the Earth (Fig. 7); in either case the degree of coupling between tectosphere and mesosphere via the ductile asthenosphere can only be recognised by evaluating the reference frames for each. McKenzie (1972) proposed a method for determining TPW by calculating weighted vectors representing the horizontal displacements of all points on the Earth's surface derived from the palaeomagnetic data. McElhinny (1973) used this method to estimate the magnitude of TPW since Eocene times; the resultant vector derived from the plate velocity vectors was shown to be small and essentially random, thus implying that there has been no significant TPW during Cenozoic times.

Jurdy and Van der Voo (1974) developed a more refined method which decomposes the displacement field of a plate into a part recording rigid rotation reflecting movement of the total lithosphere, and a part resulting from relative plate motions. It is

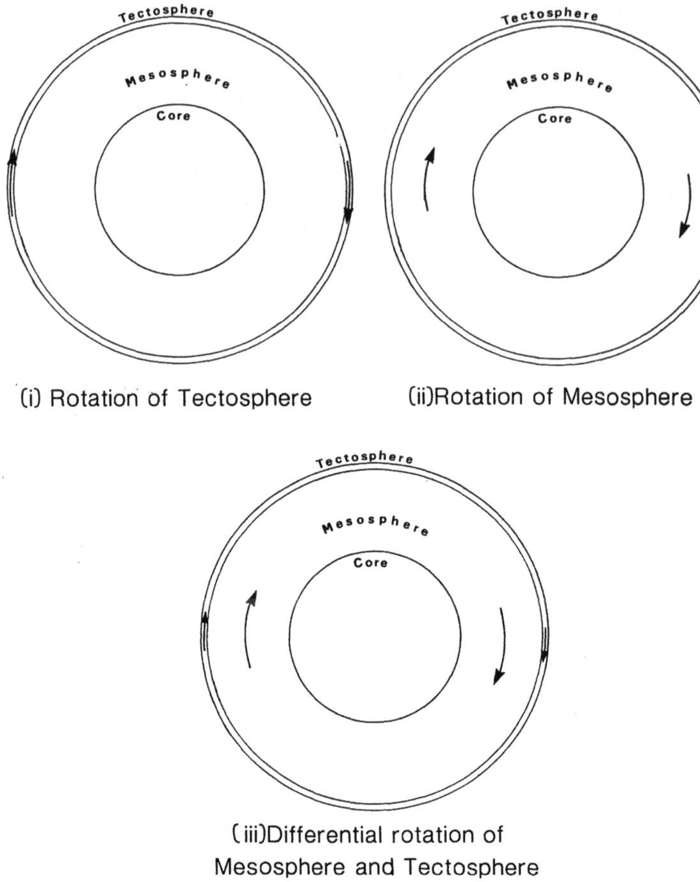

Fig. 7. Models for True Polar Wander modified after Jurdy (1981). Note that the Tectosphere is not to scale and in the real Earth is separated from the Mesosphere by the Asthenosphere

based on the best fit of displacements all over the Earth, rather than merely at the poles as in the case of McKenzie's method. Using reconstructions developed from sea floor spreading data, Jurdy and Van der Voo (1975) determined a cumulative TPW of 6.4°, slightly less than the radius of 95 percent confidence about the mean, and suggesting, like the result of McElhinny, that little motion of the combined lithosphere with respect to the spin axis has occurred since the latter part of Mesozoic times. This result is probably an artefact of the present plate geometry. Davies and Solomon (1985) have analysed the net contributions of ridge-push and trench-pull forces, and they conclude that these forces are at present almost equally and oppositely directed. Residual forces in the lithosphere shell may therefore be inadequate to promote appreciable TPW at the present time.

More recent analyses confirm that motion of the global lithosphere with respect to the pole has been small during the last part of geological times, but suggest nevertheless that it has taken place. Courtillot and Besse (1987) argue for a total motion of 5° during the last 65 Ma; they note that the 60 Ma poles for both lithosphere and mesosphere TPW (discussed below) happen to be very close and account for the early view that no motion of the total lithosphere has taken place with respect to the pole.

Relative movements between the lithospheric shell and the mantle below emerge when the mesosphere reference frame is compared with the palaeomagnetic reference axis. The mesosphere frame is recognised at the surface by buoyant plumes, either coupled to lower mantle convection or directly related to the surface of the core, which rise to thermally and mechanically impinge on the overlying lithosphere to be expressed as the hotspots. The derivative magmas result from deep partial melting and tend to be chemically primitive: alkalic basalts and their associated derivatives can be produced in addition to tholeiitic basalts of shallower origin. The thermal anomaly produced by the plume is responsible for updoming and fracturing the lithosphere, often at locations remote from plate margins. As the plates pass over the plumes, linear island chains are built up on the oceanic crust and migrating volcanic provinces are produced on the continents. To provide a reliable reference frame it is necessary to be sure that the plumes do not move relative to one another. This is generally believed to be the case: Morgan (1972) calculated that relative movements of no more than 0.5 cm/year could explain the divergence of Pacific island chains, and more recent work tends to support the view that relative movements between the plumes are unlikely to be more than a few mm/year; no detectable motions are evident over the last 10 Ma (Chase 1979; Minster and Jordan 1978) and probably over the last 100 Ma (Duncan 1981). Most recent evidence and numerical modelling suggest that the plumes probably undergo small and essentially random movements with respect to each other as they rise through, and interact with, the surrounding mesosphere and asthenosphere.

Since palaeomagnetic data allow the motion of a plate to be tracked with respect to the spin axis while the plume traces yield the motion of the plate over the mantle, the kinematic data sets from the two frames can be combined to determine the motion of the mantle with respect to the rotational axis. Several analyses combining palaeomagnetic, kinematic and hotspot data have been undertaken over the past decade. They have yielded a range of results partly because different palaeomagnetic and kinematic data sets have been employed, and partly because many of the hotspots are still poorly defined and dated.

Harrison and Lindh (1982) compared the geomagnetic and hot spot reference frames to show a progressive divergence of 5°–7° between 20 and 70 Ma and rising to a value of 18° between 70 and 180 Ma. These values were greater than the errors involved and they concluded that the mantle layer generating the hotspots must be drifting slowly but irregularly with respect to the spin axis. Morgan (1981) identified a somewhat larger movement of about 30° between the two reference frames during the past 180 Ma, and Jurdy (1981), whilst finding less than 5° of relative motion between the total lithosphere and spin axis since the beginning of Cenozoic times, found a larger motion (10°–12°) of the hotspot reference frame. Comparable values were resolved by Andrews (1985) and by Livermore et al. (1984) who identified between 17° and 19° of motion between the hotspots and the geomagnetic axis during the past 90 Ma.

Courtillot and Besse (1987) used a kinematic model for the relative motions between the Eurasian, African, North American and Indian plates to subtract relative motions between them and derive a composite APW path from the collective palaeomagnetic data. The resultant path shows a hairpin 140–120 Ma ago, after a stillstand period between 200 and 170 Ma followed by a progressive migration to the present position. Although the general shapes are similar, there are significant contrasts between this path and the path of plumes with respect to the palaeomagnetic pole as determined by Morgan (1983). The difference between them is the record of relative motions between the mesosphere and tectosphere. Fast TPW is recognised between 200 and 170 Ma ago to be followed by a long 60 Ma period of stillstand.

There is, as yet, no unanimity on the significance of TPW relative to the mesosphere hotspot frame. This is largely caused by uncertainties in the dating and definition of the hot spots: in addition to questions surrounding the reliability of many of the age determinations themselves, it is only the youngest rocks lying on the top of the hotspot volcanic province which can usually be sampled and dated; the initiation of plume activity may therefore be quite uncertain. In addition, hotspot traces are notoriously difficult to trace within the continental crust where surface expression is limited by the thick lithosphere. Plumes may then be unable to cause melting, or only appear as rare highly alkalic magmas such as kimberlites (Crough et al. 1980). Since the lithosphere moved only slightly with respect to the spin axis during the last part of geological time while the plumes have moved by larger amounts relative to the tectosphere, it is to be anticipated that the plumes, and hence the mesosphere host, will illustrate significant movement relative to the palaeomagnetic reference frame. Again, if the hotspots remain fixed relative to each other, movements cannot be everywhere the same because they take place over a spherical shell. Instead they should illustrate a rolling motion. Duncan et al. (1972) examined the temporal migration of volcanism in the Faeroe-Iceland sector of the North Atlantic and in the Central European volcanic province. They identified a collective 23° movement since the beginning of Cenozoic times. Hargraves and Duncan (1973) subsequently examined the traces of eight plumes over the last 50 Ma and recognised variable rotational movements with respect to the spin axis. They performed a best fit analysis on these variable movements to suggest that the mesosphere is "rolling" in a clockwise sense with respect to the lithosphere. The axis of this apparent rotation lies near 30 °E, 30 °N.

The recognition of an axis of rolling motion oblique to the rotation axis, while admittedly of a tentative nature, is of especial interest because of the demonstrably close relationship between the residual geoid and the Earth's rotation (Gough 1977; Le Pichon and Huchon 1984). The apparent axis of mesosphere roll is close to the "eye" of a spiralling set of geosutures described by Neev and Hall (1982). These authors define a geosuture as a deep shear zone which affects the crust on a global scale, although it may be concealed along part of its length. Obviously the system identified (Fig. 8) by Neev and Hall must be somewhat selective because it does not incorporate certain old lineaments oblique to the fabric, but the defined lineaments all collectively converge in a counter clockwise sense into an "eye" located within the Alpine-Himalayan orogenic belt. The members belonging to this set include lineaments as old as Precambrian in age, but they continue directly into oceanic fracture zones and have analogies with the ca. 4000 Ma spiralling geosutures on the planet Mars (Neev and Hall, op. cit.). Both the axis of this rotation and the counter-clockwise sense of lithosphere spiral are compatible with the inferred sense of mantle roll. The

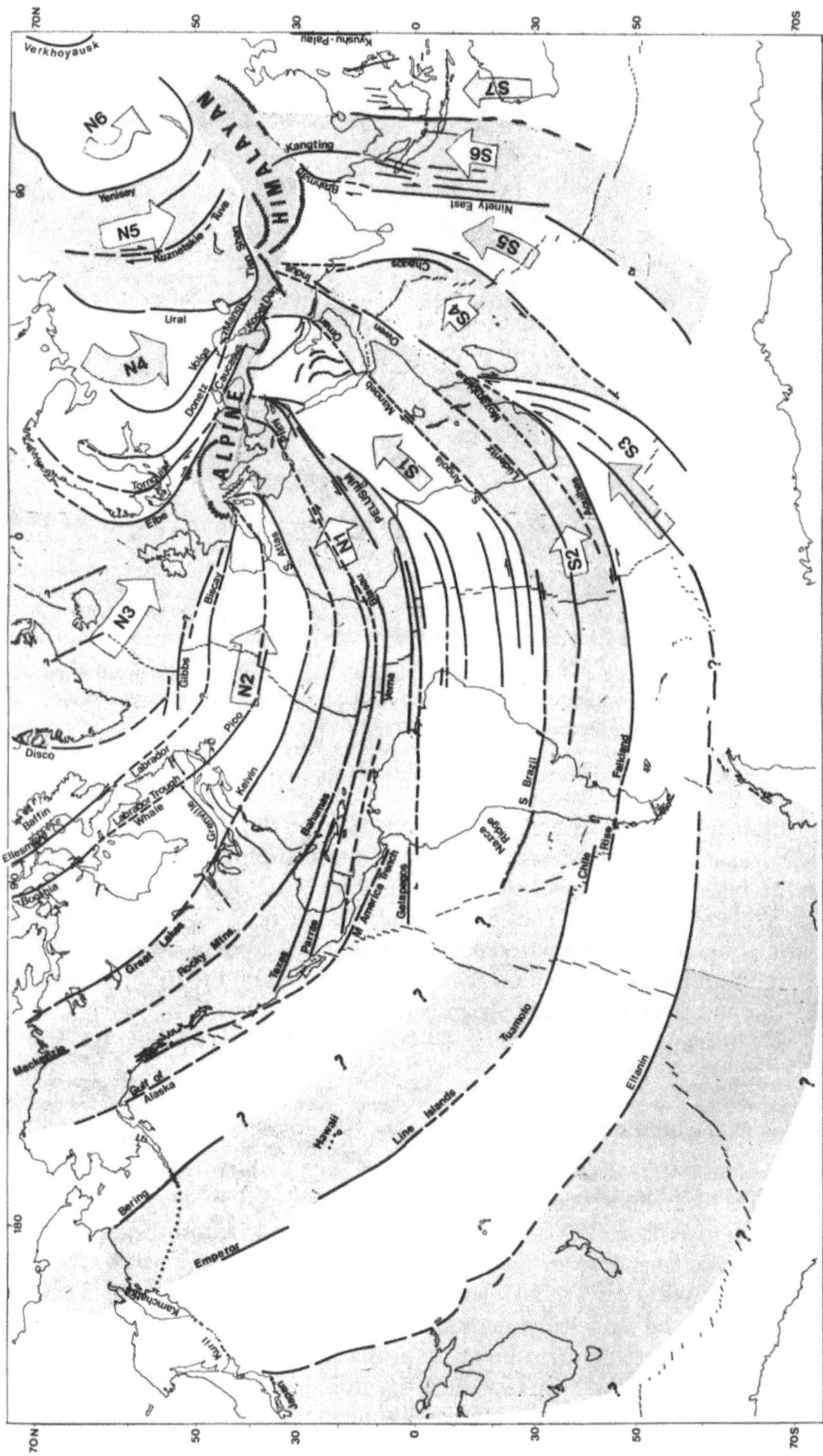

Fig. 8. The global system of counter-clockwise converging geosutures identified by Neev and Hall (1982). The segments of crust between the sutures are alternatively *stippled* and *blank* and are interpreted to move in a centripetal fashion at different rates to produce the spiral pattern. The relative rates of motion of the individual slices are shown by the lengths of the *thick arrows* and the "eye" of the spiral is located along the Alpine-Himalayan belt near the 40°N parallel. (After Neev and Hall 1982), and reproduced by permission of the authors

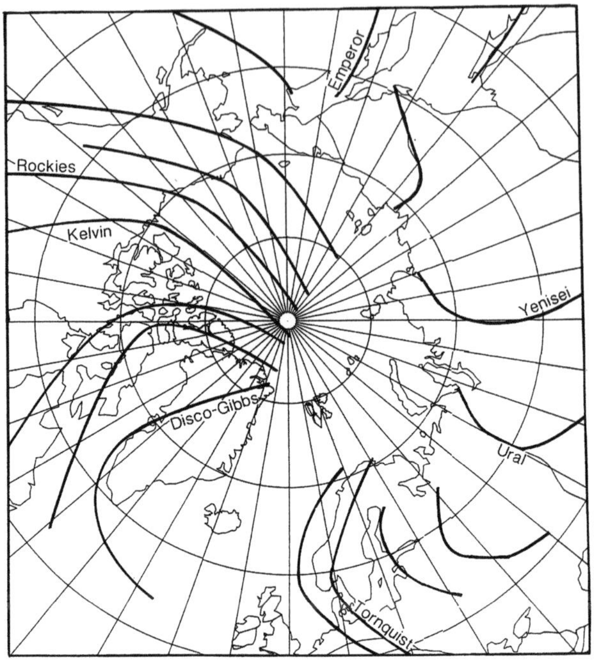

Fig. 9. The distribution of the geosutures recognised by Neev and Hall (1982) in the north polar regions. Note the convolving clockwise spiral pattern formed by the geosutures with an "eye" located on the Eurasian side of the North Pole

geosutures might then owe their development, or enhancement (in the case of the ancient examples), to a coupling between mesosphere and tectosphere, and hence to the drag between the faster mesosphere and the slower tectosphere. According to Neev and Hall, there is also a second and subsidiary spiral of geosutures with an apparent link to the Earth's rotation. The pattern described in Figure 8 converges towards the North Pole (Fig. 9) into a clockwise convolving spiral centred on the Eurasian side of the North Pole and close to the 80 °N parallel.

4 The Geomagnetic Field Source and the Core-Mantle Boundary Zone

The first indication of second order terms in the time-averaged geomagnetic field came from the work of Wilson (1970, 1971). At that time there were insufficient data to carry out a spherical harmonic analysis. He supposed that the second-order term likely to be of most importance was a zonal quadrupole which could be considered together with the dipole and modelled as a dipole offset along the rotational axis; this is simply a way of modelling the field observed at the surface and does not have a physical meaning. Wilson and McElhinny (1974) identified a progressive decrease in this offset dipole between 25 Ma and the present, although this does not seem to have been recognised in more recent work. Livermore et al. (1984) identify a small axial quadrupole of the same sign as the dipole which may have persisted throughout Cenozoic times and would be equivalent to this northward offset of the axial dipole field; during Cretaceaous times this appears to have changed sign with respect to the dipole to yield a southward offset. A third order effect noted by Wilson (1971) is for poles to plot to the right of the geographic pole with respect to the observer. This may be an artefact of biasing produced by the general westward drifting of non-dipole anomalies in a data set which, of necessity, works with only the direction and not the intensity of the field (Wilson, personal communication).

Marton (1986) found that the time-averaged geomagnetic field approximates most closely to a non-axial offset dipole. By analysing the Cretaceous to Permian palaeomagnetic data and using published kinematic models, he considers that this has been a long-term feature of the time-averaged field. Throughout this time the direction of dipole offset has been stationary in the equatorial western Pacific, a feature which applies also to the Quaternary-Recent fields (Adam et al. 1975; Creer 1977). All these dipole offsets are disposed close to a conspicuous positive zone in the residual geoid (Fig. 10) and suggest the existence of a long-term heterogeneity responsible for distorting the axial symmetry of the time-averaged field. The correlation between the contemporary non-dipole field and the geoid is also evidence for a link between the magnetic and gravitational fields (Hide and Malin 1970). Marton (op. cit.) has suggested that, provided the equatorial positive belt is a long-term feature of the geoid (see, for example, Chase and Sprowl 1983; Le Pichon and Huchon 1984), then the palaeomagnetic frame supplies a means for locating the equatorial positive bulge

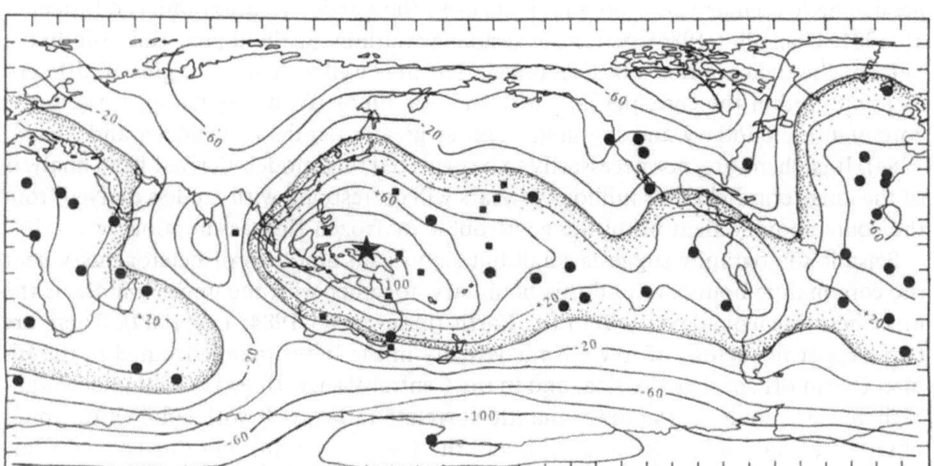

Fig. 10. The geoid based on the global gravity model GEM 10B after Chase and Sprowl (1983); the uphill side of the zero-metre contour is *stippled* and a 20-metre contour interval is employed. The *circles* are the locations of the hotspots recognised by Crough and Jurdy (1980). The *star* is the location of the mean offset geomagnetic centre recognised from Permian, Triassic, Jurassic and Cretaceous palaeomagnetic data, while the *small squares* are the locations of the magnetic centres deduced by several workers from a range of Carboniferous to Recent data and compiled by Marton (1986)

during the geological past, and that the latter is therefore a core reference frame. Whilst there is a strong correlation between the residual geoid and the locations of most of the hotspots (Crough and Jurdy 1980), the distribution of the latter is not uniform. The correlation with the geoid is most apparent in the Atlantic-African sector and the central Pacific, whilst the western Pacific, correlating with the offsets identified by Marton and others, is largely devoid of hotspots. This suggests that the source of the geoid positive belt may actually be composite and result from the gravitational signature of the hotspots and heterogeneities linked to on offset dipole source (Marton, op. cit). By relocating the geoid data with respect to the hotspot frame, Marton demonstrates a systematic counter-clockwise rotation of the mantle with respect to a rotation pole at ca. 37 °E, 0 °N since Permian times. Since there are large uncertainties involved in this estimation, the proximity of this pole to the pole of mantle roll inferred by Hargraves and Duncan (1973) is noteworthy. According to Marton it was accompanied by a similar "core roll", indicated by the offset dipole location, in roughly the opposite direction.

If such long-term asymmetry of the average geomagnetic field source can be established by refinement of the geomagnetic and kinematic constraints, this is of particular interest, not least because the coupling between the core and mantle can influence the Earth's rotation. On the assumption that flow in the liquid outer core is large-scale and geostrophic, Gire (1985) has inverted secular variation data to model the motion of the fluid iron at the core–mantle boundary. He demonstrates a simple axisymmetrical pattern with an ascending jet on the equator south of India and an antipodal descending flow near the coast of Peru. The overpressures caused by geostrophic flow at the boundary appear to have an order of magnitude and a geometry that can affect the inertia of the mantle. In this way they may excite the Chandler wobble (Le Mouel et al. 1985) and produce the apparent links between geomagnetic secular variation and changes in the Earth's rotation rate (Le Mouel et al. 1981; Courtillot 1986). Since the secular variations of the dipole and non-dipole parts of the field appear to be decoupled, they may well have their origins at different locations within the core, perhaps with the non-dipole field originating close to the core mantle boundary and the main field at greater depths (McFadden and Merrill 1984). It is, therefore, not necessarily to be expected that models derived from analysis of the long-term field over millions of years will correspond with models derived from the contemporary field using the geostrophic or frozen-flux assumptions.

Seismic tomography suggests small but significant amounts of heterogeneity near the core-mantle boundary and this boundary, together with the upper mantle, is the most heterogeneous layer within the Earth (Dziewonski 1984; Lay 1989). There are two large-scale regions of low seismic velocity in the lower mantle located in the Indian Ocean off southern Africa, and in the Central Pacific (Fig. 11). Complementary high velocity zones in the lower mantle correlate rather well with subduction zones observed at the surface (central Asia and the Pacific margin) and this has led some workers to postulate that subducted slabs continue to sink down into the lower mantle; the return of cold slabs to this zone may be the major way in which this part of the Earth is cooled (Hager 1984). This would also imply that subduction zones have an important control on mantle convection; it is complemented by the observation that the present geoid corresponds to the position of the subduction zones at ca. 125 Ma (Chase and Sprowl 1983) when, as is argued above, that lower mantle convection was more strongly coupled to plate motions (Le Pichon et al. 1985).

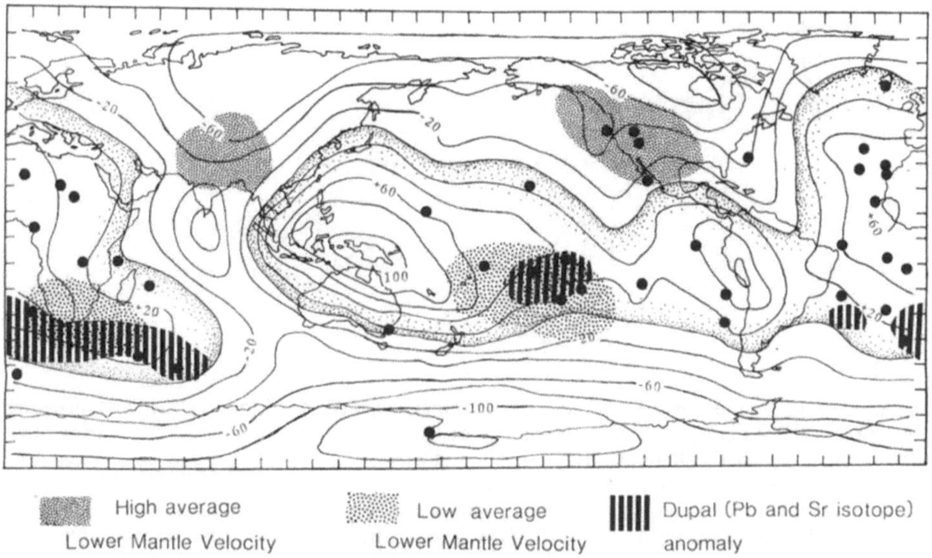

Fig. 11. The geoid and hotspots as defined in Figure 10 with the locations of the zone of high and low average lower mantle velocities with the regional distribution of the "Dupal anomaly" defined by anomalously high $^{87}Sr/^{86}Sr$, $^{207}Pb/^{204}Pb$ and $^{208}Pb/^{204}Pb$ ratios in basaltic lavas from the ocean crust. (Adapted from Castillo 1988)

The low velocity zones in the lower mantle correlate with active hotspots (see Fig. 11) but they are displaced to the west of the zones of poloidal core flow inferred from contemporary secular variation, while the Pacific centre lies to the east of the centre of long-term geomagnetic asymmetry identified by Marton (1986). Nevertheless the lower mantle zones delineated by seismology correlate with isotopic anomalies in oceanic basalts (Dupré and Allègre 1983). This correlation suggests a link between lower mantle processes and magmatic activity at the Earth's surface (Castillo 1988) which is implicit in the plume models for contemporary lithosphere tectonics. The most specific of these models by Loper and his colleagues (e.g., 1988) postulates that the D″ thermal boundary layer (see Lay 1989 for a recent summary of knowledge concerning this layer) at the base of the Earth's mantle undergoes cyclic changes linked to heat supply from the core. As a result, it progressively thickens and ultimately becomes unstable; hot material is then erupted and rises via mantle plumes to the surface. It should be possible to monitor such a process by study of the geomagnetic field both because the rate of reversals of the field is believed to be related to the rate of supply of energy to the dynamo (McFadden and Merrill 1984), and because the magnitude of palaeosecular variation is apparently greater when the energy supply is larger. Uncertainties in the thermal diffusivity and the thickness of the D″ layer at the time of breakdown limit the accuracy with which this process can be modelled; cycles of 30 and 200 Ma plume ejection have been proposed (Courtillot and Besse 1987; Loper et al. 1988).

Whilst the significance of a 30 Ma cycle is debatable, a cycle of Earth behaviour of the order of 200 Ma in duration is supported by the correlation of a range of

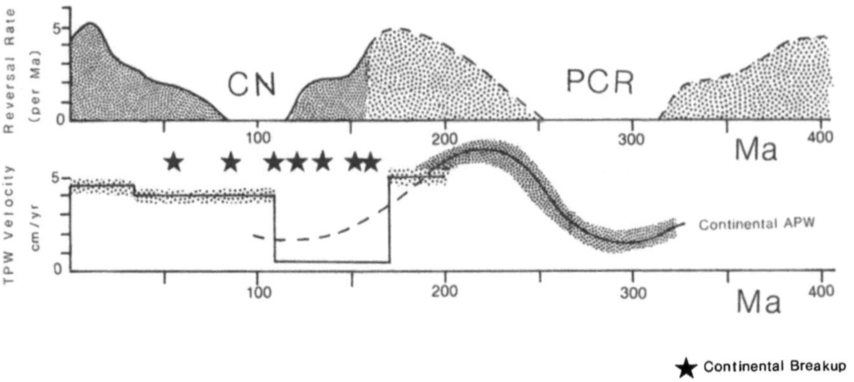

Fig. 12. Changes in the frequency of reversals of the geomagnetic field (*number* per million years) over the last 400 Ma compared with the velocity of true polar wander deduced from a cumulative APW path for the continental crust by Courtillot and Besse (1987). TPW can only be deduced over the time interval represented by the present ocean crust, and prior to this the rate of APW recorded by the continental crust of Pangaea (after Piper 1987) is used to suggest that TPW may have also peaked at ca. 200 Ma and been low at ca. 300 Ma if (as has been the case over the past 200 Ma) continental motions follow mesosphere movements in a more subdued form. (*CN* is the Cretaceous Normal Superchron and *PCR* is the Permo-Carboniferous Superchron). Note that times of decaying reversal rate appear to be periods of little or no TPW, whereas periods of growing reversal rate are periods of rapid TPW and that there may be a time-lag between TPW and the rise in reversal rate. The correlation of continental breakup with the interval of low TPW supports the view that the grouping of the continental crust to form Pangaea led to mantle insulation, heating and to hotspot development (Anderson 1982), and ultimately to breakup and decoupling of mesosphere and tectosphere. (Modified from Courtillot and Besse 1987)

phenomena of shallow and deep origins (Fig. 12 and Piper 1987). Fast TPW (ca. 5 cm/year) occurred from 200 to 170 Ma and was followed by a 60 Ma period of standstill before resuming at 110 Ma (at rates ca. 4 cm/year) up to the present day (Courtillot and Besse 1987). A similar pattern emerges from ocean floor spreading rates which showed a two- to threefold increase at 110 Ma in the Atlantic and Pacific to be followed by a gradual decline since then. The absence of sea floor older than about 200 Ma does not permit backward extrapolation of these estimates but a secondary indication is provided by the older APW paths. Pangaea underwent a rapid movement at velocities of up to 6 cm/year between 250 and 180 Ma before velocities seem to have fallen to 2–4 cm/year when this supercontinent began to break up (Piper 1987). Over the same interval of time the rate of geomagnetic field reversal increased following the long Permo-Carboniferous reversed superchron (PCR, 320–250 Ma) to a peak in Jurassic times (Piper 1987) followed by a progressive decline to the Cretaceous Normal Superchron (CN, 110–83 Ma). Since then the rate of reversals has built up in a smooth monotonic way probably to attain a peak about 10 Ma ago (McFadden and Merrill 1984). The collective pattern is suggestive of a long-term cycle (Piper 1987 and Fig. 12). The correlation with TPW and plate velocities suggests that times of decaying reversal rate are also periods of little or no TPW, whereas periods of growing reversal rate are periods of fast TPW (Fig. 12); there is however, a possible time-lag between an increase in TPW and the onset of the reversal cycle because the onset of renewed TPW 110 Ma ago occurs early in the CN superchron. The simple models of episodic plume release from the core–mantle boundary postulate a short but uncertain delay in their expression at the surface in terms of an outburst of new hotspots and continental breakup. A wide range of environmental consequences may result which do not form a subject of this review but are expressed in the geological record (Morris et al. 1979; Piper 1987 Fig. 12.14), in some cases with an apparent cycle comparable with the ca. 150–200 Ma cycle evident in Figure 12.

The data of Figure 12 provide evidence for long-term variations over the past 400 Ma in these indicators of lithosphere, mesosphere and core dynamics. Whilst it is clear that processes in the core (as reflected in the geomagnetic dynamo) influence the mantle (as seen in plumes and TPW) and ultimately the tectosphere (as seen in tectonic and magmatic activity and the motion of the plates), the collective system is a complex one with many variables and current models cannot accommodate them all with much precision. The most popular current view advocated by Courtillot and Besse (1987) and Loper et al. (1988) is that the D″ thermal boundary layer episodically becomes unstable and releases thermal plumes to a viscous and cold lower mantle. These in turn are assumed to promote thermals in the upper mantle, as expressed in the hotspots, and to interact with asthenosphere convection and subducting slabs, so influencing relative plate motions and lithosphere TPW. It has been proposed that the eruption of the Deccan traps in India at the Cretaceous–Tertiary boundary represents one of the largest single plume releases during the last 200 Ma (Courtillot and Besse 1987). This surface release occurred during the peak in continental velocities between Campanian and Palaeocene times which carried Eurasia and North America out of the Pangaean great circle frame (Le Pichon et al. 1985). The presently active end of the hotspot trace responsible for this event is now located in the Indian Ocean at Reunion Island close to isotopic and seismic anomalies in the lower mantle discussed above, and in the vicinity of a proposed thermal anomaly under Southern Africa to which Gubbins (1987) attributes the linear increase in reversal frequency since the CN superchron. This catastrophic event also occurred close to the end of the CN superchron and may be linked to the profound geological and biological changes characterising Cretaceous–Tertiary boundary (Courtillot and Besse, op. cit.).

References

Adam NV, Benkova NP, Khramov AN, Cherevko TN (1975) Spherical harmonic analysis of the geomagnetic field of the Bruhnes epoch. Studia Geophys Geol 19:141–149

Anderson DL (1982) Hot spots, polar wander, Mesozoic convection and the geoid. Nature (Lond) 297:391–393

Andrews JA (1985) True polar wander: an analysis of Cenozoic and Mesozoic palaeomagnetic poles. J Geophys Res 89:7737–7750

Bond GC, Christie-Blick N, Kominz MA (1984) Break-up of a supercontinent between 625 Ma and 555 Ma: new evidence and implications for continental histories. Earth Planet Sci Lett 70:325–345

Castillo P (1988) The Dupal anomaly as a trace of the upwelling lower mantle. Nature (Lond) 336:667–670

Chase CG (1979) Subduction, the geoid and lower mantle convection. Nature (Lond) 282:464–468

Chase CG, Sprowl DR (1983) The modern geoid and ancient plate boundaries. Earth Planet Sci Lett 62:314–320

Christensen U (1983) A numerical model of coupled subcontinental ocean convection. Tectonophysics 95:1–23
Courtillot V (1986) A French program to study the Earth's core. EOS (Trans Am Geophys Un) 67:811–812
Courtillot V, Besse J (1987) Magnetic field reversals, polar wander, and core-mantle coupling. Science 237:1140–1147
Creer KM (1977) Geomagnetic secular variations during the last 25 000 years: an interpretation of data obtained from rapidly deposited sediments. Geophys JR Astron Soc 48:91–109
Crough ST, Jurdy DM (1980) Subducted lithosphere, hot spots and the geoid. Earth Planet Sci Lett 48:15–22
Crough ST, Morgan WJ, Hargraves RB (1980) Kimberlites: their relation to mantle hot spots. Earth Planet Sci Lett 50:260–274
Davies DM, Solomon SC (1985) True polar wander and plate-driving forces. J Geophys Res 90:1837–1841
Duncan RA (1981) Hotspots in the southern oceans – an absolute frame of reference for motions of the Gondwana continents. Tectonophysics 74:29–42
Duncan RA, Petersen N, Hargraves RB (1972) Mantle plumes, movement of the European Plate and polar wandering. Nature (Lond) 239:82–86
Dupré B, Allègre CJ (1983) Pb-Sr isotope variations in Indian Ocean basalts and mixing phenomena. Nature (Lond) 303:142–146
Dziewonski AM (1984) Mapping the lower mantle by determination of lateral heterogeneity in P velocity up to degree and order 6. J Geophys Res 89:5929–5952
Evans ME (1976) Test of the dipolar nature of the geomagnetic field throughout Phanerozoic time. Nature (Lond) 262:676–677
Fleck RJ, Greenwood WR, Hadley DG, Anderson RE, Schmidt DL (1980) Age and evolution of the southern part of the Arabian Shield. In: Evolution and Mineralisation of the Arabian-Nubian Shield. Inst App Geol (Jeddah) Bull 3:1–17
Gire C (1985) Sur la variation séculaire du champ magnétique terrestre et les mouvements des couches externes du noyau fluide. Ph. D. thesis, Univ Paris VII
Goldreich P, Toomre A (1969) Some remarks on polar wandering. J Geophys Res 74:2555–2569
Gough DI (1977) The geoid and single cell mantle convection. Earth Planet Sci Lett 34:360–364
Gubbins D (1987) Mechanism for geomagnetic polarity reversals. Nature (Lond) 326:167–169
Hager BH (1984) Subducted slabs and the geoid: constraints on mantle rheology and flow. J Geophys Res 89:6003–6015
Hargraves RB, Duncan RA (1973) Does the mantle roll? Nature (Lond) 245:361–363
Harrison CG, Lindh T (1982) Comparison between the hot spot and geomagnetic field reference frames. Nature (Lond) 300:251–252
Hide R, Malin SRC (1970) Novel correlations between global features of Earth's gravitational and magnetic fields. Nature (Lond) 225:605–609
Irving E, McGlynn JC (1976) Proterozoic magnetostratigraphy and the tectonic evolution of Laurentia. Phil Trans Soc Lond A 280:243–265
Jurdy DM (1981) True polar wander. Tectonophysics 74:1–16
Jurdy DM, Voo van der R (1974) A method for the separation of true polar wander and continental drift including results for the last 55 m.y. J Geophys Res 79:2945–2952
Jurdy DM, Voo van der R (1975) True polar wander since the early Cretaceous. Science 187: 1193–1196
Lay T (1989) Structure of the Core-Mantle transitional zone. Trans Am Geophys Un (EOS) 60(4):49–58
Le Mouel JL, Madden TR, Decruix J, Courtillot V (1981) Decade fluctuations in geomagnetic westward drift and earth rotation. Nature (Lond) 290:763–765
Le Mouel JL, Gire C, Hinderer J (1985) Sur l'excitation possible de l'oscillation chandlerienne par les mouvements à la surface du noyau. CR Acad Sci Paris 301:27–38
Le Pichon X, Gaulier JM (1986) Pangée une phénomène répétitif. CR Acad Sci Paris 303:737–741
Le Pichon X, Huchon P (1984) Geoid, Pangaea and convection. Earth Planet Sci Lett 67:123–135
Le Pichon X, Huchon P, Barrier E (1985) Pangaea, Geoid and the evolution of the western margin of the Pacific Ocean. In: Nasu N (ed) Formation of Active Ocean Margins. Terrapub, Tokyo, pp 3–45

Livermore RA, Vine FJ, Smith AG (1984) Plate motions and the geomagnetic field, II Jurassic to Tertiary. Geophys J R Astron Soc 79:939–961
Loper DE, McCartney K, Buzyna G (1988) A model of correlated episodicity in magnetic field reversals, climate and mass extinctions. J Geol 96:1–15
Marton P (1986) Models for true polar wander from palaeomagnetism, geoid and hot spot studies. Tectonophysics 121:331–344
McElhinny MW (1973) Palaeomagnetism and Plate Tectonics. Univ Press, Cambridge, 358 p
McFadden PL, Merrill RT (1984) Lower mantle convection and geomagnetism. J Geophys Res 89:3354–3362
McKenzie DP (1972) Plate tectonics. In: Robertson E (ed) The Nature of the Solid Earth. McGraw-Hill, New York, pp 323–360
Minster JB, Jordan TH (1978) Present day plate motions. J Geophys Res 83:5331–5354
Molnar R, Tapponnier P (1977) Relation of the tectonics of eastern China to the India-Eurasian collision: application of strike-line field theory to large scale continental tectonics. Geophysics 5:212–216
Morgan WJ (1972) Plate motions and deep mantle convection. Geol Soc Am Mem 132:7–22
Morgan WJ (1981) Hot spot tracks and the opening of the Atlantic and Indian Oceans. In: Emiliani C (ed) The Sea. Wiley, New York, pp 443–487
Morgan WJ (1983) Hotspot tracts and the early rifting of the Atlantic. Tectonophysics 94:123–139
Morris WA, Schmidt PW, Roy JL (1979) A graphical approach to polar paths: palaeomagnetic cycles and global tectonics. Phys Earth Planet Int 19:85–99
Nairn AEM (ed) (1964) Problems in Palaeoclimatology. Interscience, London, 720 p
Neev D, Hall JK (1982) A global system of spiralling geosutures. J Geophys Res 87:10689–10708
Patchett PJ, Bylund G, Upton BGJ (1978) Palaeomagnetism and the Grenville orogeny, new Rb-Sr ages from dolerites in Canada and Greenland. Earth Planet Sci Lett 40:349–364
Piper JDA (1975) Proterozoic supercontinent: time duration and the Grenville problem. Nature (Lond) 256:519–520
Piper JDA (1976) Definition of pre-2000 m.y. apparent polar wander movements. Earth Planet Sci Lett 28:470–478
Piper JDA (1978) Geological and geophysical evidence relating to continental growth and dynamics and the hydrosphere in Precambrian times: a review and analysis. In: Brosche P, Sündermann J (eds) Tidal Friction and the Earth's Rotation I. Springer, Berlin Heidelberg New York, pp 197–241
Piper JDA (1982) Movements of the continental crust and lithosphere-asthenosphere systems in Precambrian times. In: Brosche P, Sündermann J (eds) Tidal Friction and the Earth's Rotation II. Springer, Berlin Heidelberg New York Tokyo, pp 253–315
Piper JDA (1987) Palaeomagnetism and the Continental Crust. Open Univ Press, Milton Keynes, 434 p
Piper JDA (1989) The unique signature of the Proterozoic eon. Tectonophysics (in press)
Piper JDA, Grant S (1989) A palaeomagnetic test of the axial dipole assumption and implications for continental distribution through geological time. Phys Earth Planet Int
Piper JDA, Briden JC, Lomax K (1973) Precambrian Africa and South America as a single continent. Nature (Lond) 245:244–248
Rikitake T (1966) Electromagnetism and the Earth's Interior. Devel Solid Earth Geophys 2, Elsevier, Amsterdam, 308 p
Sears JW, Price RA (1978) The Siberian Connection: a case for Precambrian separation of the North American and Siberian cratons. Geology 6:267–270
Tarling DH (1983) Palaeomagnetism, Principles and Applications in Geology, Geophysics and Archaeology. Chapman and Hall, London, pp 379
Wilson RL (1970) Permanent aspects of the Earth's non-dipole magnetic field over Upper Tertiary times. Geophys J R Astron Soc 19:417–437
Wilson RL (1971) Dipole-offset – the time average palaeomagnetic field over the past 25 million years. Geophys J R Astron Soc 22:491–504
Wilson RL, McElhinny MW (1974) Investigation of the large scale palaeomagnetic field over the past 25 million years. Eastward shift of the Icelandic spreading ridge. Geophys J R Astron Soc 39:570–586
Zhang Huimin, Zhang Wenghi (1985) Palaeomagnetic data, late Precambrian magnetostratigraphy and tectonic evolution of eastern China. Precambrian Res 29:65–75

MIX
Papier aus verantwortungsvollen Quellen
Paper from responsible sources
FSC® C105338

If you have any concerns about our products,
you can contact us on
ProductSafety@springernature.com

In case Publisher is established outside the EU,
the EU authorized representative is:
Springer Nature Customer Service Center GmbH
Europaplatz 3, 69115 Heidelberg, Germany

Printed by Libri Plureos GmbH
in Hamburg, Germany